Fine Particles in Medicine and Pharmacy

Egon Matijević
Editor

Fine Particles in Medicine and Pharmacy

 Springer

Editor
Egon Matijević
Center for Advanced Materials Processing
Clarkson University
Potsdam, New York 13699-5184,
USA
ematijevic@clarkson.edu

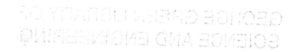

ISBN 978-1-4614-0378-4   e-ISBN 978-1-4614-0379-1
DOI 10.1007/978-1-4614-0379-1
Springer New York Dordrecht Heidelberg London

Library of Congress Control Number: 2011937020

Printed on acid-free paper

Springer is part of Springer Science+Business Media (www.springer.com)

Preface

There are several areas in medicine and pharmacy in which finely dispersed matter plays a significant role. Drugs in solid state can be produced as particles of diverse shapes and structures, in modal sizes ranging from a few nanometers to a few micrometers. Such medication is commonly combined with inactive diluents, while the pills themselves are often coated with layers which protect them from the effects of the environment, such as humidity. Both these chemically inert components in the delivery systems (diluents and shells) may also control the release of the active component.

Thus, there are many physical aspects of the medication which can affect its functionality. The first deals simply with the size of the active substance to be delivered. For example, in recent years much emphasis has been placed on the use of nanosize active materials. A recent issue of ACS NANO published several articles on the subject, including "Impact of Nanotechnology on Drug Delivery", by O.C. Farukhzad and R. Langer (3, 2009, 16–20), and "Virtual Issue on Nanomaterials for Drug Delivery" by P.T. Hammond (5, 2011, 681–684).

Another significant aspect of the drug in the pill form is the morphology of active molecules, which affects many properties of a medication, including its stability, solubility, and release. Most of the medically active compounds tend to form polymorphs, i.e., the same molecules being differently packed in the solid state, which determines their functionality. This aspect of drugs was dealt with in great detail by A.M. Rouhi in *Chemical and Engineering News* (American Chemical Society), February 24, 2003 issue, under the title "The Right Stuff: From research and development to the clinic, getting drug crystals right is full of pitfalls." It is, therefore, of great importance to produce drug delivery systems (e.g. pills) that contain the active compound in the stable state, and assure its controlled delivery.

Fine particles of different size and other properties (optical, magnetic, adhesive, etc.) play an essential role in the diagnostics, such as barium sulfate slurries in the X-ray of intestines, or well defined magnetic particles used as a biosensor, or nanodispersed gold used in bioimaging, to mention a few. Such specific uses of fine particles are described in several chapters. A short description of the contents of individual chapters is given below.

In the first chapter Vladimir Privman addresses the advancement of modeling approaches aimed at explaining morphological and geometrical features of fine particles. Specifically, discussed are certain aspects of particle shape selection and size uniformity, emerging as results of kinetics involving diffusional transport of matter in solution synthesis of nanocrystals and colloids. Processes ranging from nucleation to growth by aggregation, and mechanisms of uniform shape development are reviewed, with selected results outlined in some detail.

In the second chapter, Egon Matijević demonstrates that uniform drug dispersions can be prepared by precipitation in solutions. Indeed, in some cases, the same substance is obtained as particles of different, but uniform shapes, by altering the experimental conditions, or by varying additives. Furthermore, it is possible to coat so prepared drugs with an inorganic layer of alumina or silica, thus altering the surface reactivity and charge of the resulting particles. Such layers protect the cores and may promote specific reactions within the body.

The chapter by Silvana Andreescu, Maryna Onatska, Joseph Erlichman, Ana Estevez, and J.C. Leiter focuses on the interactions of the most widely used nanoparticles of metal oxides with cells and tissues in relation to the physico-chemical properties, biocompatibility, and cytotoxic reflexes in model biological systems, and selected biomedical applications. New and emerging uses of these particles as neuroprotective and therapeutic agents in the treatment of medical diseases related to reactive oxygen species, such as spinal cord repair, stroke, and degenerative retinal disorders are discussed. Furthermore, issues related to biocompatibility and toxicity of these nanoparticles for *in vivo* biomedical applications are dealt with in some detail.

Dan Goia and Tapan Sau contribute a comprehensive review of uniform colloidal gold, as applied in medicine and biology. Specifically, they describe how functionalized gold particles are used in bioimaging (optical, immunostaining, computed tomography, magnetic resonance, phagokinetic tracking), biosensing (optical and electrochemical), drug delivery, and therapeutic applications. Also described are additives for the preparation of highly dispersed active nanogold, including the complex (core–shell) and hierarchical structures, involving both inorganic and organic phases.

In their chapter Evgeny Katz and Marcos Pita deal with magnetic particles (microspheres, nanospheres, and ferrofluids), which are extensively used as labeling units and immobilization platforms in various biosensing schemes, mainly for immunosensing and DNA analysis, as well as in environmental monitoring. Biomolecule-functionalized magnetic particles generally exist in a 'core–shell' configuration through organic linkers, often organized as a polymeric 'shell' around the core. The state-of-the-art in the preparation, characterization, and application of biomolecule-functionalized magnetic particles and other related micro/nano-objects allows for efficient performance of various *in vitro* and *in vivo* biosensors, many of which are directed to biomedical applications.

The focus of the chapter by Devon Shipp and Broden Rutherglen is on the degradable polymer particles in drug delivery applications, based on their architectural design. Specifically, the authors consider polyanhydrides, which have the unusual property of undergoing surface erosion, and to predictable therapeutic agent release rates of approximate zero-order kinetics.

Artem Melman, in his contribution, describes an innovative method for the preparation of uniform nanoproteins, which involves their growth on monodispersed protein templates. This process is extensively involved in biomineralization in a multitude of living organisms, providing structures of exceptional complexity and uniformity. Current availability of pure recombinant cage shaped proteins and viruses offer limitless possibilities for their modification, and for targeted delivery on nanoparticles.

The chapter by Philip K. Hopke and Zuocheng Wang deals with the delivery and the effectiveness of medicine dosages deposited in the respiratory tract. Their study was originally driven by the concern regarding the effects of radioactive particles in this application. Empirical studies in animals and physical models of human airways have provided data which allows the prediction of regional deposition roles.

The chapter by Maria Hepel and Magdalena Stobiecka describes new bioanalytical sensing platforms, based on functionalized nanoparticles, for the detection of biomarkers of oxidative stress. These biomarkers and biomolecules indicate the diminished capacity of a biological system to counteract an invasion (or overproduction) of reactive oxygen species and other radicals. The oxidative stress has been implicated in a number of diseases, including diabetes, cancer, Alzheimer's, autism, and others. The detection methods for the oxidative stress biomarkers, such as glutathione, homocysteine, and cysteine, presented in this chapter, are based on their interactions with monolayer-protected gold nanoparticles. Such functionalized particles have also been shown to amplify the analytical signal in molecularly-templated conductive polymer sensors for the detection of biomolecules, and novel designs of molecularly-imprinted poly(orthophenylenediamine) sensor films.

In his chapter Sergiy Minko discusses the synthesis and applications of multifunctional hierarchically organized, multilevel structured, active hybrid colloidal particles, uniform in size and shape. Such particles are capable of programmed and controlled responses to changes in the environment or to external signals. Furthermore, various core–shell structures were synthesized in two steps consisting of metals, oxides, or polymers of different sizes and shapes, and functionalized with stimuli-responsive polymers. Specifically, deposition, precipitation on colloidal templates, grafting to the surface of particles, and self-assembly of amphiphilic block-copolymers, were extensively used for the synthesis of the core–shell colloids. A properly engineered combination of sensitivity to external stimuli with resulting changes in the particles' properties is critically important for drug delivery capsules, capsules for diagnostics and, particles-biosensors. The development of these stimuli-responsive colloids is driven by several important applications: including, biosensors that respond to changes in the chemical and biological environment, stimuli-responsive capsules that can release the cargo upon external stimuli for delivery of drugs and contrasting agents, and biocomposite materials that can adapt to living tissue.

In the final chapter, Richard Partch, Adrienne Stamper, Evon Ford, Abeer Al Bawab and Fadwa Odeh, deal with the incidence of overdoses of chemicals into the body, causing either serious injury to organs or even death. The latter is more common than what is generally believed to be the case. Among such chemicals are prescription therapeutics, illicit derivatives, biotoxins, and those found in beverages and food, leached from packaging. In this chapter it is demonstrated that both oil–water microemulsions and functionalized carrier nanoparticles are capable of removing overdosed concentrations of several of the problem chemicals from liquids including blood, both *in vitro* and *in vivo*.

Egon Matijević

Contents

Contributors

Prof. Silvana Andreescu Department of Chemistry & Biomolecular Science, Clarkson University, 8 Clarkson Avenue, SC, Box 5810, Potsdam, NY 13699, USA
e-mail: eandrees@clarkson.edu

Abeer Al Bawab Faculty of Science, and Director of Nanotechnology Center, University of Jordan, Amman, Jordan

Joseph S. Erlichman Department of Biology, St. Lawrence University, Canton, NY 13617, USA

Ana Estevez Department of Biology, St. Lawrence University, Canton, NY 13617, USA

Prof. Dan V. Goia Center for Advanced Materials Processing, Department of Chemistry & Biomolecular Science, Clarkson University, 8 Clarkson Avenue, CAMP, Box 5814, Potsdam, NY 13699, USA
e-mail: goiadanv@clarkson.edu

Prof. Maria Hepel Department of Chemistry, State University of New York, Potsdam, NY 13676, USA
e-mail: hepelmr@potsdam.edu

Prof. Philip Hopke Department of Chemical & Biomolecular Engineering, Clarkson University, 8 Clarkson Avenue, CAMP Annex, Box 5708, Potsdam, NY 13699, USA
e-mail: phopke@clarkson.edu

Center for Air Resources Engineering and Science, Clarkson University, Box 5708, Potsdam, NY 13699, USA

Prof. Evgeny Katz Department of Chemistry & Biomolecular Science, Clarkson University, Potsdam, NY, USA
e-mail: ekatz@clarkson.edu

J. C. Leiter Department of Physiology, Dartmouth Medical School, Lebanon, NH 03756, USA

Prof. Egon Matijević Department of Chemistry & Biomolecular Science, Clarkson University, 8 Clarkson Avenue, CAMP, Box 5814, Potsdam, NY 13699, USA
e-mail: matiegon@clarkson.edu

Prof. Artem Melman Department of Chemistry & Biomolecular Science, Clarkson University, 8 Clarkson Avenue, SC, Box 5810, Potsdam, NY 13699, USA
e-mail: amelman@clarkson.edu

Prof. Sergiy Minko Department of Chemistry and Biomolecular Science, Clarkson University, 8 Clarkson Avenue, SC, Box 5810, Potsdam, NY 13699, USA
e-mail: sminko@clarkson.edu

Fadwa Odeh Chemistry Department, University of Jordan, Amman, Jordan

Maryna Ornatska Department of Chemistry and Biomolecular Science, Clarkson University, 8 Clarkson Avenue, SC, Box 5810, Potsdam, NY 13699, USA

Prof. Richard Partch Chemistry Department and Center for Advanced Materials Processing, Clarkson University, Potsdam 13699-5814, NY, USA
e-mail: partch@clarkson.edu

Evon Ford Chemistry Department and Center for Advanced Materials Processing, Clarkson University, 503 Queensgate Rd, Baltimore, MD 21229, USA

Baltimore, MD, USA

Marcos Pita Instituto de Catálisis y Petroleoquímica, CSIC. C/Marie Curie 2, 28049 Madrid, Spain
e-mail: marcospita@icp.csic.es

Prof. Vladimir Privman Department of Physics, Center for Advanced Materials Processing, Clarkson University, 8 Clarkson Avenue, CAMP, Box 5721, Potsdam, NY 13699, USA
e-mail: privman@clarkson.edu

Tapan K. Sau International Institute of Information Technology, Gachibowli, Hyderabad 500032, India

Prof. Devon Shipp Department of Chemistry & Biomolecular Science, Clarkson University, 8 Clarkson Avenue, Box 5810, Potsdam, NY 13699, USA
e-mail: dshipp@clarkson.edu

Mannkind Corp., North Avenue Paine, Valencia, CA 91355, USA

Adrienne Stamper Chemistry Department and Center for Advanced Materials Processing, Clarkson University, Mannkind Corporation, 1 Casper St., Danbury, CT 06810, USA

Magdalena Stobiecka Department of Chemistry, State University of New York, Potsdam, NY 13676, USA

Zuocheng Wang Department of Chemical & Biomolecular Engineering, Clarkson University, 8 Clarkson Avenue, CAMP Annex, Box 5708, Potsdam, NY 13699, USA

Center for Air Resources Engineering and Science, Clarkson University, Box 5708, Potsdam, NY 13699, USA

Chapter 1
Models of Size and Shape Control in Synthesis of Uniform Colloids and Nanocrystals

Vladimir Privman

Abstract We review approaches to explain mechanisms of control of uniformity (narrow distribution) of sizes and shapes in solution synthesis of nanosize crystals and colloid particles. We address aspects of modeling of geometrical features and morphology selection, emerging as results of kinetic processes involving diffusional transport of matter, ranging from nucleation to growth by aggregation and to mechanisms of formation of well-defined shapes.

Keywords Aggregation • Cluster • Colloid • Crystal • Deposition • Detachment • Diffusion•Growth•Morphology•Nanocrystal•Nanoparticle•Nanosize•Nucleation• Symmetry

1.1 Introduction

Many applications of synthetic microscopic particles require them to be uniform. Kinetic mechanisms of formation of particles of narrow size and shape distributions in solutions, differ for various types of the particles: Here we refer to colloids as suspensions of few-micron down to sub-micron size particles, whereas nanoparticles and nanostructures are objects of smaller sizes, typically under 0.01 μm (10 nm). More generally, synthesis of well-defined products has to aim at uniformity of composition, internal structure/morphology, and surface properties.

Theoretical modeling approaches have to identify key mechanisms of particle size and shape selection. Indeed, frequently the actual modeling approach is limited by the computational difficulties because numerous multi-scale kinetic processes are involved: nucleation, growth, aggregation, and surface interactions of atoms/ molecules/ions (including their chemical reactions with each other and with the solution species), of nanosize building blocks, and of the forming particles. Therefore, a realistic modeling approach typically singles out a subset of those kinetic processes that can explain size and/or shape uniformity for situations of experimental relevance. Here we review aspects of several approaches and results [1–17], includ-

V. Privman (✉)
Department of Physics, Center for Advanced Materials Processing, Clarkson University,
8 Clarkson Avenue, CAMP, Box 5721, Potsdam, NY 13699, USA
e-mail: privman@clarkson.edu

E. Matijević (ed.), *Fine Particles in Medicine and Pharmacy*,
DOI 10.1007/978-1-4614-0379-1_1, © Springer Science+Business Media, LLC 2012

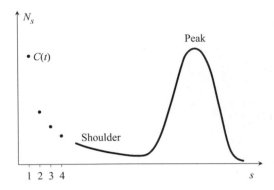

Fig. 1.1 A desirable particle size distribution at large time, t, peaked at the large cluster sizes. The peak forms and evolves due to the consumption of singlets, supplied/nucleated/maintained at the concentration $C(t)$, by a kinetic process which keeps the relative width of the peak small. The depiction for $s = 1, 2, 3, 4$ emphasizes that the particle sizes, s, are actually discrete, even though for larger s, the distribution can be treated as continuous, with $\infty > s \geq 0$

ing studies of burst-nucleation of crystalline nanoparticles in solution, the accompanying process of diffusional aggregation of these nanoparticles to form uniform polycrystalline colloids, and shape selection in nonequilibrium growth.

In applications, colloids dispersions are typically regarded as "monodispersed" for particle size distributions of relative spread up to 6–12%. At the nanoscale, we expect that most nanotechnology device applications will necessitate even stricter control: "uniform" size and shape requiring particles/structures to be "atomically identical." Thus, methodologies for uniform particle synthesis, which have a long history in colloid chemistry [1, 18, 19], have drawn renewed interest, but also faced new challenges with the advent of nanotechnology.

Here we consider those systems and synthesis processes which involve both the particles and "building blocks" from which they are formed, suspended typically in an aqueous solution of controlled chemical conditions. The transport of matter in the system, at all scales, is assumed to be diffusional. The "building blocks," termed monomers or singlets, in nanoparticle/nanostructure synthesis are solute species: atoms, ions, molecules. For colloids, the singlets are in many cases the nanosize, typically nanocrystalline precursor "primary" particles. The supply of the latter is controlled by their own burst nucleation. In both cases the monomers can also be introduced externally as a means to control the process kinetics.

Particle (cluster) size, s, distribution with a peak at large sizes, is schematically drawn in Fig. 1.1. Most processes that make the peak mean-s value grow also broaden it, for example, cluster-cluster aggregation or cluster ripening due to exchange of monomers. Therefore, they cannot yield a relatively narrow peak. This occurs because larger particles have bigger surface area for capturing small clusters/monomers, as well as on average less surface curvature, resulting in slightly better binding and thus less detachment of monomers. As a result, the larger-particle side of the peak "runs away" from the smaller particle side (see Fig. 1.1) resulting in peak broadening as the clusters grow.

Approaches to obtain a narrow size (and shape) distribution include blocking the growth of the "right side" of the peak (see Fig. 1.1) by synthesizing the particles inside nanoporous structures or objects such as micelles or inverse micelles, e.g., [20, 21]. This technique has a disadvantage that additional chemicals then remain part of the formed particles. Another approach has been by seeding, i.e., growth on top of separately/earlier prepared/synthesized smaller uniform size and shape template particles, e.g., [22].

In Sect. 1.2, we consider [9–11] the process of burst nucleation: rapid growth of particles forming in a supersaturated solution of constituent atoms, molecules or ions. This process exemplifies a size selection mechanism whereby the left side of the peak (Fig. 1.1) is eroded fast enough as compared to the peak broadening due to its growth by consumption of monomers. As a result, narrow size distribution is obtained. In practice, additional coarsening processes broaden the distribution after the initial nucleation burst, typically limiting this mechanism to nanosize crystal growth stage.

In Sect. 1.3, we describe a two-stage colloid growth mechanism [1, 6, 10, 11] yielding particle size distributions narrow on a relative scale. This involves a large supply of primary particle (precursor nanocrystal) monomers/singlets, of concentration $C(t)$, see Fig. 1.1. Availability of these monomers allows the peak to grow to large sizes in a process fast enough that the central peak is not significantly broadened. At the same time, a proper control of $C(t)$ is needed in order to avoid buildup of a significant "shoulder" always present for such growth processes at small cluster sizes (see Fig. 1.1). The monomer building blocks for such a process yielding uniform polycrystalline colloids, are actually the burst-nucleated nanocrystalline precursors (primary particles). For nanoparticle growth, there have also been studies of stepwise processes [23, 24], with batches of atomic-size monomers added for further growth of the earlier formed nanoparticles.

Let $N_s(t)$ denote the density of particles containing s singlets, at time t. Except for the small s values, see Fig. 1.1, the distribution can be treated as a function of continuous size variable. However, according to the preceding discussion the singlet concentration,

$$C(t) \equiv N_1(t), \qquad (1.1)$$

has to be separately controlled in some situations. They can be supplied as one or more batches at specific times, or generated by another process at the rate $\rho(t)$ (per unit volume). They are depleted as a result of processes involving the emergence of small clusters (the "shoulder" in Fig. 1.1), and also consumed by the growing large clusters in the main peak.

For nanoparticles, a mechanism of the early formation of the peak is by burst nucleation: nuclei of sizes larger than the critical, form from smaller "embryos" by growing over the nucleation barrier. For colloid synthesis, the initial peak formation can be facilitated by cluster-cluster aggregation of at the early growth stages. In Sect. 1.4, we generally discuss some of the issues important for improving models of uniform colloid growth, in the framework of elaboration of the two-stage model of Sect. 1.3. Specifically, we address the role of cluster-cluster aggregation. Seeding is another approach to initiating the peaked size distribution both for colloid and nanoparticle growth.

Section 1.5 addresses the problem of particle shape distribution and its control aimed at attaining uniformity. More generally, particle structure ranging from internal microscopic morphology and defects, to surface properties and to overall shapes, in both nanoparticle and colloid synthesis, are as important in applications as is particle size. Several mechanisms for particle shape control in fast, nonequilibrium growth are possible, and likely some or most apply on the case by case basis, depending on the details of the system and its kinetics. For uniform-shape growth, we have advanced a model [14], outlined in Sect. 1.5, suggesting that fast growth without development of large internal defect structures can lead to shape selection with non-spherical particle "faces" similar to those obtained in equilibrium crystal structures, but of different aspect ratios. The latter ideas have also been successfully applied [17] to explain shapes of certain growing nanostructures on surfaces, of interest in catalysis. Emergence of a relatively uniform distribution of surface structures evolving from nanoclusters to nanopyramids and then to nanopillars, was modeled [17] (not reviewed here). Finally, Sect. 1.6 offers concluding comments.

1.2   Growth of Nanoparticles by Burst Nucleation

Burst nucleation [9–11, 25, 26] is a model for growth dominated by large supply of monomers in solution. The formed embryos are assumed to be small enough that they can practically instantaneously thermally equilibrate. Therefore, the model can at best be used for growth of nanosize particles, consisting of n monomers. Indeed, a cutoff is assumed as one of the model's approximations, such that particles with $n > n_c$, where n_c is the critical cluster size (to be defined shortly), irreversibly capture diffusing solutes (monomers): atoms, ions or molecules. Whereas the dynamics in the shoulder, for $n < n_c$, see Fig. 1.2, is such that the subcritical ($n < n_c$) embryos, are instantaneously thermalized.

Burst nucleation in a supersaturated solution is driven by externally supplied or, more commonly, chemical-reaction produced supply of monomers, of concentration, $c(t)$, well over the equilibrium value c_0. In nucleation theory approaches, thermal fluctuations are assumed to cause formation of the embryos. Their surface free energy results in a free-energy barrier peaked at n_c. The actual dynamics of few-atom embryos involves complicated transitions between various sizes, shapes, and internal restructuring, and is not well understood. However, these processes are so fast that the $n < n_c$ embryo sizes are assumed approximately thermally distributed according to the Gibbs free energy of an n-monomer cluster,

$$\Delta G(n, c) = -(n - 1)kT \ln (c/c_0) + 4\pi a^2 (n^{2/3} - 1)\sigma. \qquad (1.2)$$

Fig. 1.2 The large-time nanoparticle size distribution in the burst nucleation model is sketched in the top panel. The actual distribution, depicted by the *dotted line*, if steep but continuous near n_c. The time dependence of the critical cluster size, n_c, is shown in the bottom panel. A short induction period is followed by the "burst," and then linear growth but with a negligibly small slope

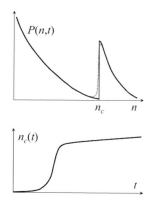

Here k is the Boltzmann constant, T is the temperature, and σ is the effective surface tension. The effective solute radius, a, is defined in such a way that the radius of an n-solute embryo is $an^{1/3}$. It can be estimated by requiring that $4\pi a^3/3$ equals the unit-cell volume per singlet (including the surrounding void volume) in the bulk material. This free-energy attains maximum (the nucleation barrier) at n_c,

$$n_c(c) = \left[\frac{8\pi a^2 \sigma}{3kT \ln (c/c_0)}\right]^3. \tag{1.3}$$

The first term in Eq. (1.2) is due to the bulk of the n-monomer cluster and is negative (since $c > c_o$), favoring growth of clusters. The logarithmic dependence on the monomer concentration derives from the entropy of mixing of noninteracting solutes. The second term represents the surface free-energy, proportional to the area, $\sim n^{2/3}$, and positive, thus suppressing growth of clusters. This term dominates for $n < n_c$, and results in the nucleation barrier. Thus, clusters with $n < n_c$ are assumed instantaneously thermally distributed. However, the kinetics of larger clusters, $n > n_c$, is assumed to correspond to fast, irreversible capture of monomers. These assumptions are typical for homogeneous nucleation. The unique aspect of burst-nucleation in solution is that the bulk free energy term in Eq. (1.2), is dependent on the monomer concentration and therefore varies with time. As a result, the critical cluster size, $n_c((c(t))$, as well as the height of the nucleation barrier, are time-dependent.

A standard assumption in the nucleation theory has been that, for approximate estimates and understanding of the growth of the cluster sizes, the distribution of their shapes can be ignored. A representative particle is assumed spherical in the calculation of its surface area and the monomer transport rate to it. There are many possible effect of the actual particle shape distribution and surface structure at various faces (for crystals) on the transport of the surrounding solute/suspension matter and on other properties and parameters. For example, effective surface tension of spherical particles depends on their radius via surface curvature. All these geometry- and structure-dependent modifications are usually ignored not just because of the computational difficulties of treating multi-parameter distributions but primarily owing to the fact that effective surface tension of nanoparticles and nanosize surface features is presently only understood to a very limited extent, e.g., [27]. Thus, σ in Eqs. (1.2) and (1.3) has either been assumed [1, 5, 7, 8] close to σ_{bulk} (which might not always be correct for particles smaller than 5–10 nm) or taken as an adjustable parameter.

During the initial burst of nucleation, c/c_0 decreases from the initial value $c(0)/c_0 \gg 1$ towards its asymptotic large-time equilibrium value 1. The late-stage form [9–11] of the particle size distribution is sketched in Fig. 1.2. Recall that embryos smaller than n_c are thermalized on time scales faster than other dynamical processes, with size distribution

$$P(n < n_c, t) = c(t) \exp\left[\frac{-\Delta G(n, c(t))}{kT}\right]. \tag{1.4}$$

Here $P(n,t)dn$ gives the number concentration per unit volume of embryos with sizes in dn.

Note that $n_c = n_c(c(t))$ is time-dependent, and that the approximate (but of course not the actual) functional form of the particle-size distribution is discontinuous at n_c (see Fig. 1.2). The production of supercritical ($n > n_c$) clusters "over the barrier" at $n = n_c$, occurs at the rate (per unit time, per unit volume) $\rho(t)$, which is modeled [1] as follows,

$$\rho(t) = K_{n_c} c P(n_c, t) = K_{n_c} c^2 \exp\left[\frac{-\Delta G(n_c, c)}{kT}\right], \tag{1.5}$$

where

$$K_n = 4\pi a n^{1/3} D, \tag{1.6}$$

is the Smoluchowski rate constant [2, 28] for the irreversible capture of diffusing solutes by growing spherical clusters of sizes $n \geq n_c \gg 1$. The quantity D is the diffusion coefficient for monomers in a dilute solution of viscosity η; D can be estimated as $\sim kT/6\pi\eta a$ (up to geometrical factors relating the effective radius a to the hydrodynamic radius).

Irreversible growth of the $n > n_c$ clusters can be described [9] by the kinetic equation

$$\frac{\partial P(n, t)}{\partial t} = (c(t) - c_0)(K_{n-1} P(n - 1, t) - K_n P(n, t)), \tag{1.7}$$

where the difference $c(t) - c_0$ is introduced in place of $c(t)$ as a multiplier of the Smoluchowski "irreversible attachment" rate constant, in order to ensure that the growth stops as $c(t)$ approaches c_0. As derived in [2], this approximately accounts for the detachment of matter if we ignore curvature and similar effects. As already mentioned, the latter include variation of the surface tension with cluster radius, resulting in variable effective "equilibrium concentration" and thus, once monomer detachment is considered, in Ostwald ripening [29] driven by exchange of monomers between the clusters. This and other possible coarsening processes, such as cluster-cluster aggregation, e.g., [30, 31], are ignored because they are typically slower than burst nucleation [1, 9–11]. The large-time linear increase [9–11] of $n_c(t)$, alluded to in bottom plot in Fig. 1.2, has a very small slope [32], and thus the particle growth would practically stop if it were only due to burst-nucleation. However, for later times the other coarsening processes will broaden, as well as further grow, the particle size distribution as compared to burst-nucleation alone.

In addition to growth (or shrinkage) of particles by various dynamical processes involving the surrounding matter, they also undergo internal restructuring, understanding of which for nanosize clusters is not well developed [33, 34]. Without restructuring, clusters would grow as fractals [30, 31], whereas density measurements and X-ray diffraction data for colloidal particles aggregated from burst-nucleated nanocrystals indicate that while they have polycrystalline structure, their density is close to that the bulk material [1, 18]. There is both experimental and indirect modeling evidence [1, 4, 5, 7, 8] that, in such irreversible-growth colloid synthesis internal restructuring leads to compact particles with smooth surfaces.

Particles in the supercritical distribution, for $n > n_c(t)$, in burst nucleation are assumed to grow irreversibly by capturing monomers. At the same time, the sizes of the subcritical, $n < n_c(t)$, particles are instantaneously redistributed by fast thermalization. The function $n_c(t)$ is increasing monotonically. This sharp distinction between two types of dynamics is an approximation of the nucleation theory. The short-time form of the supercritical distribution depends on the initial conditions. However, at large times [9–11] the particle size distribution will attain its maximum at $n = n_c$, and the shape of a truncated Gaussian: of its "right-hand side slope" (see Fig. 1.2), whereas the peak of the full Gaussian curve (not shown in Fig. 1.2), is actually well to the left of n_c.

These properties were derived [9–11] and confirmed by large-scale numerical modeling of time-dependent distributions, for a selection of parameters and initial conditions, utilizing a novel efficient numerical integration scheme; see [9]. Here we survey analytical results for large times. The kinetic equation then has an asymptotic solution of the Gaussian form

$$P_G(n,t) = \zeta(t)c_0 \exp\left[-(\alpha(t))^2(n - M(t))^2\right], \qquad (1.8)$$

for $n > n_c(t)$ (and large t), with the "peak offset" $n_c(t) - M(t)$ a positive quantity. The derivation involves a continuous-n form of Eq. (1.7), keeping terms up to the second derivative in order to capture the diffusive nature of the peak broadening,

$$\frac{\partial P}{\partial t} = (c - c_0)\left[\left(\frac{1}{2}\frac{\partial^2}{\partial n^2} - \frac{\partial}{\partial n}\right)(K_n P)\right]. \qquad (1.9)$$

Irreversible growth of supercritical clusters results in $P(n,t)$ taking on appreciable values only over a narrow range. Thus, we can further approximate

$$K_n \approx K_{n_c} = \kappa (n_c(t))^{1/3} / c_0, \tag{1.10}$$

where

$$\kappa \equiv 4\pi c_0 a D, \tag{1.11}$$

and then use Eqs. (1.3) and (1.6) to show that the product of the coefficients, $(c - c_0)K_{n_c}$, in Eq. (1.9) becomes a constant in the limit of interest,

$$\frac{\partial P}{\partial t} = \frac{z^2}{2} \left(\frac{1}{2} \frac{\partial^2}{\partial n^2} - \frac{\partial}{\partial n} \right) P, \tag{1.12}$$

where we defined

$$z^2 \equiv \frac{64\pi^2 a^3 \sigma c_0 D}{3kT}. \tag{1.13}$$

Equation (1.12) implies that the solution is indeed a Gaussian, with

$$\alpha(t) \approx 1/\sqrt{z^2 t}, \quad M(t) \approx z^2 t / 2, \quad \zeta(t) \approx \Omega / \sqrt{z^2 t}. \tag{1.14}$$

The prefactor Ω cannot be determined from this analysis, because the overall height of the distribution depends on the initial conditions. Such quantities have to be determined from the conservation of matter. Rather complicated mathematical considerations [9], result in the expression $n_c(t) - M(t) \propto \sqrt{t \ln t}$ (with a positive coefficient) for the "peak offset." Since $M(t)$ is linear in time, the $\sqrt{t \ln t}$ "offset" is sub-leading, and we get

$$n_c(t) \approx z^2 t / 2. \tag{1.15}$$

The width of the truncated Gaussian is proportional to $1/\alpha \propto \sqrt{t}$. Thus the *relative* width decreases according to $\sim t^{-1/2}$. One can also show [9] that the difference $c(t) - c_0$ approaches zero ($\sim t^{-1/3}$) as expected.

The Gaussian shape has provided a good fit not only for large times but also for intermediate times in numerical modeling for various initial conditions, including those describing initially seeded distributions; see [9]. Numerical simulations have also confirmed the other expected features, summarized in Fig. 1.2. The initial induction period is followed by the "burst," and then the asymptotically linear growth. Experimentally, it has been challenging to quantify distributions of nucleated nanocrystals, because of their non-spherical shapes and tendency to aggregate. The distribution is usually two-sided around the peak, and the final particles stop growing after a certain time. Both of these properties are at odds with the predictions of the burst-nucleation model, and the discrepancy can be attributed to the assumption of instantaneous thermalization of the clusters below the critical size and to the role of other growth mechanisms.

Fig. 1.3 Two-stage synthesis of uniform colloids as aggregates of precursor nanocrystal particles burst-nucleating in supersaturated solution

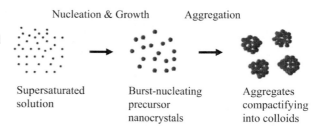

Nucleation & Growth   Aggregation

Supersaturated solution   Burst-nucleating precursor nanocrystals   Aggregates compactifying into colloids

Very small clusters, of sizes tentatively estimated [6–8, 35–37] to correspond to $n_{th} \approx 15$–25 "monomers" (atoms, ions, molecules, sub-clusters), can evolve very rapidly, so that the assumption of fast thermalization/restructuring is justified. Larger clusters then develop a bulk-like core and their internal restructuring slows down: once $n_c(t) > n_{th}$, the "classical" nucleation model should be regarded as approximate. Modifications of the model have been contemplated [9, 38, 39]. This, however, requires introduction of additional dynamical parameters which are not as well understood as those of the "minimal" model.

1.3   Colloid Synthesis by Two-stage Growth

Since the diffusion constants in aqueous environment are roughly proportional to radii of the diffusing entities, there is a drop in the rates of diffusional-transport driven processes by a factor of ~100 from single atoms to, for instance, 10 nm nanoparticles. Another ~100 drop in rates is then obtained when diffusional motion of 1 μm colloids are involved. Burst nucleation, while ideally yielding narrow size distributions, at best approximately describes formation of particles up to several tens of nm in diameter (usually much smaller). Size distribution of particles is then broadened as they further grow by other mechanisms. However, one exception has been identified: the two-stage mechanism [1] whereby the nanosized primary particles, while burst-nucleating and further growing, become the singlets for their aggregation to form uniform secondary particles of colloid sizes. This is summarized in Fig. 1.3.

Synthesis of dispersions of uniform colloid particles of various chemical compositions and shapes have been reported [1, 7–8, 18, 40–63], with particle structural properties consistent with the two-stage mechanism. Specifically, precipitated spherical particles had polycrystalline X-ray characteristics, including ZnS [42], CdS [7, 8, 41], Fe_2O_3 [40], Au, Ag, and other metals [1, 26, 57–59, 61, 63]. Experimentals have confirmed that many monodispersed inorganic colloids consist of nanocrystalline subunits [1, 7–8, 18, 40–59, 61–63], and these subunits were observed [1, 55, 57] to have the same sizes as the average diameter of the precursor singlets formed in solutions. The singlets were of sizes of approximately up to a couple of 10 nm. Such a composite structure has also been identified for certain synthesized suspensions of uniform nonspherical colloid particles [40, 49, 51, 53,

62], but these findings are presently not definitive enough to single out the two-stage growth mechanism presented in Fig. 1.3.

In this section, we outline a model that involves the coupled primary and secondary processes in the simplest formulation invoking several approximations that allow us to avoid introduction of unknown microscopic parameters. In the next section, we describe improvements that allow for a better agreement with experimental observations. The latter approach involves large-scale numerical simulations. Additional information, results and examples of parameter value fitting, etc., can be found in [1, 5, 7, 8, 12, 13].

The secondary particles are assumed here to grow by irreversible capture of singlets. This is appropriate as a description of the evolution of the already well developed peak, see Fig. 1.1, with the role of the few particles in the "shoulder" (Fig. 1.1) minimal. (The emergence of the peak is addressed later.) We use rate equations with Γ_s denoting the rate constants for singlet capture by the $s \geq 1$ aggregates, with all the other notation defined in Sect. 1.1, see Eq. (1.1) and also Fig. 1.1,

$$\frac{dN_s}{dt} = (\Gamma_{s-1}N_{s-1} - \Gamma_s N_s)C, \quad s > 2, \tag{1.16}$$

$$\frac{dN_2}{dt} = \left(\frac{1}{2}\Gamma_1 C - \Gamma_2 N_2\right)C, \tag{1.17}$$

$$\frac{dC}{dt} = \rho - \sum_{s=2}^{\infty} s\frac{dN_s}{dt} = \rho - \Gamma_1 C^2 - C\sum_{s=2}^{\infty} \Gamma_s N_s. \tag{1.18}$$

Here we ignore cluster–cluster aggregation, assuming that the only process involving the $s > 1$ aggregates is that of capturing singlets at the rate proportional to the concentration of the latter, $\Gamma_s C$, see Eq. (1.16). This assumption is generally accepted in the literature, e.g., [1, 5, 6, 64–66]. As noted, we will describe elaborations later: Indeed, processes such as cluster-cluster aggregation [30, 31], detachment [2, 4] and exchange of singlets (ripening), etc., also contribute to and modify the pattern of particle growth, and most broaden the particle size distribution. However, in the two-stage uniform colloid synthesis they are slower than the singlet-consumption driven growth.

Equations (1.16–1.18) do not account for possible particle shape and morphology distribution. This issue is not well understood and difficult to model on par with particle size distribution. However, experimentally it has been well established [1, 43–45, 50, 52, 57] that, the growing aggregates rapidly restructure into compact bulk-like particles of an approximately fixed shape, typically, though not always, spherical for two-stage aggregated colloids. Without such restructuring, the aggregates would be fractal [31, 67]. We address shape selection in Sect. 1.5.

For the model of the singlet-supply-driven particle growth, if the singlets were supplied/available constantly, then the size distribution would develop a large shoulder at small aggregates, with no pronounced peak at $s \gg 1$; cf. Fig. 1.1. If the

Fig. 1.4 Calculated colloid-particle size distribution as a function of the particle radius. The parameters were for spherical gold colloids. [1, 5]

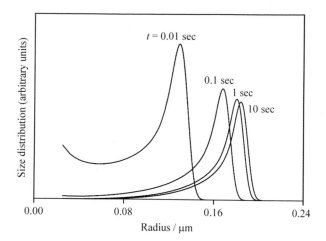

supply of singlets is limited, then only small aggregates will be formed (no growth). A key discovery in studies of colloid synthesis [1, 6] has been that there exist protocols of singlet supply, at the rate $\rho(t)$ which is a properly decreasing function of time, that yield narrow-peaked (at large sizes, s) distributions for large times. Furthermore, the primary process, that of burst-nucleated nanocrystalline precursors growing past the nucleation barrier and further coarsening, naturally "feeds" the secondary process of these precursor nanoparticles aggregating to form colloids, just at a rate like this.

Diffusional growth of the secondary (colloid) particles in particular, must be facilitated by the appropriate chemical conditions in the system, such as the ionic strength and pH. Indeed, surface potential should be close to zero (the isoelectric point) or the electrostatic screening significant enough to avoid electrostatic barriers, in order to promote fast irreversible primary particle attachment [1, 43–45, 50, 52, 57]. Clusters of sizes s are then aggregated in solution with the volume densities $N_{s=1,2,3,...}(t)$. They are identified by the number of the included primary particles (nanocrystalline domains); thus there sizes may somewhat vary. For Eqs. (1.16–1.18), we take the initial conditions $N_{s=1,2,3,...}(0) = 0$. The simplest choice for the rate constants is the Smoluchowski expression, cf. Eq. (1.6),

$$\Gamma_s \approx 4\pi R_p D_p s^{1/3}. \tag{1.19}$$

Here R_p, D_p are the effective primary particle radius and diffusion constant, the choice of which will be discussed shortly. The approximate sign is used because several possible improvement to this simplest formula can be offered. A numerical calculation result for a model of the type outlined here is shown in Fig. 1.4. It illustrates a key feature of "size selection," i.e., the practical "freezing" of the growth even for exponentially increasing time intervals (here in steps $\times 10$).

For $\rho(t)$ in Eq. (1.18), we use the rate of production of the supercritical clusters, Eq (1.5), the calculation of which requires $c(t)$. We will use the following [1] approximate relation,

$$\frac{dc}{dt} = -n_c\rho, \tag{1.20}$$

combined with Eqs. (1.3), (1.5) and (1.6). We already mentioned [9] that complicated steps are required to derive the expression (not shown) for dc/dt in burst nucleation alone, without the secondary aggregation process. When the burst-nucleated, growing supercritical particles are also consumed by the secondary aggregation, even more complicated considerations are involved. Indeed, the solute species (present with concentration c in the dilute, supersaturated solution) are also partly stored in the $n>1$ subcritical embryos, as well as in the supercritical primary particles and in the secondary aggregates. They can be captured, as assumed in our model of burst nucleation, but they can also detach back into the solution. The main advantage of the approximate Eq. (1.20) is tractability. It ignores the effect of the possible rebalancing of the "recoverable" stored solute species in various part of the particle distributions. Rather, it focuses on the loss of their availability due to the mostly unrecoverable storage in secondary particles of sizes $s = 1,2,3...$ (where the $s = 1$ particles are the "singlet" nucleated supercritical nuclei, whereas $s>2$ corresponds to their aggregates). The form of the right-hand side of Eq. (1.20), when used with Eqs. (1.3), (1.5) and (1.6), also ignores further capture by and detachment from larger particles. The resulting closed equations allow calculating the rate $\rho(t)$ of the supply of singlets for the secondary aggregation, starting with the initial supercritical concentration $c(0) \gg c_0$ of solutes, via the relations

$$\frac{dc}{dt} = -\frac{2^{14}\pi^5 a^9 \sigma^4 D_a c^2}{(3kT)^4[\ln(c/c_0)]^4} \exp\left\{-\frac{2^8\pi^3 a^6 \sigma^3}{(3kT)^3[\ln(c/c_0)]^2}\right\}, \tag{1.21}$$

$$\rho(t) = \frac{2^5\pi^2 a^3 \sigma D_a c^2}{3kT \ln(c/c_0)} \exp\left\{-\frac{2^8\pi^3 a^6 \sigma^3}{(3kT)^3[\ln(c/c_0)]^2}\right\}. \tag{1.22}$$

Here we denoted the diffusion constant of the solutes by D_a, in order to distinguish it from D_p of the primary particles.

We now discuss the choice of parameters in light of some of the simplifying assumptions made. We will also consider, in the next section, possible modifications of the model. In fact, Fig. 1.4, based on one of the sets of the parameter values used for modeling formation of uniform spherical Au particles, already includes some of those modification [5]. First, we note that if the assumption $s \gg 1$ is not made, the full Smoluchowski rate expression [2, 28] should be used, which, for aggregation of particles of sizes s_1 and s_2, on encounters due to their diffusional motion, is

$$\Gamma_{s_1,s_2 \to s_1+s_2} \simeq 4\pi\left[R_p\left(s_1^{1/3} + s_2^{1/3}\right)\right]\left[D_p\left(s_1^{-1/3} + s_2^{-1/3}\right)\right], \tag{1.23}$$

where for singlet capture $s_1 = s$ and $s_2 = 1$. This relation can introduce nontrivial factors for small particle sizes, as compared to Eq. (1.19), and also relies on the assumption that the diffusion constant of s-singlet, dense particles is inversely proportional to the radius, i.e., to $s^{-1/3}$, which might not be accurate for very small, few-singlet aggregates.

Another assumption in Eqs. (1.19) and (1.23), is that the radius of representative s-singlet, dense particles can be estimated as $R_p s^{1/3}$. However, primary particles actually have a distribution of radii, and they can also age (grow/coarsen) before their capture by and incorporation into the structure of the secondary particles. In order to partially compensate for this approximations, the following arguments can be used. Regarding the size distribution of the singlets, it has been argued that since their capture rate by the larger aggregates is approximately proportional to their radius times their diffusion constant, this rate will not be that sensitive to the particle size and size distribution, because the diffusion constant for each particle is inversely proportional to its radius. Thus, the product is well approximated by a single typical value.

The assumption of ignoring the primary particle ageing, can be circumvented by using the experimentally determined typical primary particle linear size, $2R_{exp}$, instead of attempting to estimate it as a function of time during the two-stage growth process. In fact, for the radius of the s-singlet particle, the expressions in the first factor in Eq. (1.23), which represents the sum of such terms, $R_p s^{1/3}$, should be then recalculated with the replacement

$$R_p s^{1/3} \rightarrow 1.2 R_{exp} s^{1/3}. \tag{1.24}$$

The added factor is $(0.58)^{-1/3} \approx 1.2$, where 0.58 is the filling factor of a random loose packing of spheres [68], introduced to approximately account for that as the growing secondary particle compactifies by restructuring, not all its volume will be crystalline. A fraction will consists of amorphous "bridging regions" between the nanocrystalline subunits.

At the end of the model computations, inaccuracies due to the approximations entailed in using Eq. (1.20), and the use of the uniform singlet radii, see Eq. (1.24), both possibly leading to nonconservation of the total amount of matter, can be partly compensated for [1], by renormalizing the final distributions so that the formed secondary particles contain the correct amount of matter. This effect seems not to play a significant role in the dynamics. Some additional technical issues can be found in [1, 3, 5, 7, 8, 12, 13].

1.4 Improved Models for Uniform Colloid Growth

The model of Sect. 1.3 was used for a semi-quantitative description (without adjustable parameters) of the processes of formation of spherical colloids of metals Au [1, 3, 5, 7, 12, 13, 69] and Ag [12, 13], salt CdS [7, 8], as well as argued to qualitatively explain the synthesis of monodispersed microspheres of Insulin [60]. In this section, additional elaborations will be outlined which allow to improve the two-stage model for quantitative agreement with experimental results, explored for CdS [7, 8], Au [69], and Ag [12, 13]. The CdS spherical colloid particle distribution was measured for several times during the process and for different protocols of feeding the solutes into the system, not just for the practically instantaneous supply by a fast chemical reaction as was done for the two metals. When ions (or atoms/molecules) are not released as an instantaneous "batch" or otherwise externally supplied, we have to include in the model the rate equations for their production in chemical reactions. This is in itself a rather challenging problem because for many "cookbook" colloid synthesis processes, experimental identification and more so modeling of the kinetics of various solute species involved, are not well studied.

Numerical simulations have lead to the conclusion that the key physical properties of the primary nucleation process: the effective surface tension and equilibrium concentration, when varied as adjustable parameters, mostly affect the time scales of the onset of "freezing" of the secondary particle growth, seen in Fig. 1.4. The use of the measured bulk values for these parameters yields results consistent with the experimentally observed "freezing" times. The parameters of the kinetics of the secondary aggregation primarily control the size of the final particles. Sizes numerically calculated within the "minimal" model [1, 3, 5, 7, 8, 12, 13, 69], while of the correct order of magnitude, were smaller than the experimentally observed values. This suggests that the assumed kinetics for the secondary aggregation results in too many secondary particles which then grow to sizes smaller than those experimentally observed.

Two approaches to improving the model have been considered. The first [5, 12] argues that for very small "secondary" aggregates, those consisting of one or few primary particles, the spherical-particle diffusional expressions for the rates, which are anyway ambiguous for tiny clusters, should be modified. In order to avoid introduction of many adjustable parameters, the rate $\Gamma_{1,1\to2}$, cf. Eq. (1.23), was multiplied by a "bottleneck" factor, $f<1$. Indeed, merging of two singlets (and other very small aggregates) may involve substantial restructuring, thus reducing the rate of successful formation of a bi-crystalline entity. The two nanocrystals may instead unbind and diffuse apart, or merge into a single larger nanocrystal, effectively contributing to a new process, $\Gamma_{1,1\to1}$, not in the original model. Data fits [5, 7, 12] yield values of order 10^{-3} or smaller for f. Note that both for the original model and one just described, numerical simulations require substantial computational resources. As a result, simulation speed-up techniques valid for the kinetics of larger clusters have been devised [6, 12, 13].

The second approach to improving the model [7, 8, 69], starts with the observation that the "minimal" model already assumes a bottleneck for particle merger, because only singlet capture is included; the rates in Eq. (1.23) with both $s_1>1$ and $s_2>1$, are set to zero. This corresponds to the experimental observation that larger particles were never seen to pair-wise "merge" in solution. The conjecture was adopted that the restructuring processes that cause the observed rapid compactification of the growing secondary particles, mediate the incorporation of primary particles,

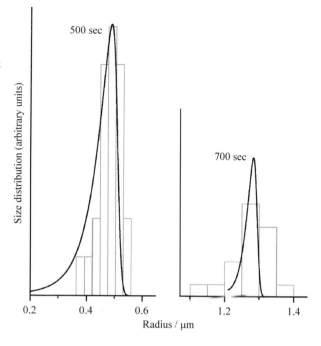

Fig. 1.5 Calculated particle size distribution (*solid lines*) are compared to the experimentally measured ones (histograms) for two different times during the growth. The parameters were for the model of synthesis of spherical CdS colloids [8], with $s_{max} = 25$

but not of larger aggregates, in the evolving structure. The incorporated particles mostly retain their crystalline core, to yield the final polycrystalline colloids. The improved model allows small secondary clusters, up to certain sizes, $s_{max} > 1$, to also be dynamically rapidly incorporated into growing aggregates on diffusional encounters. The model equations now allow $s_{1,2} > 1$ cluster-cluster aggregation with rates given by Eq. (1.23), but only as long as at least one of the sizes, s_1 or s_2 does not exceed a certain value s_{max}, see [7, 8] for details. This sharp cutoff is an approximation, but it offers the convenience of a single new adjustable parameter. Indeed, data fits for CdS and Au spherical particles, have yielded quantitative agreement with experiments, as illustrated Fig. 1.5, for fitted values of s_{max} ranging [7, 8, 69] from ~ 15 for Au, to ~ 25 for CdS. These values are intuitively reasonable as defining "small" aggregates, and they also remind us of a similar concept of the cutoff value n_{th}, discussed in Sect. 1.2, beyond which atomistic aggregates develop a well formed "bulk-like" core. Indeed, the available numerical estimate of such a quantity in solution [37], for AgBr nanoaggregates, suggests that n_{th} is comparable to or somewhat larger than ~ 18. Finally, the added cluster-cluster aggregation at small sizes, offers a mechanism for the formation of the initial peak in the secondary-particle distribution.

The singlet-capture-only, the added bottleneck-factor, and the small-cluster-aggregation approaches are all modifications of the rates of the diffusional-transport-driven irreversible-capture expression for the aggregation rate constants, Eq. (1.23). Allowing for cluster-cluster aggregation has required large-scale numerical effort and consideration of efficient algorithmic techniques for simulations, not reviewed here, including conversion of the discrete-s equations to continuum ones, with the adaptive-grid (re)discretization both in the time, t, and cluster size, s; see [7, 8].

1.5   Shape Selection in Particle Synthesis

Both colloids and nanosize particles, when synthesized according to various "cookbook" protocols to get uniform products, display a plethora of shapes and morphologies many of which are useful in applications. The morphology can range from single crystals to polycrystalline entities, as well as to amorphous or compound structures. Specifically, there is substantial experimental evidence [22, 23, 40, 49, 52, 53, 61, 62, 70–76] for growth of uniform size and shape nonspherical particles under properly chosen conditions. The challenge of explaining uniformity of shape and morphology has remained largely unanswered until recently [14]. An exception have been the "imperfect-oriented attachment" mechanism [77–80]: persistence in successive nanocrystal attachments has been argued to explain the formation of uniform short-chain aggregates, as well as mediate growth of other shapes [18, 80] for a certain range of aggregate sizes. We observe that for many growth conditions, microscopic particles are simply not sufficiently large—do not contain enough constituent monomers—to develop shape-destabilizing stable surface morphologies such as "dendritic instability" of growing side branches, then branches-on-branches, etc. In an imprecise language we can state that such synthetic particles simply don't have enough "phase space" to explore the full range of surface morphological fluctuations.

Several processes and their competition determine the resulting particle structure and specifically shape. Diffusional transport of matter results in attachment of atoms (ions, molecules) to form nanocrystals, or of nanocrystalline building blocks to grow colloids. These "monomers" can also detach/reattach, as well as move/roll on the surface. Furthermore, nanoparticles as building blocks, can restructure and even further grow on-surface by capturing solute species. Modeling all these processes at once would be a formidable numerical challenge. Therefore, one has to seek simplifying principles and capture the key mechanism(s) that would in some cases allow us to reproduce shape selection by tractable numerical modeling. Empirical experimental evidence [1] available primarily for spherical-colloid synthesis, have suggested that the building-block nanocrystals eventually get "cemented" in a dense structure, but can retain their unique crystalline cores. Diffusional transport with attachment without such restructuring would yield a fractal structure [30, 67]. On the experimental side, quantitative data on the time dependent kinetics have been rather limited [8, 81]. This represents a problem for modeling, because numerical results can presently only be compared to the measured distributions of the final particles and to data obtained from their final-configuration structural analysis.

Particle synthesis processes are usually initiated at large supersaturations leading to fast kinetics. Shape selection is then not an equilibrium growth, even though the actual shapes frequently display some of the crystallographic faces of the underlying material. One of the difficulties in modeling particle shapes numerically [30, 82, 83], has been to describe the establishment and structure of the crystalline (for nanoparticles) or compact (for colloids) stable "core" on top of which the growth of the structure then continues. Indeed, as described in Sect. 1.4, the core is formed during the early, fast growth, which is the least understood stage of the particle structure emergence in multi-cluster processes. On the other hand, at the later stages of the growth, the clusters are sufficiently dilute to treat each in isolation, capturing "monomer" matter from its environment.

In the recently reported [14] kinetic Monte Carlo study, the formed seed was taken as a compact particle, which was assumed to be approximately spherical and have already formed a well-defined internal crystalline order without any large, size-spanning defects. This initial core captures diffusing "atoms" which can only be attached in positions locally defined by the lattice symmetry of the structure. This approach is motivated by nanocrystal growth, but can also shed some light on the formation of those colloids the faces of which follow the underlying material symmetry. Indeed, it is possible that the main singlet nanocrystal in the seed dominates the emergence of the surface faces. A more likely scenario for such, typically nonspherical colloids, suggested by recent preliminary experimental observations [81] based on dark-field and bright-field TEM, for cubic-shaped polycrystalline neighborite (NaMgF$_3$) particles, is that the leading crystal structure is formed by the process of the outer shell (outer constituent singlets) of the particle recrystallizing to become effectively continuous, single-crystal, on top of a polycrystalline core. Finally, in protein crystallization [84, 85], the growth stage, from $\sim 10^2$ to $\sim 10^8$ molecules per crystal, after the initial small-cluster formation but before the onset of the really macroscopic growth modes, is also consistent with such a single-ordered-core approach.

While still requiring substantial numerical resources [14], consideration of particle growth in this regime has the flexibility of including the processes of atoms "rolling" on the surface and their detachment/reattachment, by using thermal-type, (free-)energy-barrier rules. The diffusional transport occurs in the three-dimensional (3D) space, without any lattice. However, the "registered" lattice-attachment rule starting from the seed, prevents the growing, moderate-size clusters from developing macroscopic (size-spanning) defects and ensures the maintenance of the crystal symmetry imposed by the core. We can then focus on the emergence of the surface and shape morphological features. The results [14] have allowed to identify three regimes of particle growth.

The first regime corresponds to slow growth, for instance, when the concentration of externally supplied, diffusionally transported building blocks, to be termed "atoms," is low. The time scale, τ_d, of motion—hopping to neighbor sites and detachment/reattachment, which all can be effectively viewed as surface diffusion—of the on-surface atoms is much smaller than the time scale of the addition of new monolayers, τ_{layer}. The shape of the growing cluster is then close to, but not identical with the Wulff-construction configuration [86–89]. The second regime is that of fast growth, $\tau_{layer} \ll \tau_d$, and corresponds to the formation of surface instabilities. The dynamics of the particle shape is then driven by the local diffusional fluxes, which are maximal near the highest-curvature "sharp" surface features. Small-scale random fluctuations of the surface are amplified, and the cluster assumes a shape of a random clump.

The third regime corresponds to $\tau_d \sim \tau_{layer}$. It was found [14] that in this *nonequilibrium* growth, particles can maintain an even-shaped form with well-defined faces corresponding to the underlying crystal structure imposed by the seed and atom-attachment rules. This numerically identified shape-selection is the key finding highlighted here, and it was only observed for a certain range of particle sizes. There is indeed the "persistence" effect alluded to earlier, but only for particles which have a certain measured amount of "atoms" in them. As particles grow beyond such sizes, growth modes involving bulges, dendritic structures, and other irregularities can be supported and are realized. This interesting new pattern of shape-selection in the nonequilibrium growth regime have been explored [14] for the simple cubic (SC), body-centered cubic (BCC), face-centered cubic (FCC), and hexagonal close-packed (HCP) crystal lattices. Several possible shapes were found for each symmetry, with their selection determined by the growth parameters.

Selective illustrative results [14] are presented here, for the 3D SC lattice. We first offer comments on the steady-state regime, followed by results for nonequilibrium growth. Based on preliminary studies, the seed was defined by lattice cells within a sphere with radius of 15 lattice constants. The seed atoms were fully immobile. The latter assumption was made to save computing time, based on preliminary observations that the seed rarely evolved much from its original structure. Thus, only the atoms later adsorbed at the growing structure, underwent the dynamical motion.

We first outline results for the "steady state" regime ($\tau_{layer} \gg \tau_d$) for the SC symmetry. Each atom attached to the cluster can have up to six bonds pointing to nearest neighbors, described by the set $\{\vec{e}_{int}\}$ of six lattice displacements of the type (100). The set of displacements/detachments for surface atoms, $\{\vec{e}_{mov}\}$, was defined in two different ways. Case A: $\{\vec{e}_{mov}\}_A$ included both the set $\{\vec{e}_{int}\}$ and also the 12 next-nearest-neighbor displacements of the type (110), of length $\sqrt{2}$; case B: $\{\vec{e}_{mov}\}_B = \{\vec{e}_{int}\}$. Thus, for variant B, the dynamics of the surface atoms is slower.

Figure 1.6 illustrates the resulting steady-state particle shape for the variant A of the SC simulation. We also show a schematic which illustrates the cluster shape formed with the type (100), (110), (111) lattice planes, which happen to be those dense-packed, low-index faces that dominate the low-temperature Wulff construction for the SC lattice [86–88]. The superficial similarity with this equilibrium shape is misleading. Indeed, our system's dynamical rules do not correspond to true thermal equilibration. The resulting shape is thus dependent on the dynamics. For example, Fig. 1.6 also shows the shape obtained for the same system but with variant B for the displacements/detachments (a slower surface dynamics). We conclude that the particle shape is not universal even in the steady state, in the sense expected [90] of many processes that yield macroscopic behavior in Statistical Mechanics: The microscopic details of the dynamical rules do matter. In practical terms this makes it unlikely that particle shapes can be predicted based on arguments such as minimization of some free-energy like quantity.

Fig. 1.6 The top particle exemplifies results of the steady state SC lattice simulations for variant *A* of the displacements/detachments for surface atoms. The shown particle shape is a cluster of 3.8×10^5 atoms, in a steady state with a dilute solution of freely diffusing atoms. (The *white lines* were added for guiding the eye.) Also shown is the projection of the cluster shape onto the *xy* plane, as well as the shape formed by lattice planes of the types (100), (110), (111) for equilibrium Wulff construction assuming that all these faces have equal interfacial free energies. The bottom particle exemplifies the steady state for variant *B* of the surface dynamics, and the projection of this shape onto the *xy* plane. The details and parameter values are given in [14]

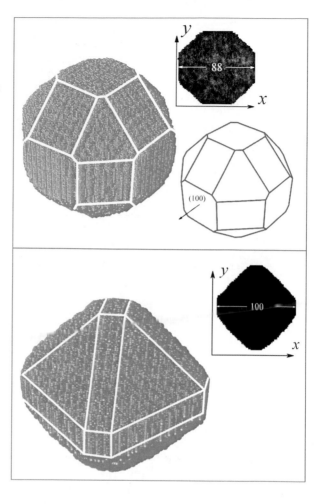

We emphasize that while well-defined particle shapes can be obtained in the steady-state regime within the present model, this does not offer a predictable and well-defined particle shape selection mechanisms in practical synthesis situations. This occurs because the model itself is not fully valid. Indeed, we observed [14] strong sensitivity of the results to the density of the monomer matter and its redistribution by transport to and from the surrounding medium. Therefore, the isolated-cluster assumption breaks down. Other clusters (particles) compete for the "atoms" (solutes) in the dilute solution, and as a result additional growth mechanisms [29] that involve exchange of matter between clusters (Ostwald ripening) cannot really be ignored (as is done in our approximate model).

The main difference between the nonequilibrium and steady-state regimes is that the former corresponds to a fast growth process fully dominated by capture of singlet matter from a dilute solution. Other processes, such as those involving exchange of matter with other clusters, or the on-surface diffusion, are slower.

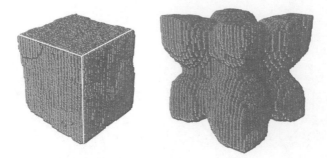

Fig. 1.7 Examples of nonequilibrium SC lattice cluster shapes (for the kinetics of type A). The cubic shape on the left, emerges already at short times, $t \approx 3 \times 10^5$, persisting for growing clusters, here shown for $t = 2.5 \times 10^6$, containing 4.5×10^5 atoms, with the cube edge length 77. (The *white lines* were added for guiding the eye.) The larger cluster on the right, of spanning size 125, was grown with different parameter values, and is here shown at time $t = 5.2 \times 10^6$, containing 1.8×10^6 atoms. The parameters of the simulations, details of the kinetics, and definition of the time units, are given in [14]

Fig. 1.8 Examples of shapes obtained in nonequilibrium growth for BCC, FCC and HCP lattice symmetries. (The *white lines* were added for guiding the eye)

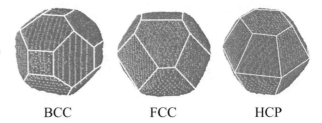

BCC   FCC   HCP

For nonequilibrium growth, the cluster shapes can be quite different. For example, for the SC lattice, a cubic shape, illustrated in Fig. 1.7, was only found in the nonequilibrium regime [14] with the kinetic transition rates, detailed in [14], for atom intake vs. surface dynamics corresponding to $\tau_d \sim \tau_{layer}$. Other, less symmetrical shapes have also been found, as illustrated in Fig. 1.7. Regular shapes obtained for nonequilibrium growth with lattice symmetries other than SC, are exemplified in Fig. 1.8. Several shapes obtained, for different symmetries, are catalogued in [14], as are examples of unstable growth and other interesting growth modes, further exploration of which has been limited by the demands on numerical resources required for simulating large particles.

The present model captures the key ingredients required for well-defined shape selection in the nonequilibrium growth regime. It avoids formation of macroscopically persisting defect structures. The dynamics of the growing particle's faces is then not controlled by extended defects—which is a well known mechanism [85, 89] that can determine growth modes in nonequilibrium crystallization. The evolving surface overwhelms small imperfections, at least as long as the particles remain not too large, even for colloids that are formed from aggregating nanocrystalline subunits. The growing cluster faces result in well-defined particle shapes and proportions. In fact, the densest-packed, low-index crystal-symmetry faces, which dominate the equilibrium crystal shapes, also emerge in this nonequilibrium growth, but actual shapes, planar faces and other surfaces present, and their proportions are not the same as in equilibrium.

1.6   Conclusion

We reviewed approaches to modeling particle size and shape selection in colloid and nanoparticles synthesis. The reviewed theories require numerical simulations to obtain results to compare with experiments and gain qualitative insight into the model predictions. The known approaches are presently limited and in most cases semi-quantitative. Furthermore, most of the experimental data are limited to observation of the final products, whereas results for time dependent, kinetic processes, as well as detailed morphological data are needed to advance our understanding of the fine-particle design, which would benefit diverse applications. Thus, notwithstanding the recent successes, we consider the status of the theoretical understanding of the kinetics of fine-particle synthesis, and the theory-experiment synergy, as preliminary. The field is widely open to future research.

Acknowledgments The author thanks his colleagues D. Goia, V. Gorshkov, I. Halaciuga, S. Libert, E. Matijević, D. Mozyrsky, J. Park, D. Robb, I. Sevonkaev, and O. Zavalov for rewarding scientific interactions and collaboration, and acknowledges funding by the US ARO under grant W911NF-05-1-0339.

References

1. V. Privman, D. V. Goia, J. Park and E. Matijević, J. Colloid Interface Sci. **213** (1999) 36-45, online at http://dx.doi.org/10.1006/jcis.1999.6106
2. D. Mozyrsky and V. Privman, J. Chem. Phys. **110** (1999) 9254-9258, online at http://dx.doi.org/10.1063/1.478849
3. J. Park and V. Privman, Recent Res. Devel. Stat. Phys. **1** (2000) 1-17, online as e-print at http://arxiv.org/abs/cond-mat/0003029
4. V. Privman and J. Park, in *Proc. Fourth International Workshop on Materials Processing at High Gravity*, edited by L. L. Regel and W. R. Wilcox (Kluwer, New York, 2001), pages 141-147, online as e-print at http://arxiv.org/abs/cond-mat/0008097
5. J. Park, V. Privman and E. Matijević, J. Phys. Chem. **B 105** (2001) 11630-11635, online at http://dx.doi.org/10.1021/jp011306a
6. V. Privman, Mat. Res. Soc. Symp. Proc. **703** (2002) Article T3.3, pages 577-585, online at http://www.mrs.org/s_mrs/bin.asp?CID=2508&DID=58023
7. S. Libert, V. Gorshkov, V. Privman, D. Goia and E. Matijević, Adv. Colloid Interface Sci. **100-102** (2003) 169-183, online at http://dx.doi.org/10.1016/S0001-8686(02)00056-8
8. S. Libert, V. Gorshkov, D. Goia, E. Matijević and V. Privman, Langmuir **19** (2003) 10679-10683, online at http://dx.doi.org/10.1021/la0302044
9. D. T. Robb and V. Privman, Langmuir **24** (2008) 26-35, online at http://dx.doi.org/10.1021/la702097g
10. V. Privman, Ann. N.Y. Acad. Sci. **1161** (2009) 508-525, online at http://dx.doi.org/10.1111/j.1749-6632.2008.04323.x

11. V. Privman, J. Optoelectron. Adv. Mater. **10** (2008) 2827-2839, online at http://joam.inoe.ro/download.php?idu = 1708
12. D. T. Robb, I. Halaciuga, V. Privman and D. V. Goia, J. Chem. Phys. **129** (2008) Article 184705, 11 pages, online at http://dx.doi.org/10.1063/1.3009625
13. I. Halaciuga, D. T. Robb, V. Privman and D. V. Goia, in *Proc. Conf. ICQNM 2009*, edited by D. Avis, C. Kollmitzer and V. Privman (IEEE Comp. Soc. Conf. Publ. Serv., Los Alamitos, California, 2009), pages 141-146, online at http://dx.doi.org/10.1109/ICQNM.2009.7
14. V. Gorshkov, A. Zavalov and V. Privman, Langmuir **25** (2009) 7940-7953, online at http://dx.doi.org/10.1021/la900613p
15. I. Sevonkaev and V. Privman, *Shape Selection in Synthesis of Colloids and Nanoparticles*, World J. Eng. **6**, Article P909, 2 pages (2009), online at http://wjoe.hebeu.edu.cn/ICCE-17%20proceedings%20Hawaii%20USA/Sevonkaev,%20Igor%20(Clarkson%20U.,%20Potsdam,%20NY)%20%20909.pdf
16. V. Gorshkov and V. Privman, Physica **E 43** (2010) 1-12, online at http://dx.doi.org/10.1016/j.physe.2010.07.006
17. V. Gorshkov, O. Zavalov, P. B. Atanassov and V. Privman, Langmuir **27** (2011) 8554-8561, online at http://dx.doi.org/10.1021/la103113f
18. E. Matijević, Chem. Mater. **5** (1993) 412-426, online at http://dx.doi.org/10.1021/cm00028a004
19. E. Matijević and D. Goia, Croat. Chem. Acta **80** (2007) 485-491, online at http://hrcak.srce.hr/file/29077
20. *Nanoparticles and Nanostructured Films: Preparation, Characterization and Applications*, edited by J. H. Fendler (Wiley-VCH, Weinheim, 1998), online at http://dx.doi.org/10.1002/9783527612079
21. J. Eastoe, M. J. Hollamby and L. Hudson, Adv. Colloid Interface Sci. **128-130** (2006) 5-15, online at http://dx.doi.org/10.1016/j.cis.2006.11.009
22. Y. Xia, Y. Xiong, B. Lim and S. E. Skrabalak, Angew. Chem. Int. Ed. **48** (2009) 60-103, online at http://dx.doi.org/10.1002/anie.200802248
23. G. Schmid, Chem. Rev. **92** (1992) 1709-1727, online at http://dx.doi.org/10.1021/cr00016a002
24. T. Teranishi, M. Hosoe, T. Tanaka and M. Miyake, J. Phys. Chem. **B 103** (1999) 3818-3827, online at http://dx.doi.org/10.1021/jp983478m
25. V. K. LaMer and R. H. Dinegar, J. Amer. Chem. Soc. **72** (1950) 4847-4854, online at http://dx.doi.org/10.1021/ja01167a001
26. V. K. LaMer, Ind. Eng. Chem. **44** (1952) 1270-1277, online at http://dx.doi.org/10.1021/ie50510a027
27. S. F. Chernov, Y. V. Fedorov and V. N. Zakharov, J. Phys. Chem. Solids **54** (1993) 963-966, online at http://dx.doi.org/10.1016/0022-3697(93)90225-G
28. G. H. Weiss, J. Statist. Phys. **42** (1986) 3-36, online at http://dx.doi.org/10.1007/BF01010838
29. P. W. Voorhees, J. Statist. Phys. **38** (1985) 231-252, online at http://dx.doi.org/10.1007/BF01017860
30. *Solids Far from Equilibrium*, edited by C. Godreche (Cambridge Univ. Press, 1992), online at http://dx.doi.org/10.2277/052141170X
31. F. Family and T. Vicsek, *Dynamics of Fractal Surfaces* (World Scientific, Singapore, 1991), online at http://www.worldscibooks.com/chaos/1452.html
32. I. Sevonkaev, private communication (2008)
33. L. J. Lewis, P. Jensen and J.-L. Barrat, Phys. Rev. **B 56** (1997) 2248-2257, online at http://dx.doi.org/10.1103/PhysRevB.56.2248
34. F. Baletto and R. Ferrando, Rev. Mod. Phys. **77** (2005) 371-423, see Section 5c, online at http://dx.doi.org/10.1103/RevModPhys.77.371
35. K. F. Kelton and A. L. Greer, Phys. Rev. **B 38** (1988) 10089-10092, online at http://dx.doi.org/10.1103/PhysRevB.38.10089
36. D. V. Goia, private communication (1999)

37. J. D. Shore, D. Perchak and Y. Shnidman, J. Chem. Phys. **113** (2000) 6276-6284, online at http://dx.doi.org/10.1063/1.1308517
38. K. F. Kelton, A. L. Greer and C. V. Thompson, J. Chem. Phys. **79** (1983) 6261-6276, online at http://dx.doi.org/10.1063/1.445731
39. F.-P. Ludwig and J. Schmelzer, J. Colloid Interface Sci. **181** (1996) 503-510, online at http://dx.doi.org/10.1006/jcis.1996.0407
40. E. Matijević and P. Scheiner, J. Colloid Interface Sci. **63** (1978) 509-524, online at http://dx.doi.org/10.1016/S0021-9797(78)80011-3
41. E. Matijević and D. Murphy Wilhelmy, J. Colloid Interface Sci. **86**, 476-484 (1982), online at http://dx.doi.org/10.1016/0021-9797(82)90093-5
42. D. Murphy Wilhelmy and E. Matijević, J. Chem. Soc. Faraday Trans. I **80** (1984) 563-570, online at http://dx.doi.org/10.1039/F19848000563
43. E. Matijević, Ann. Rev. Mater. Sci. **15** (1985) 483-516, online at http://dx.doi.org/10.1146/annurev.ms.15.080185.002411
44. M. Haruta and B. Delmon, J. Chim. Phys. **83** (1986) 859-868
45. T. Sugimoto, Adv. Colloid Interface Sci. **28** (1987) 65-108, online at http://dx.doi.org/10.1016/0001-8686(87)80009-X
46. L. H. Edelson and A. M. Glaeser, J. Am. Ceram. Soc. **71** (1988) 225-235, online at http://dx.doi.org/10.1111/j.1151-2916.1988.tb05852.x
47. U. P. Hsu, L. Rönnquist and E. Matijević, Langmuir **4** (1988) 31-37, online at http://dx.doi.org/10.1021/la00079a005
48. M. Ocaña and E. Matijević, J. Mater. Res. **5** (1990) 1083-1091, online at http://dx.doi.org/10.1557/JMR.1990.1083
49. M. P. Morales, T. Gonzáles-Carreno and C. J. Serna, J. Mater. Res. **7** (1992) 2538-2545, online at http://dx.doi.org/10.1557/JMR.1992.2538
50. T. Sugimoto, J. Colloid Interface Sci. **150** (1992) 208-225, online at http://dx.doi.org/10.1016/0021-9797(92)90282-Q
51. J. K. Bailey, C. J. Brinker and M. L. Mecartney, J. Colloid Interface Sci. **157** (1993) 1-13, online at http://dx.doi.org/10.1006/jcis.1993.1150
52. E. Matijević, Langmuir **10** (1994) 8-16, online at http://dx.doi.org/10.1021/la00013a003
53. M. Ocaña, M. P. Morales and C. J. Serna, J. Colloid Interface Sci. **171** (1995) 85-91, online at http://dx.doi.org/10.1006/jcis.1995.1153
54. M. Ocaña, C. J. Serna and E. Matijević, Colloid Polymer Sci. **273** (1995) 681-686, online at http://dx.doi.org/10.1007/BF00652261
55. S.-H. Lee, Y.-S. Her and E. Matijević, J. Colloid Interface Sci. **186** (1997) 193-202, online at http://dx.doi.org/10.1006/jcis.1996.4638
56. C. Goia and E. Matijević, J. Colloid Interface Sci. **206** (1998) 583-591, online at http://dx.doi.org/10.1006/jcis.1998.5730
57. D. V. Goia and E. Matijević, Colloids Surf. A **146** (1999) 139-152, online at http://dx.doi.org/10.1016/S0927-7757(98)00790-0
58. Z. Crnjak Orel, E. Matijević and D. V. Goia, Colloid Polymer Sci. **281** (2003) 754-759, online at http://dx.doi.org/10.1007/s00396-002-0841-6
59. I. Sondi, D. V. Goia and E. Matijević, J. Colloid Interface Sci. **260** (2003) 75-81, online at http://dx.doi.org/10.1016/S0021-9797(02)00205-9
60. L. Bromberg, J. Rashba-Step and T. Scott, Biophys. J. **89** (2005) 3424-3433, online at http://dx.doi.org/10.1529/biophysj.105.062802
61. M. Jitianu and D. V. Goia, J. Colloid Interface Sci. **309** (2007) 78-85, online at http://dx.doi.org/10.1016/j.jcis.2006.12.020
62. I. Sevonkaev, D. V. Goia and E. Matijević, J. Colloid Interface Sci. **317** (2008) 130-136, online at http://dx.doi.org/10.1016/j.jcis.2007.09.036
63. I. Halaciuga and D. V. Goia, J. Mater. Res. **23** (2008) 1776-17874, online at http://dx.doi.org/10.1557/JMR.2008.0219
64. J. A. Dirksen, S. Benjelloun and T. A. Ring, Colloid Polymer Sci. **268** (1990) 864-876, online at http://dx.doi.org/10.1007/BF01410964

65. P. G. J. van Dongen and M. H. Ernst, J. Statist. Phys. **37** (1984) 301-324, online at http://dx.doi.org/10.1007/BF01011836

66. N. V. Brilliantov and P. L. Krapivsky, J. Phys. A **24** (1991) 4787-4803, online at http://dx.doi.org/10.1088/0305-4470/24/20/014

67. D. W. Schaefer, J. E. Martin, P. Wiltzius and D. S. Cannell, Phys. Rev. Lett. **52** (1984) 2371-2374, online at http://dx.doi.org/10.1103/PhysRevLett.52.2371

68. R. M. German, *Particle Packing Characteristics* (Metal Powder Industries Federation, Princeton, 1989).

69. V. Gorshkov, S. Libert and V. Privman, unpublished (2003).

70. T. Sugimoto, *Monodispersed Particles* (Elsevier: Amsterdam, 2001), online at http://www.sciencedirect.com/science/book/9780444895691

71. *Fine Particles, Synthesis, Characterization, and Mechanisms of Growth*, edited by T. Sugimoto (Marcel Dekker: New York, 2000)

72. A. P. Alivisatos, Science **271** (1996) 933-937, online at http://dx.doi.org/10.1126/science.271.5251.933

73. V. Uskokovic and E. Matijević, J. Colloid Interface Sci. **315** (2007) 500-511, online at http://dx.doi.org/10.1016/j.jcis.2007.07.010

74. P. S. Nair, K. P. Fritz and G. D. Scholes, Small **3**, 481-487 (2007), online at http://dx.doi.org/10.1002/smll.200600558

75. C. J. Murphy, T. K. Sau, A. M. Gole, C. J. Orendorff, J. X. Gao, L. Gou, S. E. Hunyadi and T. Li, J. Phys. Chem. **B 109** (2005) 13857-13870, online at http://dx.doi.org/10.1021/jp0516846

76. D. V. Goia, J. Mater. Chem. **14** (2004) 451-458, online at http://dx.doi.org/10.1039/b311076a

77. A. S. Barnard, H. F. Xu, X. C. Li, N. Pradhan and X. G. Peng, Nanotechnology **17**, 5707 (2006), online at http://dx.doi.org/10.1088/0957-4484/17/22/029

78. Q. F. Lu, H. B. Zeng, Z. Y. Wang, X. L. Cao and L. D. Zhang, Nanotechnology **17** (2006) 2098-2104, online at http://dx.doi.org/10.1088/0957-4484/17/9/004

79. N. Pradhan, H. F. Xu and X. G. Peng, Nano Lett. **6**, 720-724 (2006), online at http://dx.doi.org/10.1021/nl052497m

80. F. Huang, H. Z. Zhang and J. F. Banfield, J. Phys. Chem. **B 107** (2003) 10470-10475, online at http://dx.doi.org/10.1021/jp035518e

81. I. Sevonkaev, *Size and Shape of Uniform Particles Precipitated in Homogeneous Solutions*, Ph.D. Thesis (Clarkson University, Potsdam, 2009), online at http://gradworks.umi.com/34/01/3401449.html

82. F. Baletto and R. Ferrando, Rev. Mod. Phys. **77** (2005) 371-423, online at http://dx.doi.org/10.1103/RevModPhys.77.371

83. D. J. Wales, *Energy Landscapes. With Applications to Clusters, Biomolecules and Glasses* (Cambridge University Press, Cambridge, 2004), online at http://dx.doi.org/10.2277/0521814154

84. C. N. Nanev, Crystal Growth Design **7** (2007) 1533-1540, online at http://dx.doi.org/10.1021/cg0780807

85. D. Winn and M. F. Doherty, AIChE J. **46** (2000) 1348-1367, online at http://dx.doi.org/10.1002/aic.690460709

86. C. Herring, in *Structure and Properties of Solid Surfaces*, edited by R. Gomer and C. S. Smith (University of Chicago Press, Chicago, 1953), Chapter 1, pages 5-81.

87. J. K. Mackenzie, A. J. W. Moore and J. F.Nicholas, J. Phys. Chem. Solids **23** (1962) 185-196, online at http://dx.doi.org/10.1016/0022-3697(62)90001-X

88. J. A. Venables, *Introduction to Surface and Thin Film Processes* (Cambridge University Press, Cambridge, 2000), online at http://dx.doi.org/10.1017/CBO9780511755651

89. J. E. Taylor, J. W. Cahn and C. A. Handwerker, Acta Metall. Mater. **40** (1992) 1443-1474, online at http://dx.doi.org/10.1016/0956-7151(92)90090-2

90. V. Privman, in *Encyclopedia of Applied Physics* **23**, edited by J. L. Trigg (American Institute of Physics, New York, 1998), pages 31-46, online at http://dx.doi.org/10.1002/3527600434.eap553

Chapter 2
Preparation of Uniform Drug Particles

Egon Matijević

Abstract A number of commercially available drugs, used in different medical and pharmaceutical applications, have been prepared as colloid dispersions consisting of particles uniform in shape and size. In all cases the same compounds could be obtained in different morphologies, with each particle consisting of smaller sub-units. Furthermore, it is shown that the drugs molecules can be attached in uniform layers onto inert cores of desired shape and size. Finally, the drug particles could be coated with inorganic layers (e.g. alumina or silica) of different thickness, while keeping the solids fully dispersed.

Keywords Alumina • Barium Naproxenate • Budesonide • Calcium Naproxenate • Cyclosporine • Danazol • Diatrizoic Acid Ethyl Ester • Ketoprofen • Loratadine • Naproxen • Silica

2.1 Introduction

An ideal drug delivery system is expected to transport a selected medication through a specific path in a live body (human or animal) to reach an organ and react there at a desired rate. Under real conditions it is very difficult to achieve such a perfect result with the active substance, but there are continuous efforts to improve every single step of the process. It is now generally recognized that the physical state of the drug is one of many properties needing to be controlled in order to achieve the required effects in a reproducible manner.

The efficiency of the medication depends not only on its chemical composition, but also on the method by which the drug is distributed in the body. There are several general avenues of approach to administer an active substance (drug) to treat an ailment or to alleviate pain. If the drug is liquid, one can inhale the vapor, or if it is in a solution it can be delivered as a spray of finely dispersed droplets (aerosols). However, the liquids or solutions containing the active compound are most commonly taken orally or injected into veins or muscles at controlled concentrations, pH, ionic strength, etc.

E. Matijević (✉)
Department of Chemistry & Biomolecular Science, Clarkson University, 8 Clarkson Avenue, CAMP, Box 5814, Potsdam, NY 13699, USA
e-mail: ematijevic@clarkson.edu, matiegon@clarkson.edu

E. Matijević (ed.), *Fine Particles in Medicine and Pharmacy,*
DOI 10.1007/978-1-4614-0379-1_2, © Springer Science+Business Media, LLC 2012

The prevailing method for taking medication is in the solid state, as pills, designed to dispense the active ingredient at an optimum rate. While this last method is widely employed, the complexity of the proper formulation of such delivery units is often not fully appreciated. The drugs differ in their solubility, toxicity, hydrophobicity or hydrophilicity, molecular shape, and numerous other properties, affecting their desired effects. Furthermore, the active compound is usually not the only component of the pill, but incorporated in an inert substrate or coated. The function of the latter is to protect the drug from the effects of the environment and to control the rate of its release.

The use of ultrafine particles in drug delivery has shown certain advantages, because they more readily dissolve and are easily absorbed due to their small size and, consequently, high specific surface area. Such particles also better resist the hydrodynamic shear forces after adhesion to various vessel walls in the body. The present methods for drug delivery focus either on attaching the active molecules onto inert colloidal particles, or on their incorporation into ultrafine particle or polymer carriers, which are then used to produce pills [1–4]. Such systems may beneficially affect the administration of the active compounds, but they have several disadvantages. Adsorbing a drug onto colloidal particles, such as latexes, is prone to phagocytosis, which is the entrapment of foreign materials by cellular species. The accumulation of particles in various parts of the body may cause infections or other health problems. The incorporation of nanoparticles or microspheres requires non-toxic, biodegradable polymers that will be inert when decomposed into the initial monomeric species. The binding of drugs within these particles can also cause problems with their release. The preparation procedures for nanoparticles or microspheres, such as emulsion polymerization, use special solvents or reagents, which may remain in the resulting solids as toxic impurities. Finally, described methods can increase the required drug dosages since the composite particles consist mainly of the inert carrier compounds. One major improvement could be achieved by the preparation of ultrafine pure solid pharmaceutical compounds, which would maximize the usage of the drug dosage, and possibly lower the latter, without introduction of foreign materials into the body.

Some of the specific aspects and problems of the drug systems in the solid state were addressed in an article in C&EN (*Chemical & Engineering News*, American Chemical Society **2003**, 81, 32) under the title "*The Right Stuff. From research and development to the clinic, getting drug crystals right is full of pitfalls*" [5]. One of the aspects discussed in that review deals with various polymorphic forms of the same active substance, which exhibit different "bioavailability", solubility, dissolution rate, physical and chemical stability, melting point, filterability, and some other properties. For example, without giving any details, it was cited that in some toxicological animal tests the same drug was beneficial when administered in the crystalline state, but deadly when used amorphous [5].

The polymorphs refer to different arrangements of the same molecules when in solid state, which can be identified by the X-ray analysis [6]. They may appear as particles of various shapes, influencing the effectiveness of the drug function. Furthermore, the stability of different polymorphs may also vary with time, temperature, etc., so that their nature and functionality may change on storage, until a final stable form is established, which then need not be of the highest efficiency. The existence of multiple polymorphic crystal forms and amorphous solids in drug materials can, therefore, significantly complicate the formulation development. Obviously, it is desirable to incorporate the drug of the most thermodynamically stable solid-state form into the dosage so that any transformational processes are unlikely to occur during the shelf-life of the product [7].

Drug particle morphologies, which are governed by the internal crystal structures, can also be of importance in the formulation development and manufacturing. Drugs which crystallize into plates or needles may have poor flow properties, which present challenges in their blending and compaction. For example, the morphology of drug particles intended for pulmonary administration as dry powders can significantly impact the product performance [8].

The crystal form requires a careful consideration in deciding dosages of poorly water-soluble drugs, because their oral bioavailability is often dissolution rate limited. Since different crystal forms of the same drug influence their thermodynamic solubilities and deliveries, it is essential for the solid-state properties of these compounds to be carefully controlled so that their pharmacokinetic performance does not change from batch to batch or during the shelf life of the product.

The application of fine particle technology to dosage can play a major role in improving the drug product performance. For example, the oral bioavailability of poorly water-soluble drugs can be enhanced by increasing the ratio of the surface area of the substance to its mass, which increases dissolution rates [9]. This phenomenon is especially true of compounds that have a narrow absorption window in the gastrointestinal tract.

Some major problems encountered with regard to the particle size of the active compounds were specifically addressed in several patents [10–14]. For example, certain polymer forms may enhance the efficiency of treatments, such as in killing the intracellular micro-organisms. While some of these patents claim uniformity of drug particles, no micrographs are given to view their shape and to assess the modal size and size distribution.

The conventional method used in the pharmaceutical industry to produce finely dispersed drugs consists in reducing the particle size of bulk materials through dry or wet milling in the presence of surfactants [15–17]. The particle size distribution of milled solids tends to be rather broad and batch to batch variations can influence the performance of the final product. Needless to say, it is impossible to control the particle shape by such treatment of solid drugs.

There is no doubt that much work is being done in the pharmaceutical industry on the behavior of drugs as dependent on their internal and external states, when delivered as solids. However, little published information is available on the subject, especially with regard to physical properties of active substances. It is obvious that the first step to evaluate these effects is to have well defined uniform drug particles in terms of their size, shape, and structure.

A significant number of studies have been reported on "monodispersed" inorganic materials of simple and composite natures, especially when prepared by precipitation in homogeneous solutions, as described in two volumes by Sugimoto [18, 19] and in some review articles [20–22]. As one would expect, the chemical and physical properties of the resulting particles depend in a sensitive way on the experimental conditions (concentration of reactants, pH, temperature, etc). For example, in the precipitation of a metal oxide in aqueous solutions the cation can be present in a variety of hydrozied complexes, which may also include the accompanying anion. All these systems are, as a rule, unstable and change with the usual parameters, but also with some other ones, such as the age of the reactant media, exposure to light, etc. The reactant solutions often contain additives (different accompanying ions, mixed solvents, stabilizers, etc.), all affecting the complexity of the entire system and of the final product.

In contrast, one may expect most physical parameters to affect less the precipitation of pure organic compounds, especially those which do not dissociate. It would also be reasonable to assume that the shape of the resulting particles may be influenced by steric aspects of the constituent molecules. However, studies reviewed in this article will show that solids of the same pure molecular organic compounds (in this case drugs) can be obtained as uniform particles of different morphologies by the precipitation in solutions by varying some physical parameters, or chemically inert additives. The reasons for this behavior are still not fully explained.

Thus, it should be of no surprise that manufacturers of pharmaceutical products have a strong interest in developing a better understanding of the effects of physical properties of given drugs on their therapeutic and other applications. Obviously, to achieve this goal it is necessary to produce active substances as uniform particles in terms of their size, shape, internal structure, etc, so that their medical efficiencies can be evaluated as a function of the physical parameters. Surprisingly, precious few articles on dispersions of uniform drug particles are found in the published communications.

2.2   Preparation Methods for Uniform Drug Particles

While some effects of finely dispersed matter in terms of the particle size, shape, and structure were reported since ancient times, extensive studies of such systems began in the last century. Obviously, to evaluate optical, magnetic, adsorptive, and other properties of dispersions as a function of the physical nature of constituent particles, it was necessary to produce well defined matter of desired morphological characteristics [18–22]. However, except for polymer latexes, only a few dispersions of uniform organic particles have been described in the literature [23–31].

There are several procedures available to obtain "monodispersed" particles in the micrometer or smaller size range. The simplest would be the aerosol process in which uniform droplets of a liquid or a solution are first produced in a convenient generator [32]. Each droplet can function as an individual "reactor" for the production of a single particle, either by physical or chemical means. When using solutions of a pure product, the solvent needs to evaporate, yielding the solid (e.g. drug) in a finely dispersed state, as was demonstrated on the example of NaCl [33]. Alternatively, the droplet can consist of an unsaturated compound, which on exposure to a reacting vapor causes the saturation of the bonds, resulting in an active solid particle of spherical shape, as was done by exposing octadecene droplets to bromine vapor [34]. The aerosol process offers some significant advantages over other techniques, because it can be carried out without any byproducts in a highly clean environment. The shortcoming is that there have been no efforts to design efficient generators for the production of uniform droplets of a desired size in large quantities.

From a practical point of view reactions in solutions are preferable avenues to obtain uniform particles. There are several different ways to achieve such dispersions by precipitation. Some cases described in this review, such as the controlled double jet precipitation (CDJP) technique, were used to generate uniform drug particles. This technique was first developed to obtain photographic materials [35], but was later successfully applied in the production of various well-defined inorganic particles [36–39]. The solutions of reactants are simultaneously injected at a controlled rate and temperature into the suspending liquid (in most cases water), which may also contain additives (e.g. stabilizers). For better control of the process, agitation can be produced by suspending the CDJP generator in an ultrasonic bath.

Several other avenues of approach were taken to prepare drug particles of different compositions and shapes based on precipitation. The employed procedures given in more detail are as follows:

1. Forming a sparingly soluble salt of an ionizable drug in an aqueous solution.
2. Precipitating the drug in a solution by changing the pH.
3. Adding a miscible non-solvent to the drug in a solvent (solvent shifting).
4. Vaporizing the more volatile solvent of the drug in a solution of two miscible liquids, one being a nonsolvent.

In all cases additives (surfactants, polymers) may be admixed to control the size, shape, and structure of the resulting dispersed solids. A number of uniform drug particles of different chemical compositions, structures, morphologies, and modal sizes, obtained by the described procedures are reviewed in sections that follow.

2.3   Precipitation of Sparingly Soluble Salts

2.3.1   Naproxen [40, 41]

Naproxen ($CH_3OC_{10}H_6CH(CH_3)CO_2H$) is a non-steroidal anti-inflammatory drug present in several over-the-counter products [42]. The compound is a carboxylic acid of the following composition:

Naproxen

and the crystal structure reported by Kim et al. [43].

In principle, finely dispersed particles of an ionizable drug can be obtained by precipitating a sparingly soluble salt of its active part. If the latter is an anion, the salt should form with the cation of an added electrolyte. In the specific case in which the carboxylic group is involved, just changing the pH may also cause the precipitation of the drug.

Both these approaches were employed in the case of naproxen. The drug can be obtained in solid state by lowering the pH of the compound in aqueous solutions [40], or by forming a metal carboxylate [41]. Clinical studies showed that the derived metal salts of naproxen possess the same anti-inflammatory activity as the free carboxylic acid of naproxen [44]. Particles of sparingly soluble salts, as well as of insoluble complexes of naproxen with rare earth metals, have been reported in the literature, but no attempt was made to control their size and morphology. Here it is shown that the double jet precipitation (CDJP) technique can be used to produce uniform micrometer-size calcium and barium naproxenate particles.

2.3.1.1   The pH Effect

To obtain uniform particles of naproxen of different morphologies the solid carboxylate powder was dissolved in an aqueous solution at pH 12 and subsequently acidified with HCl in the presence of the stabilizer polyvinylpyrrolidone (PVP, MW 10,000) either at room temperature or at 60°C. At pH>6, only small amounts of solids were observed.

Figure 2.1 displays naproxen formed at pH 6, which initially consists of relatively uniform spheroidal particles of ~ 10 μm in diameter (Fig. 2.1a). Under this condition less than 10% of the dissolved naproxen is precipitated. When the acid content reaches 0.8–1.25 mol of H^+ per mole of naproxen, resulting in the pH ~ 4–5, the particles are irregular in shape of ~ 1–4 μm in size (Fig. 2.1b). A closer examination of the two micrographs indicates that both kinds of these solids are aggregates of smaller subunits, which are clearly visualized in Fig. 2.1c. On heating these solutions for up to 12 h at 60°C or ageing them for up to 2 days, the somewhat irregular precipitates disintegrate into smaller, but rather uniform, rod-like particles ~ 1 μm long and ~ 0.3–0.4 μm wide, displayed in Fig. 2.1d. Under these conditions, nearly all of the originally dissolved drug is precipitated. The XRD patterns of the powders, before and after the pertization of naproxen into the smaller particles, show these solids to be crystalline.

It is important to emphasize the composite nature of the described naproxen particles. Indeed, much evidence has become available showing that many, if not most, synthesized uniform colloids consist of smaller subunits [45–47]. A model was developed to explain how a large number of small size precursors can yield uniform colloidal spheres [48, 49].

Fig. 2.1 Scanning electron micrographs (SEM) of naproxen particles. **a** Precipitated in a 0.02 mol dm^{-3} aqueous naproxen solution containing 1 wt% of (PVP), and acidified to pH 6 with HCI. **b** Naproxen particles produced at the same conditions as in (**a**) except at pH 4. **c** Higher magnification of particles in (**b**). **d** Particles obtained after peptization of the initial dispersion shown in (**b**). [40]

2.3.2   Naproxenate Salts [41]

Naproxenate salts with alkaline earth cations (Ca^{2+} and Ba^{2+}) were precipitated by mixing aqueous solutions of sodium naproxenate with solutions of calcium or barium salts, either in the absence or in the presence of additives, including surfactants. The latter were chosen not to form salts with the same cations, but to affect the morphology of the resulting naproxenate salt particles.

2.3.2.1   Calcium Naproxenate [41]

Calcium naproxenate particles of different shapes were prepared with the CDJP process by reacting solutions of sodium naproxenate with solutions of calcium acetate in the absence or in the presence of different stabilizers. In all cases anisometric particles were obtained, the shape and uniformity of which depended on the additives and the method of agitation. Figure 2.2 displays the SEM of Ca-naproxenate particles precipitated at room temperature in the absence (Fig. 2.2a) and in the presence of arabic gum (Fig. 2.2b) or Igepal (Fig. 2.2c) [41]. The sample in Fig. 2.2d is of the same composition as in Fig. 2.2c, except the reaction generator was suspended in an ultrasonic bath. Experiments using some other surfactants (e.g. Tween, Triton-x) yielded irregular particles of less well defined shapes.

2.3.2.2   Barium Naproxenate [41]

Ba-naproxenate precipitated in the CDJP reactor by mixing solutions of sodium naproxenate and $BaCl_2$, yielded, as a rule, rather ill defined dispersions. Somewhat improved particles of different shapes were obtained using the same method in the presence of certain additives (Triton X-100 and Tween 80), as illustrated in Fig. 2.3.

It is noteworthy that the so precipitated salts have the same FTIR spectra as the Na-naproxenate as shown in Fig. 2.4.

The admixing of polyvinylalcohol (PVA) yielded rather uniform crystalline spheres of high specific surface area, having complex internal structure, shown in Fig. 2.5. In this case Ba-acetate was employed instead of $BaCl_2$.

2.4   Miscible Solvent/Nonsolvent Systems

As indicated earlier there are, in general, two approaches to produce fine particles by using miscible liquids, one being a solvent and the other a nonsolvent for the same drug. In the first approach a nonsolvent is added to the solution of the active

Fig. 2.2 Scanning electron micrographs (SEM) of calcium naproxenate prepared by adding 25 cm³ of both 0.3 mol dm⁻³ sodium naproxenate and calcium acetate solutions into the CDJP reactor containing 25 cm³ of a sodium hydroxide solution at pH ~ 8.8 in the absence of an additive (**a**) and in the presence of 6 wt% Arabic gum (**b**), or 2 wt% Igepal CA-897 (**c**). The same system as in (**c**) with the CDJP reactor immersed in an ultrasonic bath (**d**). [41]

Fig. 2.3 Scanning electron micrographs (SEM) of barium naproxenate particles prepared by admixing 25 cm³ each of 0.3 mol dm⁻³ naproxen and $BaCl_2$ solutions into 25 cm³ of sodium hydroxide solution at pH 10.5, in the CDJP reactor in the presence of 1 wt% Triton X-100 (**a**), and 1% Tween-80 (**b**). [41]

Fig. 2.4 FTIR spectra KBr pellets of (**a**) barium naproxenate prepared in the absence of a surfactant, and (**b**) of a commercial sample of Na⁻ Nanaproxenate. [41]

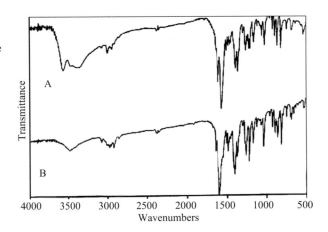

substance until the solubility limit is exceeded and the solid separates. In the second method the drug is dissolved in a mixture of the nonsolvent and solvent and then the latter is slowly removed by appropriate means, such as evaporation. In several cases both ways were employed to produce drug dispersions yielding particles of different morphologies.

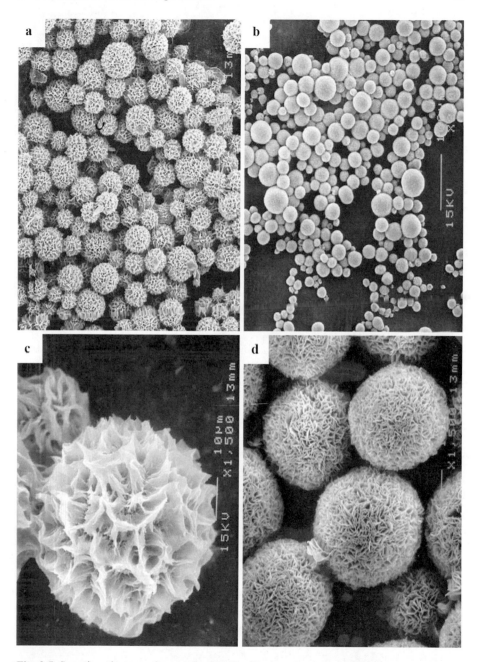

Fig. 2.5 Scanning electron micrographs (SEM) of barium naproxenate particles prepared under the same conditions as for the sample shown in Fig. 2.3a with the reactor immersed in an ultrasonic bath in the presence of 1 wt% PVA (MW = 123,000–186,000 (**a**), and 6% PVA (MW = 13,000–23,000) (**b**). Micrographs of (**c**) and (**d**) are of particles shown in (**a**) and (**b**) respectively, at higher magnifications. [41]

2.4.1   Budesonide [50]

Budesonide ($16\alpha,17\alpha$-butylidenedioxy-$11\beta,21$-1,4-diene-3,20-dione) is a hydro-cortisone steroid used to control asthma and allergies.

Budesonide

The drug of the shown structure is soluble in ethanol and acetone, but insoluble in water, which can be used to produce dispersions of uniform fine particles of different morphologies and modal sizes.

For this purpose two approaches were adopted: (a) precipitation and (b) evaporation. In the first case water is added into the solution of budesonide in ethanol to cause it to precipitate. In the second method, water is first admixed into the solution of the drug in ethanol in quantities too low to cause its separation, and then the alcohol is slowly removed by evaporation.

2.4.1.1   Addition of a Miscible Nonsolvent [50]

The nonsolvent (water) was directly introduced into the ethanol solution of budesonide in an ultrasonic bath (60 kHz), until the solids started to form, which occurred after an induction period of a few minutes. The best results in terms of the morphology and other characteristics of the resulting particles were obtained with ethanol/aqueous, acetone/aqueous, or acetone/alcane systems. Figure 2.6a illustrates a dispersion obtained by adding in a test tube 4 cm^3 of water to 2 cm^3 of a 0.035 mol dm^{-3} budesonide solution in ethanol, while sonicating. After a few seconds the system became turbid, and the process was completed in less than 10 min.

The same kind of experiments were performed with a lower budesonide concentration in ethanol, in which case a larger amount of water had to be admixed to induce precipitation of the drug. A dispersion obtained by mixing 2 cm^3 of a 0.009 mol dm^{-3} budesonide solution in ethanol with 15 cm^3 of water, while sonicating, is displayed in Fig. 2.6b. Large polydispersed crystals of rectangular shape (up to ~ 10 μm in length and width) were so obtained.

Different stabilizers, soluble in the solvent mixture, can be added to prepare well dispersed particles of other shapes. For example, Fig. 2.7a illustrates budesonide

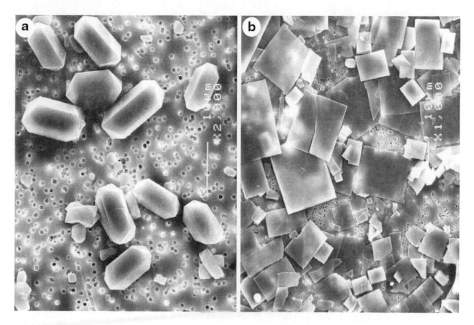

Fig. 2.6 Scanning electron micrographs (SEM) of budesonide crystals prepared by precipitation while sonicating (**a**) 0.035 mol dm^{-3} solution of the drug in ethanol to which were added 4 cm^3 of water; and (**b**) to 2 cm^3 of a 0.009 mol dm^{-3} of the drug in ethanol were added 15 cm^3 of water. [50]

particles precipitated in ethanol/water solutions containing hydrooxypropyl cellulose (HPC-SL), treated in the same manner as samples shown in Fig. 2.6.

2.4.1.2   Evaporation of Solutions with Mixed Solvents

Colloidal spherical budesonide was obtained by evaporating, either fully or partially, ethanol/water solutions of the drug in the presence of the same stabilizers as before. The electron micrograph in Fig. 2.7b displays such particles by drying on a glass slide droplets of the same composition as in Fig. 2.7a. Partial evaporation of ethanol in the same system produced rather uniform spheres when undisturbed and so do experiments carried out in larger volumes.

The evaporation of the acetone/aqueous solution system in the absence of a stabilizer yielded a solid film on the glass surface. However, spherical particles were formed when 1 wt% HPC-SL was added to the solution. Acetone/alcane systems yielded spheres of narrow size distributions in the micrometer size range.

To increase the quantity of particles and to improve the control of the kinetics of evaporation, experiments were carried out in a "rotavapor." Under optimum conditions spherical particles of narrow size distribution could be obtained [50].

It is noteworthy that the XRD spectra of all particles have the same structure, except those of spheres which are amorphous (Fig. 2.8).

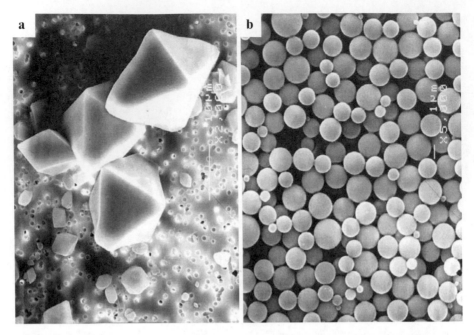

Fig. 2.7 Scanning electron micrographs (SEM) of budesonide particles; **a** prepared as in Fig. 2.6a except 1 cm³ of 0.1 wt% of hyohoxypropyl cellubse (HPC-SL) was present in the reacting mixture. **b** The same system except each droplet was dried at room temperature. [50]

Fig. 2.8 XRD spectra of budesonide: (a) raw material, and of particles illustrated in Figs. 2.6a (b), 2.7a (c) and 2.8a (d), Fig. 2.7b (e) not shown here. [50]

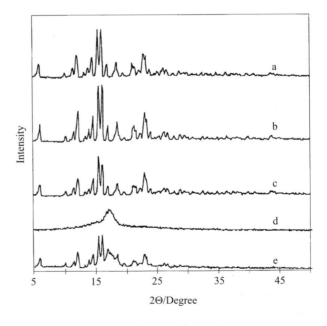

Fig. 2.9 Scanning electron micrographs (SEM) of danazol particles obtained by adding under magnetic stirring 5 cm³ of a 0.1 wt% aqueous solution of Carbowax (**a**) or Igepal (**b**) to 1 cm⁻³ of 0.000 mol dm⁻³ ethanol solution of the drug. [51]

2.4.2   Danazol [51]

Danazol is a synthetic steroid for the treatment of endometriosis. The compound is insoluble in water, but soluble in ethanol, methanol, and acetone.

Danazol

Unlike the results with other drugs described in this review, the evaporation technique applied to pure ethanol/water solutions of danazol failed to yield well defined particles. Instead, rather uniform rodlike crystallites could be obtained by precipitation in the presence of some additives. In general, 0.1 wt% aqueous surfactant solutions were admixed to stock solutions of danazol in ethanol at room temperature, while magnetically stirred. As expected, a minimum amount of the miscible non-solvent was required for the precipitation to take place. However, once this amount was added, the particles were rapidly formed (in less than 1 min). Continued aging did not affect either the shape or the size of the resulting solids. The morphology of the particles could be influenced by the surfactant and by the nonsolvent used. Thus, in the presence of Carbowax, clusters of thin needles appeared (Fig. 2.9a), while in the solution of Igepal ribbon-like particles (Fig. 2.9b) were generated [51].

2.4.3   Cyclosporine [52]

Cyclosporine is a member of a family of compounds that possess immunosuppressive activity and is commonly administered following an organ transplant to prevent rejection. The composition of the compound, using the common abbreviation for the amino acid segments, is as follows:

C_62H_{111}N_{11}O_{12}   Mol. Wt. 1202.61

Cyclosporine

To produce well defined particles advantage is taken of the much higher solubility of this drug in ethanol than in water. The commercial product, as received, consisted of aggregates of ill defined particles, illustrated in Fig. 2.10a. In contrast, rather uniform dispersions could be obtained by evaporation and precipitation techniques, using the drug dissolved in ethanol–water mixtures.

At appropriate conditions (given in Fig. 2.10), adding rapidly a sufficient amount of water to the solution of the drug in ethanol, produced uniform spheres of ~ 1 μm

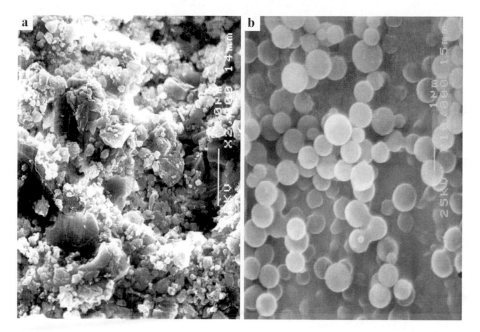

Fig. 2.10 Scanning electron micrographs (SEM) of (**a**) commercial cyclosporine powder, and (**b**) particles obtained by rapidly adding in an ultrasonic bath 4 cm^3 of water (pH 6.5) into 2 cm^3 of a 0.03 mol dm^{-3} solution of the drug in ethanol. [52]

in diameter. Similar particles could be obtained at lower cyclosporine concentrations, which required a larger amount of water. One experimental parameter of importance was the rate of adding the nonsolvent. If the latter was slow, the size distribution was broader.

Spheres were also produced by the evaporation technique. In this procedure water (nonsolvent) was slowly added to an ethanol solution of cyclosporine to yield a mixed solution still below the solubility limit of the drug. A drop of this solution was placed on a glass slide and then alcohol was allowed to evaporate, which caused cyclosporine to precipitate. The influences of (a) the initial volume ratios of alcohol to water, ranging from 10 to 1.6, (b) the added hydroxypropyl cellulose (HPC-SL), and (c) the presence of NaCl were also investigated [52]. Thus, replacing water with a 1 wt% HPC-SL solution, under otherwise the same conditions, yielded particles of the same morphology, but somewhat smaller in size, ranging between 0.25 and 1 μm in diameter. The average size could also be reduced by using aqueous electrolyte solutions in the precipitation experiments. Such spherical cyclosporine particles were not stable in terms of their physical characteristics. A powder of spherical particles aged for one month at room temperature transformed into large rectangular units, which were aggregates of the original spherical subunits.

2.4.4   Loratadine [51]

Loratadine, of the structural formula:

Loratadine

is a long acting antihistamine used to treat symptoms of seasonal allergies, such as rhinitis ("hay-fever"). This solid is well soluble in many organic solvents, including methanol, ethanol, 2-propanol, and acetone, but it is insoluble in water. The solubility in ethanol at 25°C was estimated to exceed several grams in 100 cm^3 of alcohol. As received, loratadine powder consisted of rather large (from a few μm to mm in length) irregular rod-like crystallites, the X-ray diffraction pattern of which is given in Fig. 2.11.

2.4.4.1   Particles Preparation [51]

Two different methods were used to obtain this drug as uniform particles of different morphologies. In the first process precipitates were formed by adding the nonsolvent (either water or aqueous surfactant solutions) to the ethanol solutions of the drug. The elongated particles so obtained were smaller than those of the commercial product, albeit more uniform.

In the second method the drug was introduced into the alcohol-water solution in sufficiently small amounts not to exceed the solubility. Ethanol was then evaporated using a vacuum pump, to cause solids to separate as colloidal particles. Specifically, to different volumes (1–50 cm^3 of a 0.015 mol dm^{-3} loratadine solution in ethanol), either water or surfactant solutions were added, always in a volume ratio of 2:1 (drug solution/nonsolvent).

Fig. 2.11 X-ray diffraction
patterns (XRD) of lorata-
dine particles: *a* as received
powder, *b* precipitated as
elongated particles shown in
Fig. 2.12b and *c* as spherical
particles shown in Fig. 2.12a.
[51]

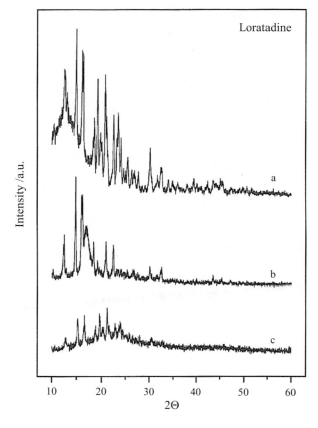

The mixtures were then cooled in an ice/water bath before applying vacuum. The duration of the treatment varied from 20 min to several hours, which depended on the volume of the sample.

It was established that the nature of the precipitates was rather sensitive to the experimental parameters. In order to attain uniform spheres of the drug, an optimized process was developed as follows. After partial evaporation of the liquid (~40% of the total volume), the same volume of the nonsolvent was added to the suspension, and then the evaporation was continued. To ensure that most of the alcohol was removed from the system, the same procedure had to be repeated at least twice. During this treatment the temperature of the dispersion fell below 5°C. The micrograph in Fig. 2.12a displays a typical example of spherical particles so obtained, the diameter of which is ~2 μm. It is noteworthy, that this sample shows the same major X-ray diffraction peaks (Fig. 2.11, pattern c) as the original powder.

Initially 2 cm^3 samples were used in this study, but scaling up the total reaction volume to 75 cm^3 gave essentially the same product. The rate of evaporation had a small effect on the characteristics of the particles. However, it was essential that the entire content of alcohol be removed by evaporation; otherwise, on aging at room

Fig. 2.12 Scanning electron micrographs (SEM) of loratadine particles: **a** spheres formed by vacuum evaporation of ethanol from the solution of the drug in alcohol/water mixture, and **b** rods transformed from the spheres shown in (**a**) by aging the dispersion with some ethanol remaining in the system. [51]

temperature for several hours, dispersed spheres transformed into a fibrillar structure, shown in Fig. 2.12b.

Due to their size, spherical particles settled on standing, but could be redispersed by shaking. Some of the drug particles tended to adhere to the container walls, which could have been caused by the particle charge.

Hoping to improve the dispersion stability of spherical drug particles, the same preparation procedure was modified by replacing water with 0.1 wt% of an aqueous surfactant solution (Tween, PVP, Arquad, or Aerosol). The so obtained particles were similar to those illustrated in Fig. 2.12b, yet more stable.

2.5   Composite Drug Particles

Medications in pills are seldom present as single pure substances. Instead, drugs are either mixed with an inert material, or coated, or both. Additives are often used as diluents to control the delivery of the active component over a period of time. Furthermore, in many cases the medication is contained in coated pills to protect it from the effects of the environment, such as humidity or oxidation. The encapsulation also provides for dosage forms which have sustained delivery properties [53, 54].

The coating of particles with layers of different compositions may also alter their surface charge properties and hydration of the pill. For example, a layer of alumina on the pill would have an isoelectric point at pH ~ 9, while coating with silica would result in an i.e.p. of ~ 2. Obviously, such coatings can be used to attach a drug to a tissue, if so desired, or to prevent the attachment so that the active component can be delivered elsewhere. For these reasons some of the drug particles described in this review were coated with medically inert layers of different compositions, or the drug was used as a coating on inactive uniform cores [55]. In the latter case, by selecting smaller cores the specific surface area of the exposed active compound could be substantially increased with lesser amounts of the drug for a more efficient delivery.

2.5.1   *Coating of Inert Nanosize Particles with Drugs [53]*

As a rule, it is rather difficult to obtain stable nanosize particles of bioactive materials. One method is to attach such drugs onto inert small cores. For example, it was earlier demonstrated that bioactive molecules, such as amino acids, can be attached onto uniform metal oxide spheres, i.e. aspartic acid on chromium hydroxide [56]. Such attachments are particularly readily achieved, if a bond can be formed between a chemical group on the active molecule and a site on the adsorbent particle [57].

Specifically, either naproxen (Sect. 3.1) or ketoprofen, of the composition:

Ketoprofen

were deposited on nanosize alumina particles [Degussa C (DGC) alumina, γ-Al$_2$O$_3$, particle size ~ 13 nm, specific surface area (SSA) 100 $\pm$ 10 m^2 g^{-1}, or Nalco alumina particle size ~ 8 nm, SSA 550 $\pm$ 20 m^2 g^{-1}], which yielded stable dispersions in aqueous and ethanol media [55].

For this purpose the stock solutions of drugs in water were prepared in concentrations of 0.05–0.15 mol dm^{-3} at pH ~ 11, and then the latter was decreased with HCl to pH 7.2. Measured volumes of dispersions of alumina, containing known amounts of solids, were first diluted as needed either with water or with ethanol. Then stock solutions of different concentrations of naproxen or ketoprofen were added under stirring to these dispersions to reach the final desired volume. In order to avoid aggregation of the alumina nanoparticles, their concentration was kept at 0.05 wt%. After equilibration for extended periods of time (several hours) the solids

Fig. 2.13 Adsorbed amounts of drugs on nanosize alumina (DGC), Γ, μmol m^{-2}) as a function of the concentration (mol dm^{-3}) for naproxen in water ($\circ$) and ketoprofen in water ($\bullet$) at pH 7.2$\pm$0.1, and naproxen in ethanol (Δ) and ketoprofen in ethanol ($\blacktriangledown$). [55]

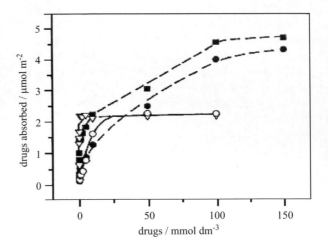

were separated by centrifugation and dried in vacuum. Alternatively, the dispersions could be freeze-dried. The so prepared powders were readily redispersed.

Figure 2.13 shows deposition data of the two drugs on finely dispersed alumina in aqueous and ethanol dispersions, which demonstrate their binding onto the cores.

Electron micrographs in Fig. 2.14 display the alumina particles used as cores and of the same adsorbent coated with naproxen. The infrared spectra in Fig. 2.15 of the cores, of naproxen, and of coated cores clearly show that the last is not a simple combination of individual components.

2.5.2 Coating of Drug Particles with Alumina [53]

It was well documented in the literature that electrostatic attraction alone does not suffice to produce the binding of solutes onto the surface of solids dispersed in a liquid (preferably water). In contrast hydrolyzed metal ions readily interact either with neutral or oppositely charged species on solid surfaces as established in much detail experimentally [58, 59].

For this reason to coat drugs with alumina, the metal ion needed to be first hydrolyzed, which was readily achieved by aging the aqueous solutions of a salt of this metal at elevated temperatures (e.g. 90°C) for extended periods of time [59, 60]. It was earlier demonstrated that this simple procedure causes aluminum ion to react with water yielding positively charged hydrolyzed complexes, which readily adsorb and/or interact with surface sites of hydrophobic solids. This reaction can either neutralize the negatively charged ionic groups on the particles or even reverse them to positive [61–63]. At sufficiently long aging times the hydrolysis of aluminum ions results in the precipitation of sparingly soluble products, which can also react with a substrate (e.g. particles) to yield a coating.

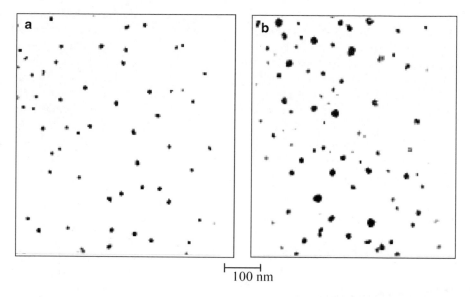

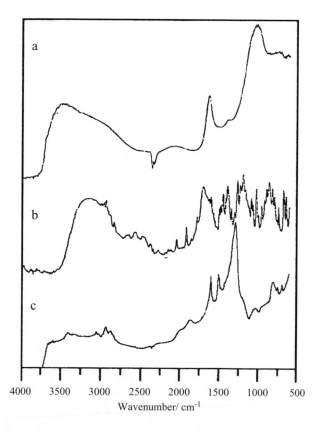

Fig. 2.14 Transmission electron micrographs (TEM) of (**a**) Nalco alumina and (**b**) the same alumina coated with naproxen. [55]

Fig. 2.15 ATR-FTIR spectra of (*a*) DGC alumina, (*b*) naproxen, and (*c*) the same alumina coated with naproxen in water. [55]

Wavenumber/ cm^{-1}

Here is described the coating of loratadine and danazol particles with alumina by the described method [51]. Since the IEP of the drug particle was at pH ~ 4 and that of aluminum (hydrous) oxide particles at pH ~ 8.5, strong electrostatic attraction may be expected at pH values between those of the cores and of the coating materials. The binding itself is then facilitated by surface interactions (e.g. condensation). The coating of particles was carried out by adding aluminum salt solutions to the dispersions of drugs and then increasing the pH with a base. For this purpose the aluminum ions were partly prehydrolyzed by aging a 0.01 mol dm^{-3} aqueous solution of either aluminum nitrate or aluminum sulfate at 70°C for 50 h. This procedure was shown previously [59, 60] to enhance the interactions of the formed solute aluminum complexes with the colloidal drug particles.

2.5.2.1 Coating of Loratadine Particles

In general, to 5 cm^3 of aqueous dispersions of rodlike loratadine particles (prepared as described in Sect. 2.4.4.1), containing 0.003 mol dm^{-3} (1.9 mg cm^{-3}) solids, were added 0.5 cm^3 of a pretreated (aged) 0.0010 mol dm^{-3} Al(NO$_3$)$_3$ or 0.0005 mol dm^{-3} Al$_2$(SO$_4$)$_3$ solution, at room temperature, while stirred. After approximately 10 min the suspensions were neutralized with a dilute NaOH solution (0.01 mol dm^{-3}) to pH 6–6.5 and kept overnight at room temperature.

A peak characteristic for aluminum by the EDS analysis (Fig. 2.16) of so treated drug particles clearly indicates a successful coating.

The electrokinetic measurements given in Fig. 2.17 show the dependence of the ζ-potential on the pH of loratadine cores (a), and of the coated particles prepared using hydrolyzed aluminum ions, obtained by aging aqueous solutions of either Al$_2$(SO$_4$)$_3$ (b), or Al(NO$_3$)$_3$ (c). There is a significant shift of the isoelectric point (IEP) of coated particles as compared to that of the pure drug, although the effect with hydrolyzed aluminum sulfate is less pronounced than that with aluminum nitrate. It has been known that the sulfate ion forms complexes with aluminum ion in aqueous solutions, which in turn affect the ζ-potential of the particles, i.e. the IEP is lower than with the hydrolyzed aluminum ion in the presence of nitrate ion.

2.5.2.2 Coating of Danazol Particles [51]

Essentially the same procedure was used in coating danazol. Specifically, to 5 cm^3 of an aqueous danazol dispersion of platelets (Sect. 4.2) containing 1.0 mg cm^{-3} of the drug, were added 0.8 cm^{-3} of a 0.0010 mol dm^{-3} aged Al(SO$_4$)$_3$ solution and after 10 min the dispersion was neutralized with a 0.010 mol dm^{-3} NaOH solution to pH 6–6.5, and then kept for 20 h at room temperature.

The SEM micrograph of coated danazol particles (Fig. 2.18a) clearly shows that a uniform surface layer was formed on the cores, which is substantiated by the EDS result given in Fig. 2.18b.

Fig. 2.16 EDS analysis of rod-like loratadine particles coated with aluminum (hydrous) oxide. [51]

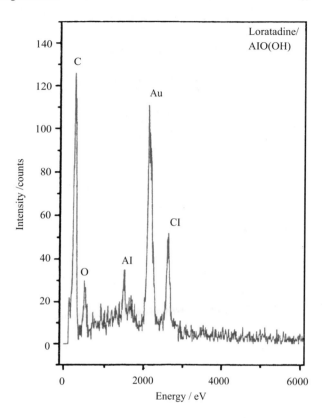

Fig. 2.17 Zeta potential as a function of the pH of (*a*) elongated loratadine particles (shown in Fig. 2.12) (*b*) the same particles coated with aluminum (hydrous) oxide using an aged $Al_2(SO_4)_3$ solution, and (*c*) the same particles using an aged $Al(NO_3)_3$ solution, as described in Sect. 2.5.2.1. [51]

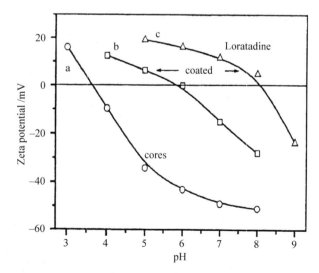

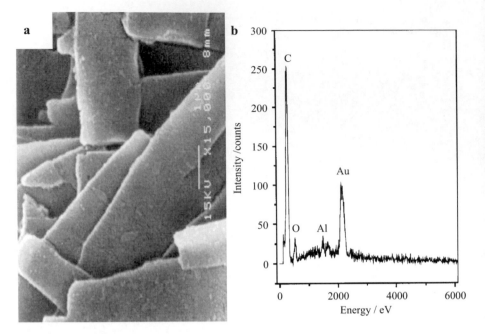

Fig. 2.18 a SEM of danzol particles coated with aluminum (hydrous) oxide, as described in Sect. 2.5.2.2, and **b** the EDS analysis of the same particle. [51]

2.5.3   Coating with Silica [6, 51]

Coating with silica affects the particles in two general ways. As cited before, the isoelectric point of silica is much lower, while the degree of hydration is increased. Both of these properties have an effect on the performance of the modified medication. For this reason two of the studied drugs, loratadine and danazol, and a finely dispersed X-ray diagnostic images agent (ethylester of diatrizoic acid, EEDA) were coated with silica layers.

Since loratadine and danazol are soluble in alcohol, the well-known methods for particle coating with silica based on tetraethylorthosilicate, (TEOS) [65–67] could not be employed. Instead, an alcohol-free method based on sodium silicate solutions, was used to coat the particles with silica.

2.5.3.1   Coating of Loratadine Particles [51]

An "active" silicic acid solution was prepared using a cationic exchanger following a procedure described by Iler [64]. Accordingly, an aqueous solution of Na_2SiO_3 was mixed with the beads of the cationic exchange resin Amberlite as follows. The original sodium silicate liquid was diluted with water in the weight ratio 1:9. To 20 cm^3 of this solution, 3 g of the resin were added and the mixture was agitated

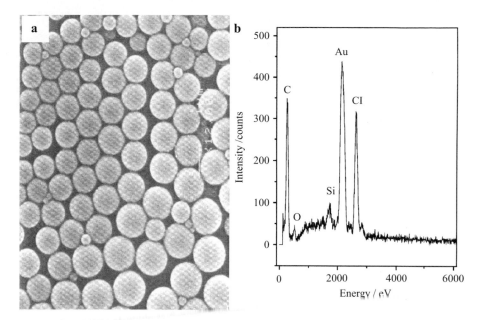

Fig. 2.19 a SEM of spherical loratadine particles coated with a silica layer, when 0.08 g of "active" silicic acid was added to 1 g of the dispersion (described in Sect. 2.4.4) and **b** the EDS analysis of the same particles. [51]

until the pH reached 9.5. The resin was then removed by filtration. The final active silicic acid solution contained ~2.9 wt% SiO_2.

To produce silica shells on spherical loratadine particles (shown in Fig. 2.19) the "active" silicic acid, in amounts ranging from 0.04 to 0.12 g, was added to 5 cm^{-3} of the aqueous dispersion containing ~7 mg cm^{-3} of the drug in water. The resulting mixture was then kept for different periods of time (up to several days) at room temperature, although the presence of silica on the particles could be already detected 2 h after combining the reactants.

One example of such coated loratadine spheres is displayed in Fig. 2.19a, while the presence of the coating was corroborated by the EDS analysis, as shown in Fig. 2.19b. Little effect was observed when different amounts of the "active silicic acid" or larger amounts of the drug (3 g) were used in the same system. Dispersions of coated loratadine were more stable than those of bare particles; i.e., the former could be easily redispersed in water by shaking.

Electrokinetic measurements revealed that the ζ-potentials of coated particles at pH 5.5 and 7.0 were 20 mV more negative than those of the cores, due to the presence of the silica shell.

2.5.3.2   Coating of Danazol Particles [51]

Somewhat different "active" silicic acid was prepared from the sodium silicate solution (3.25 SiO_2:Na_2O, H_2O), by diluting the original liquid with water in the

Fig. 2.20 SEM of danazol
particles coated with silica
layer.

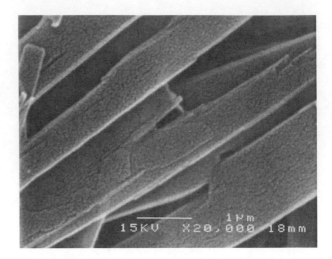

Fig. 2.21 Zeta potential as
a function of pH of danazol
particles and of the same
particles coated with a silica
layer

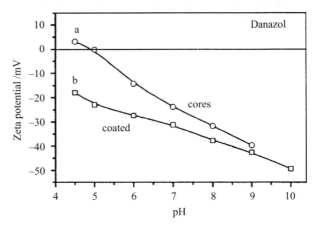

weight ratio 1–9. To 20 cm^{-3} of this solution were added 3 g of the Amberlite ion
exchange resin under agitation until pH 9.5 was reached. After removing the resin
the active silicic acid contained ~2.9 wt% SiO$_2$ [67].

To coat danazol particles, 5 cm^{-3} of the dispersion containing 1 mg cm^{-3} of solids, prepared as described in Sect. 4.2, and the "active" silica solution, were aged
for 20 h at room temperature.

The electron micrograph in Fig. 2.20 illustrates such danazol particles coated
with a silica layer. The presence of the shell is further substantiated by the electrokinetic data in Fig. 2.21 which show coated particles to be more negatively charged
than the cores over the entire pH range. The curve of ζ-potential for coated particles
yields the isoelectric point of ~2, which is typical for silica.

Fig. 2.22 Zeta potential of rod-like EEDA particles as a function of the pH (*a*), and (*b*) the same particles coated with silica. [6]

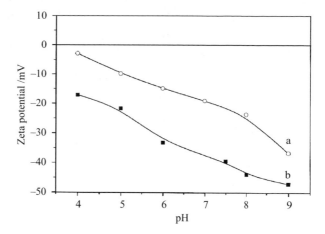

2.5.3.3   Coating of EEDA [6]

In the final example nanoparticles of diatrizoic agent (ethyl ester of diatrizoic acid, EEDA) of the composition:

which is as an X-ray contrast agent for computer topography images of the blood pool and lymphatic systems [68].

While this compound was available as finely dispersed particles of different, but uniform shapes only those consisting of rods were coated with silica, using an active silica coating solution (3.25 SiO_2:Na_2O, H_2O) described earlier [51].

Electrokinetic data in Fig. 2.22 clearly show that the particles were encapsulated by silica over the entire pH range studied.

References

1. Tomlinson, E., Davis, S.S., Eds. *Site–Specific Drug Delivery;* John Wiley and Sons; Great Britain, 1986.
2. Brindley, A., Davis, S.S., Davies, M.C., Watts, J.F. *J. Colloid Interface Sci.* **1995**, 171, 150.
3. Guiot, P., Couvreur, P., Eds. *Polymeric Nanoparticles and Microspheres;* CRC Press, 1986.

4. Rembaum, A., Tokes, Z., Eds. *Microspheres: Medical and Biological Applications;* CRC Press, 1988.
5. Rouhi, A.M., *Chem Eng News* **2003**, 81, 32.
6. Bosch, H.W., Škapin, S.D., Matijević, E. *Colloids Surf. A* **2004**, 250, 43.
7. Stowell, G.W., Behme, R.J., Denton, S.M., Pfeiffer, I., Sancilio, F.D., Whittal, L.B., Whittle, R.R. *J. Pharm. Sci.* **2002**, 91, 2481.
8. Crowder, T.M., Rosati, J.A., Schroeter, J.D., Hickey, A.J., Martonen, T.B. *Pharm. Res.* **2002**, 19, 239.
9. Hintz, R.J., Johnson, K.C., *Int. J. Pharm.* **1989**, 51,9.
10. Frank, S., Loforoth, J., Bostanian, L., U.S. Patent 5,780,063, 1998.
11. Violante, M., Fisher, H. U.S. Patennt 4,826,689, 1989.
12. Violante, M.R., Steigbigel, R.T. U.S. Patent 4,783,484, 1988.
13. Motojama, S., Sato, S., Umeda, S., Yasumi, E., Hirotsuneand, S., Takasaka, Y., Tsujino, T. U.S. Patent 4,540,602, 1985.
14. Fessi, H., Devissaguet, J., Puisieux, C., E.P. Patent 275,796 B1, 1992.
15. Liversidge, G., Cooper, E., McIntire, G., Shaw, J.M., Cundy, J., Bishop, K., Czekai, D., U.S. Patent 5,145,684, 1992.
16. Liversidge, G.G., Cundy, K.C. *Int. J. Pharm.* **1995**, 125, 91.
17. Liversidge, G.G., Conzentino, P. *Int. J. Pharm.* **1995**, 125, 309.
18. Sugimoto, T., Ed. *Fine Particles Synthesis, Characteristics and Mechanisms of Growth*; Marcel Dekker, 2000.
19. Sugimoto, T., Ed. *Monodispersed Particles*; Elsevier, 2000.
20. Matijević, E. *Langmuir* **1994**, 10, 8.
21. Matijević, E. *Chem. Mater.* **1993**, 5, 412.
22. Matijević, E. *Pure Appl. Chem.* **1992**, 64, 1703.
23. Horn, D. *Agnew. Makromol. Chem.* **1989**, 166, 139.
24. Horn, D., Luddecke, E., *NATO ASI Series, Series 3: High Technology* **1996**, 12, 761.
25. Zelenez, A., Sonnenberg, W., Matijević, E. *Colloid Polym. Sci.* **1998**,276,838.
26. Auweter, H., Haberkorn, H., Heckmann, W., Horn, D., Luddecke, E., Rieger, J., Weiss, H. *Angew. Chem. Intern. Ed* **1999**, 38, 2188.
27. Brick, M.C., Palmer, H.J., Whitesides, T.H. *Langmuir* **2003**, 19, 6367.
28. Sinha, V.R., Singla, A.K., Wadhawan, S., Kaushik, R., Kumria, R., Bansal, K., Dhawan, S. *Int. J. Pharm.* **2004**, 274, 1.
29. Bromberg, L.; Rashba-Step, J.; Scott, T. *Biophys. J.* **2005**, *89*, 3424.
30. Uskoković, V.; Matijević, E. *J. Colloid Interface Sci.* **2007**, *315*, 500.
31. Patel, P.C., Giljohann, D.A., Seferos, D.S., Mirkin, C.A., *Proc. Natl. Acad. Sci. U.S.A.* **2008**, 105, 17222.
32. Matijević, E.; Partch, R.E. Synthesis of Monodispersed Colloids by Chemical Reactions in Aerosols. In *Fine Particles: Synthesis, Characterization and Mechanism of Growth*, Sugimoto, T.; Ed., CRC Press, **2000**, pp 97-113.
33. Espenscheid, W.F., Matijević, E. Kerker, M. *J. Phys. Chem.* **1964**, 68, 2831.
34. McRae, D., Matijević, E., Davis, E.J. *J. Colloid Interface Sci.* **1975**, 53, 411.
35. Berry, C.R, Precipitation and growth of silver halide emulsion grains In *The Theory of the Photographic Process*, 4th Ed. James, T.H., Ed., MacMillan, New York, 1977, pp 88– 104.
36. Sugimoto, T. A Controlled Double-jet Process In *Fine Particles: Synthesis, Characterization and Mechanism of Growth*, T. Sugimoto, Ed., Marcel Dekker, New York, 2000, pp 280-289.
37. Wey, J. In *Preparation and Properties of Solid State Materials;* Wilcox, M. Ed.; Marcel Dekker, 1981; Vol. 6; pp 67-117.
38. Stávek, J.; Šípek, M.; Hirasawa, I.; Toyokura, K. *Chem. Mater.* **1992**, *4*, 545.
39. Her, Y.-S.; Matijević, E.; Chon, M.C. *J. Mater. Res.* **1995**, *10*, 3106.
40. Pozarnsky, G.A., Matijević, E. *Colloids Surf. A* **1997**, 125, 47.
41. Goia, C.; Matijević, E. *J. Colloid Interface Sci.* **1998**, *206*, 583.

42. Flower, R.J., Moncada, S., Vane, J.R., in Goodman, G.A., Goodman, L.S., *The Pharmacological Basis of Therapeutics,* 6th ed., Gilman, A. Ed., Macmillan Publishing Company, 1980; p 710.
43. Kim, Y.B.; Song, H.J.; Park, I.Y. *Arch. Pharm. Res.* **1987,** *10,* 232.
44. Harrison, I.T., Lewis, B., Nelson, P., Rooks, W., Roszkowski, A., Tomolonis, A., Fried, J.H. *J. Med. Chem.* **1970,** 13, 203.
45. Matijević, E.; Goia, D.V. *Croat. Chem. Acta,* **2007,** *80,* 485.
46. Libert, S.; Goia, D.V.; Matijević, E. *Langmuir* **2003,** *19,* 10673.
47. Libert, S.; Gorshkov, V.; Privman, V.; Goia, D.V.; Matijević, E. *Adv. Colloid Interface Sci.* **2003,** 100-102, 169.
48. Privman, V.; Goia, D.V.; Park, J.; Matijević, E. *J. Colloid Interface Sci.* **1999,** 213, 36.
49. Park, J.; Privman, V.; Matijević, E. *J. Phys. Chem. B* **2001,** 105, 11630.
50. Ruch, F.; Matijević, E. *J. Colloid Interface Sci.* **2000,** *229,* 207.
51. Škapin, S.; Matijević, E. *J. Colloid Interface Sci.* **2004,** *272,* 90.
52. Joguet, L.; Matijević, E. *J. Colloid Interface Sci.* **2002,** 250, 503.
53. Wurster, D. Particle-coating methods, 2nd ed., in: Libermann, H.A; Lachman, L.; Schwartz (Eds.), *Pharmaceutical Dosage Forms: Tablets, vol. 3,* Marcel Dekker, 1990, p 161.
54. Chang, J., Robinson, R.K. In *Particle Coating Methods– Pharmaceutical Dosage Forms: Tablets, 2nd Ed.;* Libermann, H., Lachman, L., Schwartz, J. Eds.; Marcel Dekker, 1990, Vol 3, p 199.
55. Joguet, L.; Sondi, I.; Matijević, E. *J. Colloid Interface Sci.* **2002,** *251,* 284.
56. Matijević, E., A Critical Review of the Electrokinetics of Monodispersed Inorganic Colloids In *Interfacial Electrokinetics and Electrophoresis,* A.V. Delgado, Ed., Marcel Dekker, NY, 2002, pp 199-218.
57. Matijević, E.; Sapieszko, R.S., Metal Oxides: Forced Hydrolysis in Homogeneous Solutions. In *Fine Particles: Synthesis, Characterization and Mechanism of Growth,* Sugimoto, T. Ed., Marcel Dekker, New York, 2000, pp 2-34.
58. Matijević, E.; Force, C.G. *Kolloid Z.Z. Polym.* 1968, *225,* 33.
59. Matijević, E.; Janauer, G.E.; Kerker, M. *J. Colloid Sci.* 1964, *19,* 333.
60. Matijević, E. Charge Reversal of Lyophobic Colloids In *Principles and Applications of Water Chemistry,* Faust, S., Hunter, J., Eds.; Wiley, 1967; pp 328-369.
61. Matijević, E. *J. Colloid Interface. Sci.* **1973,** 43, 217.
62. Matijević, E.; Kratohvil, S. *Adv. Ceramic Mater.* **1987,** *2,* 798.
63. Škapin, S.D.; Matijević, E. *Farm. Vestn.* **2003,** *54,* 511.
64. Stober, R.W.; Fink, A.; Bohn, E. *J. Colloid Interface Sci.* **1968,** *26,* 62.
65. Ohmori, M.; Matijević, E. *J. Colloid Interface Sci.* **1992,** *150,* 594.
66. Giesche, H.; Matijević, E. *J. Mater. Res.* **1994,** *9,* 436.
67. Iler, R. U.S. Patent 2,885,366, 1959.
68. Liversidge, G.L., Cooper, E.R., Shaw, J.M., Mcintire, G.L. U.S. Patent 5,318,767, 1994.

Chapter 3
Biomedical Applications of Metal Oxide Nanoparticles

**Silvana Andreescu, Maryna Ornatska, Joseph S. Erlichman,
Ana Estevez and J. C. Leiter**

Abstract Metal oxide nanoparticles have a unique structure, interesting and unusual redox and catalytic properties, high surface area, good mechanical stability and are biocompatible. For these reasons, metal oxide nanoparticles have attracted considerable interest in the field of biomedical therapeutics, bio-imaging and bio-sensing. This chapter discusses properties and biomedical applications of selected nanometer size metal oxides. These materials have become important components in medical implants, cancer diagnosis and therapy and in neurochemical monitoring. For example, titania is the material of choice in medical implants; it provides an excellent biocompatible surface for cell attachment and proliferation. Ceria-based nanoparticles, on the other hand, have recently received a great deal of attention because of their redox, auto-catalytic and antioxidant properties. Several other metal oxides have been used as gas sensing nanoprobes for cell labeling and separation, as contrast agents for magnetic resonance imaging (MRI) and as carriers for targeted drug delivery. New and emerging applications of nanoceria as neuroprotective agents possessing antioxidant/free radical scavenging properties are emerging in the biomedical field, and ceria-based nanoparticles may be used as therapeutic agents in the treatment of medical diseases related to reactive oxygen species, such as spinal cord repair, stroke and degenerative retinal disorders. Issues related to biocompatibility and toxicity of these nanoparticles for *in vivo* biomedical applications remain to be fully explored.

Keywords Metal oxide nanoparticles • Biomedical applications • Surface functionalization • Interaction of nanoparticles with proteins • Cells and tissues • Nanoparticle toxicity • Oxidative stress

3.1 Introduction

In recent years, numerous types of metal oxides in particulate, film or composite forms have become key components in catalysis, solid oxide fuel cells and sensing devices [5–9]. More recently, metal oxides nanoparticles (NPs) have been used in a

S. Andreescu (✉)
Department of Chemistry & Biomolecular Science, Clarkson University,
8 Clarkson Avenue, SC, Box 5810, Potsdam, NY 13699, USA
e-mail: eandrees@clarkson.edu

E. Matijević (ed.), *Fine Particles in Medicine and Pharmacy,*
DOI 10.1007/978-1-4614-0379-1_3, © Springer Science+Business Media, LLC 2012

variety of biomedical applications due to the high surface area and enhanced magnetic and catalytic activity [10, 11]. Interestingly, the biomedically useful characteristics of these NPs cannot be predicted by the bulk properties of the corresponding macroscopic size material. The most widely used NPs are magnetic iron oxide (Fe_3O_4), titania (TiO_2), zirconia (ZrO_2) and more recently ceria (CeO_2). These materials have interesting catalytic, antioxidant and bactericidal activities, mechanical stability and biocompatibility, making them amenable for a variety of disparate biomedical applications such as therapeutic and bio-imaging agents, components in medical implants, drug delivery systems and probes for neurochemical monitoring [12–17]. For example, titania is the material of choice in medical implants because it provides a biocompatible surface for cell attachment and proliferation [18, 19]. Magnetic iron oxides have been intensively used in cell labeling and separation, in magnetic resonance imaging (MRI) and in targeted drug delivery. Ceria has recently received a great deal of attention due to its catalytic and antioxidant capacity [20–22]. Many of the useful properties of metal oxides are dictated by the morphological and physico-chemical characteristics of these particles, but their chemical properties vary broadly depending on the preparation method and processing used in the manufacturing process.

Many research papers and reviews covering the synthesis and characterization of various metal oxides have been published [23–34]. Advances in colloid chemistry have enabled synthesis of a variety of nanometer size metal oxides [35], and a wide array of metal oxides of various sizes and compositions are commercially available. Significant progress has been made in the control of chemical composition, size, size distribution and shape [36–42]. It is now possible to obtain uniform particles with controlled shapes other than simple spheres [24, 43]. By varying the precursor, reductant, solvent, dispersant and the experimental conditions (e.g. pH, temperature), it is possible to alter the morphology and physico-chemical properties [44]. The final particle size depends on the nucleation rate, the fraction of the atoms involved in the formation of the nuclei and the extent of the aggregation. Both aqueous and non-aqueous synthetic approaches have been reported [29–33, 35, 45]. Production of well-dispersed uniform metal oxides by precipitation in aqueous solutions has been achieved by a very elegant procedure called "forced hydrolysis" developed by the senior editor of this book [34]. The polarity of the solvent and the stabilizer also affects size, charge and shape. The stabilizers are usually large molecules containing functional groups that can attach to the particle surface thereby controlling particle solubility, assembly behavior and morphology. Natural polymers like gelatin, dextrin and polysaccharides, synthetic polymers like polyvinyl alcohol and polyacrylamides and organic molecules with large alkyl chains can be used as stabilizers. While such additives improve the ability to generate stable dispersions, these coatings may change the behavior of the NPs in a biological media, induce toxicity or affect the primary biomedically useful properties of the NPs. One way to avoid the use of surfactants and produce high-purity biocompatible NPs is to use solvents that act both as reactant and dispersing agent in a 'surfactant-free' synthesis [29–33, 35, 45]. In this case, the solvent can act as a 'capping' agent binding to the particle surface to prevent agglomeration. Many biomedical applications of metal oxides involve surface modification with biocompatible coatings, fluorescence tags or biological molecules (e.g. antibodies, receptors) for targeted biological response and imaging purposes [46, 47]. Uniform surface modification with precise control of the number of functionalities on the NPs surface is an important and challenging task.

Application of the NPs in the biomedical field relies on the interactions between NPs with proteins, cells and tissues. By virtue of their small size, NPs can interact and enter biological systems through various routes; they can gain access to cellular and tissue locations (for example, mitochondria and the central nervous system) that are not accessible to larger size particles [5, 48]. Whether NPs cross cellular membranes and the blood-brain barrier, interact with cells, organs and tissues, accumulate at different locations in the body, remain inert or interfere with normal physiological processes, will determine their behavior, fate and transport in biological systems [13, 49–51]. *In vivo* biomedical applications require NPs with high purity, polydispersity and high stability, and these properties depend on the size, shape, charge and surface coverage of each NP and, in biological media, on the entry sites and size of the biological target encountered by each NP [14, 52, 53].

The interaction of metal oxides with proteins, cells and tissues is strongly related with the biological material's surface chemistry because the particles will first "see" the surface of the biological target [54–63]. Thus, NP characteristics like surface charge, surface coating, surface area, particle reactivity, composition, aggregation and dissolution may all affect cellular uptake, *in vivo* reactivity, toxicity and distribution across tissues [6, 14, 60, 64–67]. Interfacial processes at the boundary between the inorganic material and the surrounding biological environment are complex and surface specific [63]. Studies of these interactions involve, in addition to material-specific research, investigations of protein adsorption, orientation and structure at the interface [68], biophysical and biological studies at cell and tissue levels as well as *in vivo* experiments in animal models. A multidisciplinary approach, involving knowledge from material chemistry, biology, biochemistry and medicine is needed to study and understand the fundamental processes associated with these complex systems.

This chapter focuses on the most widely used metal oxides, their interaction with cells and tissue in relation to the physico-chemical properties, biocompatibility and cytotoxic response in model biological systems and selected biomedical applications. The last section of this chapter is dedicated to biomedical applications of antioxindant NPs with special emphasis on the use of cerium oxide to mitigate oxidative stress in diseases related to reactive oxygen species.

3.2   Metal Oxide Nanoparticles: Research Needs and Opportunities in the Biomedical Field

A broad range of metal oxide NPs have been investigated in the biomedical field. Table 3.1 summarizes the most widely used metal oxides, their primary characteristics and examples of biomedical applications. Further research is needed to engineer

Table 3.1 Examples of metal oxides types, properties and potential applications

Oxides	Types	Properties	Biomedical applications	Ref
Iron oxides	Fe_2O_3, Fe_3O_4, Fe_3S_4, MeO-Fe_2O_3 where M = Ni, Co, Zn,... etc)	Magnetism Catalytic properties	Molecular labels Imaging Drug delivery/ MRI Biomagnetic separations	[4, 15, 70–74]
TiO_2	Single crystalline metal oxide; Wide band gap semiconductor; Common forms are rutile and anatase	Photocatalytic activity Biocompatibility	Biomedical implants Photocatalytic damage Neurochemical monitoring Promoter for cell adhesion	[18, 75–78]
ZnO, MnO_2 ZrO_2	Used mainly in composite films	Catalytic properties Anti-inflammatory action	Antibacterial and anti-inflammatory agents	[79]
CeO_2	Present in two oxidation states Ce^{3+}/Ce^{4+}	Catalytic properties antioxidant/ radical scavenging properties Biocompatibility	Neuroprotective agents	[12, 20, 21, 80–84]

the surface and morphology of the NPs to tailor their physico-chemical properties to specific applications. Synthesis and modification of the NPs should improve biocompatibility and accessibility to target cells and tissue locations while conserving their biologically useful properties. For any given biomedical application, particles should first be characterized and their physico-chemical properties (surface composition, charge, nature of the coating, hydrophilicity, etc.) determined. Optimizing the synthetic procedure and a systematic investigation of the engineering variables that affect the dimensional, physical and chemical properties is important for the rational design of NPs with controllable structure and surface proterties [69]. However, obtaining NPs with predictable structure, size, composition and functionality, and a well-defined biological response, biodistribution and pharmacokinetic characteristics is a challenging task.

Structural design requirements can vary considerable depending on the final application of the NPs. The size of the particles can be tailored to accumulate or permeate intracellular or extracellular compartments, depending on whether they are used in biosensing applications diagnosis or tumor targeting or for drug delivery purposes. Their surface can be modified to facilitate transport and accumulation at a specific location. For example, restricted permeation with high accumulation in tumors *in vivo* was observed with 100 nm particles while 20 nm particles were found to rapidly diffuse throughout the tumor matrix [69]. Other questions that must be addressed include: How particle design can be optimized toward a specific application (e.g. increased therapeutic efficiency, controlled delivery, optimized diagnostic)? Can this be predicted? Is this related to surface reactivity, surface charge or NP composition? How NPs interact with biological systems? What is the mechanism of cellular and tissue uptake? Is it possible to control *in vivo* response through systematic modifications of the NPs surface? Can *in vitro* studies be used to predict *in vivo* behavior? How are these particles cleared once they accumulate *in vivo*? In the following sections we discuss these challenges for the most widely used metal oxides in relation to the material characteristics of the target tissue and provide examples of specific biomedical applications.

3.3   Surface Modification of Metal Oxide NPs: Biocompatibility and Surface Properties

While native NPs can have the desired properties in terms of size, shape and degree of dispersity, they might lack biocompatibility. Most biomedical applications involve NPs that have been surface-modified with biocompatible layers to provide chemical functionalities, prevent aggregation, avoid non-specific protein adsorption and minimize toxic effects in physiological conditions. Surface coating can be achieved post-synthesis or during synthesis (in situ synthesis or one-pot synthesis) in a single step reaction in the presence of biocompatible matrices such as polymeric hydrogels [4]. In the post-synthesis route, the coating materials are added to the already formed particles using chemical grafting, ligand exchange [85], or by surface-entrapment via hydrophobic and electrostatic interactions. In the in-situ synthesis, coating materials are added in the same reaction solution with the metal oxide precursors; the coating is formed simultaneously with the particles. Examples of surface modifications and NP stabilization procedures include chemical grafting of various organic functionalities [26] and addition of silica [73], gold [85] or polymeric [26, 86–89] shells onto the particle surface to form a core-shell structure. Examples of general functionalization strategies are shown in Fig. 3.1 for iron oxide NPs [1].

The outer shell in the core-shell structures can be engineered to carry multiple functions (Fig. 3.2): drug carrier, anchoring point for biomolecules (antibodies, receptors, proteins) or imaging agents (fluorophores and radionuclides), provide stability and enhanced biocompatibility and improve intracellular behavior [46]. Such multifunctional NPs can serve as imaging agents, therapeutic carriers (e.g. drug loaded magnetic particles) or the particle itself can be the drug.

To allow attachment of fluorophores or biomolecules, surface reactive groups such as carboxylic, amino and hydroxyl groups are first introduced. One of the most widely used methods is based on 1-ethyl-3(3-dimethylaminopropyl)carbodiimide (EDC) or a combination of EDC with N-hydroxy-succinimide (NHS) to produce stable carboxylic or amine-functionalized NPs [2]. Figure 3.3 shows an example of the chemical functionalization procedure for NPs with covalently bound chitosan

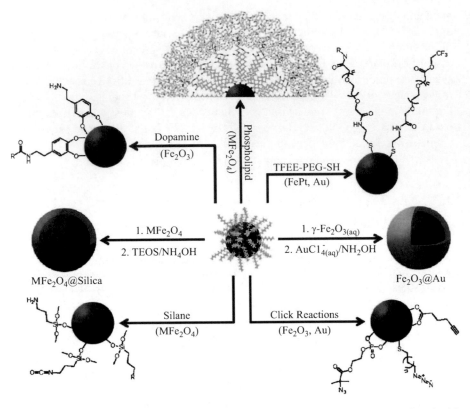

Fig. 3.1 Examples of surface functionalization of magnetic iron oxide NPs. (Reproduced with permission from ACS, Latham & Williams, Accounts of Chem. Res., 41, 3, 2008, 411–420 [1])

Fig. 3.2 Examples of surface modified NPs with multiple functionalities

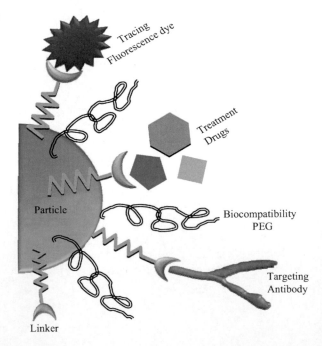

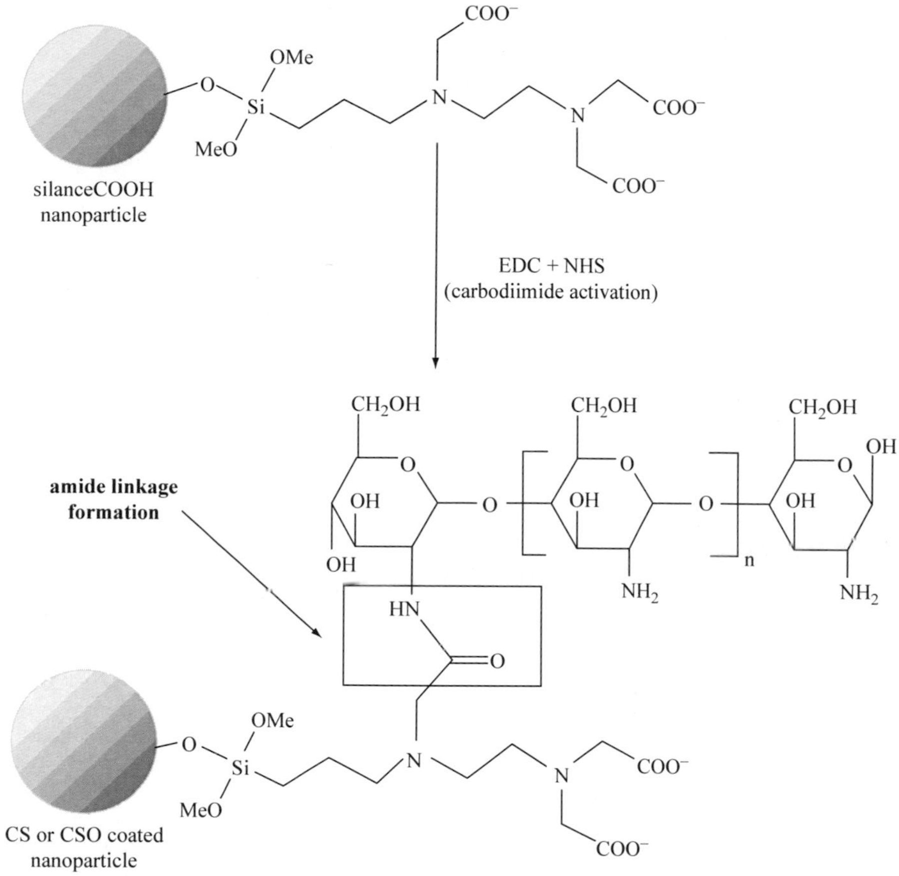

Fig. 3.3 Chemical modification of the NPs surface with the biopolymer chitosan, using the carbodiimide/NHS procedure. Chitosan binding takes place with the amide linkage formation between the carboxylic group on the NPs surface and the amino group of the chitosan. (Reproduced with permission from Lopez-Crus et al., J. Mater. Chem., 2009, 19, 6870–6876 [2])

biopolymer via the EDC/NHS chemistry [2]. Glutaraldehyde is another bifunctional reagent commonly used to attach functional labels. Glutaraldehyde binds at one end to amino groups of the NPs and at the other end to the amino groups of biomolecules *via* a Schiff reaction.

Natural or synthetic polymeric macromolecules have been used most frequently as NPs coatings for biomedical applications. These polymers can add neutral, charged, hydrophilic or amphiphilic properties onto the NPs surface. Their main function is to stabilize NPs, prevent aggregation, enhance biocompatibility and facilitate cellular uptake. Commonly used polymers include poly(ethylene glycol) (PEG; also known as polyethylene oxide, PEO), dextran, poly-lactic acid, poly-acrylic acid, hyaluronic acid, polyethylene imine (PEI), polyvinyl alcohol (PVA),

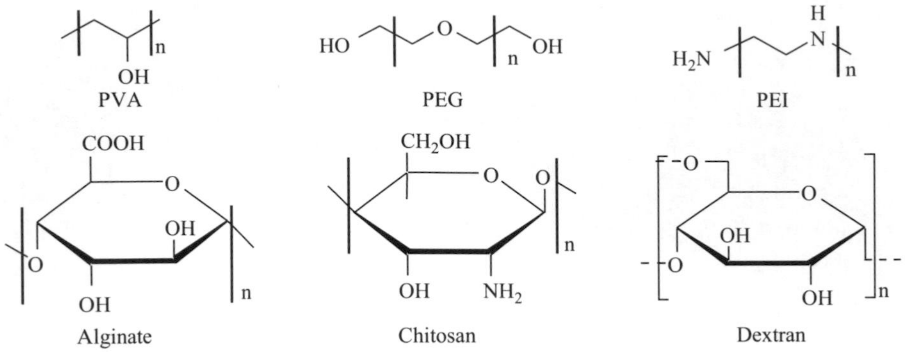

Fig. 3.4 Chemical structure of most commonly used polymers as nanoparticle coating materials: polyvinyl alcohol (PVA), poly(ethylene glycol) (PEG), polyethylene imine (PEI), alginate, chitosan and dextran

polyacrylic acid (PAA), alginate and chitosan and its derivatives (their chemical structures are shown in Fig. 3.4) [26, 86]. PEG is a hydrophilic, water soluble non-toxic FDA-approved synthetic polymer that increases biocompatibility, enhances blood circulation time and improves resistance to non-specific protein adsorption [87–89]. PEG coated iron oxide NPs showed excellent solubility and stability in aqueous solutions and physiological conditions [75]. Dextran is a glucose-derived biodegradable polysaccharide that improves biocompatibility and provides NPs with hydroxyl functionalities. Chitosan is a non-toxic hydrophilic cationic poly-saccharide commonly used in pharmaceutical formulations and in NPs used as drug delivery vehicles [90]. When used as a coating material, chitosan adds amino groups to the NP surface. Chitosan can be either adsorbed or covalently linked to the NP surface as shown in Fig. 3.3 [2]. Chitosan coated iron oxide NPs are stable in biological buffers over a wide pH range. Alginate is another polysaccharide with multiple carboxyl functionalities that produces highly stable alginate coated NPs. PEI is a synthetic cationic polymer that forms cationic complexes with NPs and enhances cell entry via endocytosis [91]. PVA is a hydrophilic biocompatible poly-mer commonly used to prevent NPs aggregation. PAA is used to produce carboxyl functionalized NPs [86]. An important variable in the functionalization procedure is the polymer length. By selecting and optimizing the length and size of the coating material, it is possible to modulate *in vivo* effects including cellular uptake, drug delivery efficiency and toxicity [91]. The length of the PEO copolymer coating NPs was, for example, inversely correlated with toxicity. NPs functionalized with polymeric chains less than or equal to 0.75 kDa were the most toxic, and NPs coated with chains equal to or longer than 15 kDa were the least toxic [92].

Surface modified NPs and the quality of the coating can be determined by mea-suring the particle charge (zeta potential deduced from measurements of electro-phoretic mobility), particle size distribution (dynamic light scattering), size of the crystal using transmission electron microscopy (TEM) and X-Ray diffraction. Fou-rier Transform Infrared Spectroscopy and x-ray photoelectron (XPS) spectra can be used to probe the presence of functional groups [73, 86]. Further improvements in surface modification techniques are needed for successful *in vivo* use of NPs. Specifically, enhancing coating efficiency while ensuring biocompatibility, controlling the number and location of functionalities and achieving high reproducibility during synthesis are several important objectives that need to be addressed in future research activities.

3.4 Interaction of Metal Oxide NPs with Biological Systems

The interaction between the NPs and the biological entities (proteins, cells, tissue) occurs first at the particle surface. Therefore the composition, charge, morphology and the available surface area are important parameters. The following sections cover aspects of the interaction between NPs and proteins, cells and tissue and the effect of physico-chemical parameters.

3.4.1 Interaction of Nanoparticles with Proteins

In a physiological environment, proteins with affinity for the NP surface attach and cover the particle, forming a "protein corona" [93, 94]. Such interactions are complex due to the large surface area available, the variability between different NPs of various compositions and coatings and the wide number and composition of natural proteins present in biological systems. The factors controlling formation of the "protein corona" are poorly understood. Further study is required to determine how binding constants, affinity and stoichiometry affect the formation of the protein coat. Assessing the lifetime of protein-NPs complexes, the association/dissociation kinetics and the competition between the various proteins for the NP surface is also needed. These are important studies since the nature of the protein coat may dictate the fate and transport of the NPs within cells and tissues, the biodistribution and clearance within the body and the biological response to the NPs [14, 95, 96]. Greater understanding of the factors involved in the creation of the protein coat may also provide insight into the design of NPs with improved drug delivery and diagnostic functions.

Both the amount and the type of protein bound onto the surface influence the transport and biodistribution of the NPs, and these are closely linked with the surface properties of the inorganic particles. Studies of NPs with plasma proteins have been carried out after incubation of the NPs with plasma followed by isolation of the protein-NPs complex in conditions that mimic the bloodstream [3, 95]. Centrifugation, microfiltration, size exclusion chromatography, affinity chromatography, magnetic separation or a combination of these methods have been used

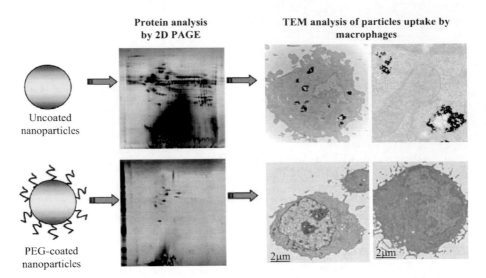

Fig. 3.5 Protein binding and macrophage uptake for PEG modified and unmodified NPs and transmission electron microscopy (TEM) demonstrating differential accumulation of the particles inside cells: PEGylated particles bind less protein and are not taken up by macrophages. (Reproduced with permission from the ACS (Dobrovolskaia et al., Mol Pharm., 2008) [3])

to separate the protein-NPs complex from the biological fluid. The amount of total protein bound to the NP can be quantified using proteomic approaches, e.g. sodium dodecyl sulfate polyacrylamide electrophoresis (SDS-PAGE), Western Blot, Mass Spectrometry (MS), High Performance Liquid Chromatography (HPLC), capillary electrophoresis, protein microarray as well as Isothermal Titration Calorimetry (ITC) and Surface Plasmon Resonance (SPR). Size and surface properties of the NPs, rather than the bulk characteristics, were found to affect the protein adsorption pattern and determine the composition of the adsorbed protein layer when NPs were exposed to human serum [93, 97]. When the particle is larger than the protein, multiple proteins bind to each particle and form extended aggregates. On the other hand, when the protein is larger than the particle, multiple particles bind each protein [98]. There are only a few examples of studies of particle-protein complex in a relevant biological environment [93, 126] and only some of these have explored the interaction between metal oxide NPs and proteins [94, 99]. As determined by ultracentrifugation, SDS-PAGE, gel electrophoresis and MS, metal oxide NPs of TiO_2, ZnO and SiO_2 bound a wide variety of plasma proteins [99]. Different shapes of TiO_2 nanostructures (nanospheres, nanorods and nanotubes) had different protein binding patterns; nanospheres bound more proteins than the other structures. The metal oxide NPs-protein complex reached equilibrium within a few minutes of incubation and agglomeration of the NPs had a large effect on protein binding [99]. One way to prevent protein binding is to functionalize the particles with a biocompatible PEG layer to create a sterically protective shell around each NP. This procedure changed the interaction of the NPs with plasma proteins and enhanced particle circulation in the blood. Figure 3.5 shows an example of the uptake by macrophages of protected and unprotected particles (citrate stabilized gold NPs); note that PEG-coated particles are not taken up as effectively [3].

3.4.2   Interaction of Nanoparticles with Cells and Tissues

NPs may alter multiple signals in the cell environment determining cell fate and development [100]. The first biologically important interaction after exposure to NPs is the rapid adsorption of proteins. The type, amount and conformation [101, 102] of the adsorbed proteins regulate cellular adherence, cell migration, proliferation, and differentiation [94, 103]. NPs can alter membranes, change gene expression, disrupt mitochondrial function, form reactive oxygen species (ROS), and decrease cell viability. Yet, the mechanisms by which extra-cellular compounds are recognized by and/or gain entry into target cells are not fully investigated [104]. Presumably, surface chemistry determines cellular binding of NPs, and these non-specific interactions with plasma proteins are likely the main factors responsible for NPs binding to endothelial cell surface [96]. Moderate levels of intracellular iron oxide NPs were found to affect neural cell functioning [48]. Moreover, NPs (e.g. ceria) can either scavenge reactive oxygen species (ROS) or actually enhance ROS formation, depending on the conditions [105]. On the other hand, the toxicity of copper NPs [106, 107] was associated with mitochondrial failure, ketogenesis, fatty acid beta-oxidation and glycolysis, while the toxicity of TiO_2 NPs was attributed to changes in the cell surface, membrane breakage and oxidative stress [108].

Cellular uptake of NPs may occur by pinocytosis, non-specific or receptor-mediated endocytosis or phagocytosis. Attachment and entry of NPs into cells is determined, like binding to proteins, by factors like electrostatic and hydrophobic interactions, specific chemical interactions as well as particle size [67, 91, 109]. The attachment of the particles to the cell membrane seems to be most affected by their surface charge. Patil et al. showed that cells accumulate NPs to a greater extent when the surface charges on the particles are negative demonstrating that electrostatic interactions are important determinants of protein adsorption and cellular uptake [110]. Cell membranes possess large negatively charged domains, which should repel negatively charged NPs. However, Wilhelm et al. [111]. suggested that the negatively charged particles bind at cationic sites that form clusters on the cell surface because of their repulsive interactions with the negatively charged domains of the cell surface. In addition, the NPs, already bound on the cell surface, present a reduced charge density that may favor adsorption of other free particles. Thus, the high cellular uptake of negatively charged NPs is related first to the non-specific adsorption on the cell membrane and second to the formation of NP clusters [110].

3.5 Main Types of Metal Oxide Nanoparticles with Potential Use in Biomedicine

Most biomedical applications of metal oxide NPs involve magnetic iron oxide NPs as molecular labels and titania as components of biomedical implants. Other particles, like ceria, have only recently been considered as potential candidates in the biomedical arena. This section provides an overview of the most promising and widely used NPs in biomedicine.

3.5.1 Magnetic Iron Oxide Nanoparticles

Magnetic iron oxide (Fe_2O_3 and Fe_3O_4) NPs have attracted considerable interest in the biomedical field for cell labeling and separation, magnetic resonance imaging (MRI), targeted drug delivery, bio-imaging, hyperthermia and biosensing [4, 15, 70–74, 112–114]. Such applications require well-dispersed, chemically stable, uniform particles with high magnetization values and sizes smaller than 100 nm. Various forms and sizes of magnetic iron oxide NPs are available and can be purchased from several companies, or they can be prepared using established synthetic procedures [1, 113, 115], recently reviewed in detail [25, 26, 46]. Hydrolytic synthetic routes and thermal decomposition methods are most common. For most biomedical applications, particles are functionalized with biocompatible coatings [4, 46]. Most iron oxide NPs currently in pre-clinical or clinical testing are dextran coated. Surface coatings (e.g. thickness, hydrophobicity, presence of ligands) and size can affect relaxivity of magnetic NPs (the efficiency with which a contrast agent can accelerate the proton relaxation rate) [112, 116–119]. Relaxation rates increase as the size of the NP increased, but increasing coating thickness decreased relaxivity [47]. The magnetism can also be affected by adding dopants (Fe^{2+} is replaced by transition metals M^{2+} (M = Mn, Ni, Co)) [114]. However, enhanced MR capabilities of doped structures may be associated with enhanced toxicity *in vivo* do to the release of the dopant (toxic species like Ni, Co, etc.).

Highly magnetized NPs are used as contrast agents for molecular and cellular imaging in MRI [47, 120]. Such applications require transport and accumulation of the contrast agent to the target site for disease detection [121]. PEG and dextran functionalized particles have been used frequently for this purpose. Formation of nanoscale aggregates of superparamagnetic NPs with sizes ranging from 60–150 nm increased the efficiency of NP contrast agents due to enhanced relaxivity. Higher detection sensitivity can be obtained by immobilizing specific ligands (e.g. antibodies, receptors) with high affinity towards the molecular targets to facilitate preferential accumulation of the NPs at the desired anatomical location [87]. Further attachment of fluorophores onto the NPs surface provided, in addition to magnetic resonance, fluorescence imaging capabilities [4]. Immuno and fluorescently labeled (e.g. Cy5, FITC, Texas Red) iron oxide NPs maintained both

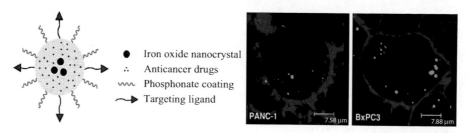

Fig. 3.6 A nanocomposite of magnetic iron oxide NPs carrying anticancer drugs and targeting ligands in a silica sol-gel matrix used as drug delivery vehicle. The particles were internalized by pancreatic cancer cells as seen by the fluorescence images. (Reproduced with permission from Liong et al. ACS Nano, 2, 2008, 889–896 [4])

immunoreactivity (and/or fluoresce) and magnetic properties. NPs with multiple functionalities are usually at least 20–30 nm in size. Most particles used in molecular imaging are between 30 and 100 nm (NPs larger than 40 nm are called small particle iron oxide—SPIO, and particles smaller then 40 nm are called ultra-small particle iron oxide—USPIO). Larger particles will not enter cells and will be taken up by macrophages and the reticulo-endothelial system [87]. Small particles less than 10 nm diameter can be easily taken up by endocytosis [63, 121] and have a longer blood circulation times [87, 113]. The same particles used for diagnosis can also be engineered to deliver drugs [73]. Loading drugs onto a mesoporous silica matrix loaded with magnetic iron oxide and functionalized with a phosphonate coating is an example of a drug delivery NP; these particles crossed the cellular membrane and were internalized by pancreatic cancer cells (Fig. 3.6) [4]. Magnetically targeted drug-delivery systems have received a great deal of attention since the location and properties of NPs can be controlled or manipulated by an external magnetic field.

Surface-functionalized superparamagnetic NPs were used for *in vivo* imaging and as nanoprobes for studying molecular interactions in living cells. Multifunctional iron oxide nanospheres with fluorescence, magnetic and bio-targeting properties were used for visual cell sorting and manipulation of apoptotic cells [122]. Clinical applications of iron oxide NPs include imaging of the liver, gastrointestinal tract, metastatic lymph nodes, uterine neoplasms, ischemia-associated inflammation in stroke lesions, atherosclerosis, multiple sclerosis, kidney disease, infection and brain tumors and to study cerebra myocardial or renal perfusion [112]. Such applications have been described in detail in several recent reviews [25, 46, 112, 114, 123].

The cellular response and toxicological effects of magnetic iron oxide NPs have been studied *in vitro* and *in vivo* [92, 124], and different cellular responses were observed for coated and noncoated NPs. Modifying magnetic NPs with polymers like dextran, chitosan and PAA was found to influence cell proliferation and cell viability to a small extent. In one study, internalization of the NPs by cells adversely affected cell behavior, and this was dependent on particle coating [125]. Dextran coated NPs induced alterations in cell behavior and morphology that were different from the reaction to bare NPs. In another study, toxicity of polyethylene oxide (PEO) functionalized magnetic NPs in *in vitro* cellular assays varied depending on the molecular weight of the polymer—shorter polymer chains induced greater toxicity [92]. NP uptake by the cells varied with the cell type [126]. In general, magnetic NPs were characterized by low toxicity. Iron oxide NPs are believed to be biodegradable; the iron is released and then metabolized through normal biochemical pathway [126, 127]. Nonetheless, iron NPs may induce release of ROS and increase oxidative stress.

3.5.2   Titania

Titania (TiO_2) is the most investigated and well characterized single crystalline metal oxide. The most common forms are anatase and rutile. Most applications of TiO_2 are in catalysis and photocatalysis, photovoltaic cells, as dielectric gate materials, and in photo-assisted degradation of organic molecules. Biomedical applications include bone implants, UV blockers for skin protection and as support for promoting cell adhesion. Nano-TiO_2 of different forms and shapes such as spheres, wires and hollow spheres, and high surface area nanostructured TiO_2 thin films have been produced and many are commercially available [75]. The surface science, electronic structure, and properties of TiO_2 have been reviewed by Diebold, 2003 [128]. Several reports discussed synthesis of uniform TiO_2 NPs with desirable surface functionalities [129–131] and discussed their applications in the biomedical field [131, 132]. The mechanical and physical properties of TiO_2 NPs can vary depending on the preparation method. Titania powders and films can be obtained by sol-gel processes via a colloidal or an alkoxide route [18], heat and chemical treatment and by electrochemical methods [76, 130]. Electrochemical methods are mainly used to form uniform titania film coatings and have the advantage of controlling film thickness by varying the electrochemical deposition parameters. Other types of structures, such as nano-shells with a Ag core and a titania shell, were fabricated using a layer-by-layer assembly technique and a negatively charged polyacrylic acid polymer [75]. Nano-shells with a 10–30 nm diameter were used to produce bioactive coatings with tunable permeability and sieving functions for dopamine and ascorbic acid for use in neurochemical monitoring. The films were compatible with nervous tissue and allowed selective detection of dopamine, an important neurotransmitter, in the presence of ascorbic acid [75]. Under physiological conditions, the negative charge of the TiO_2 NPs facilitated permeation of positively changed dopamine while restricting the access of ascorbic acid, a negatively charged compound and a major interfering compound in biological systems. The TiO_2-polyelectrolyte assembly was compatible with nerve cells (PC12 cells) and could be used in implantable neurological devices for in-vivo monitoring of brain activity. Similar core-shell structures with a ZrO_2 shell and an average diameter of ~35 nm and a shell thickness of 2–3 nm were prepared [133]. Recently, we have demonstrated that incorporation of titania, in combination with ceria NPs in enzyme sensors facilitates detection of low levels of dopamine *in vivo* (detection limits of

1 nM) in the brain of anesthetized rats [134–136]. Both ceria and titania have potential uses in bioanalytical sensors for monitoring analytes of biomedical importance (dopamine, glucose, glutamate, hydrogen peroxide, ROS. etc) [134–136]. Such systems might be useful in neuroscience research involving studies of the role of neurotransmitters on disorders like stroke, epilepsy, Alzheimer and Parkinson, and other biomedical conditions.

Biocompatibility, good mechanical stability and self-cleaning properties under UV exposure are all valuable characteristics of TiO_2. Electrochemically deposited titania films on stainless steel substrates provide a protective, biocompatible layer in biomedical implants [76]. Sol-gel derived titania with controlled porosity and composition promote protein adsorption and improve tissue attachment of implants in vitro and in vivo [77, 78]. The deposition of a titania film on medical implants facilitates strong bonding of the implant to the surrounding tissue through the formation of a calcium phosphate layer onto the surface of the implant [18]. Both anatase and rutile were tested for this purpose and no significant difference in tissue responses were found between the two forms of titania [78, 137]. Cell attachment on titania surfaces was comparable to attachment on glass. The titania coating can also be engineered to deliver drugs at the implantation site using titania nanotubes [137]. The strong protein adsorption and bone bonding characteristics of titania are related to the existence of surface hydroxyl groups, and protein adsorption can be improved by changing the chemical composition and oxide structure of the implant surface (e.g. doping with ions like Ca, F, P and Mg) [131]. A study of the influence of the nanostructured morphology and anions present in TiO_2 oxide layers on cell adhesion and proliferation demonstrated that cells are mainly influenced by adsorbed anions rather than the morphology of the surface [19]. Adsorption of proteins onto these metal oxide surfaces is also essential in studying mechanisms taking place between biological entities and medical implants. The reactions at the Ti-OH surface are related to the surface charge, which can be made either positively or negatively charged, depending on the environmental pH.

Evaluation of in vivo toxicity is needed prior to biomedical use. Several papers have reported cytotoxicity studies of titania nanostructures using in vivo and in vitro models [138]. Some results indicated the occurrence of oxidative stress and inflammatory responses after administration of titania NPs [138, 139], but few toxic effects were observed following low doses of anatase titania NPs of 5 nm diameter [139, 140]. However, caution must be taken in drawing conclusions from early nanotoxicology studies [141] due to the large number of uncontrolled experimental variables, and the variability among biological models and the different types of titania nanostructures. Toxicity considerations of metal oxide NPs are discussed in Sect. 3.7 in this chapter.

3.5.3 Ceria

Cerium oxide or ceria (CeO_2) and doped CeO_2 films have received significant attention as a result of interesting catalytic and free radical scavenging properties, which make them attractive for multiple applications in biology, medicine [82, 142], bio-sensing [134–136, 143, 144] environmental remediation [142] and catalysis [144, 145]. Cerium is a lanthanide element with a fluoride-like crystalline structure. Cerium can exist in two oxidation states: Ce^{3+} or Ce^{4+}. Under a redox environment, these two states may be interchanged [12]. Changes in the oxidation state can occur as a result of factors such as temperature, pH, presence of reactive species and partial pressure of oxygen [20, 146–148]. The many useful catalytic properties of ceria have been attributed to the presence of highly mobile lattice oxygen present at its surface and a large diffusion coefficient for the oxygen vacancies [149], which facilitates the conversion Ce^{4+} to Ce^{3+} and thus, allows oxygen to be released or stored in the crystalline structure of cerium oxide.

$$2CeO_2 \leftrightarrow Ce_2O_3 + 1/2\ O_2$$

In addition to pure ceria, mixed ceria-based metal oxides exhibit increased oxygen vacancies due to the formation of defective lattice structures, resulting in new materials with enhanced physico-chemical properties and greater catalytic activity [150]. The structure of cerium oxide may be modified by mixing or doping ceria with other rare-earth or transition metal oxides [151, 152]. For example, binary and tertiary mixtures of cerium oxide and zirconia (ZrO_2), yttria (Y_2O_3) or titania (TiO_2) have been made [150, 151, 153–155]. In these doped materials, the local oxygen environment is modified and more efficient redox processes are permitted. Most studies attribute the enhanced oxygen mobility of mixed and doped oxides to defective crystal structures with an increased number of oxygen "vacancies" and their thermodynamic equilibrium [151]. Procedures for preparing homogenous mixtures of metal oxides with increased oxygen vacancies have also been reported [21, 156, 157] for CeO_2 mixed with ZrO_2, TiO_2, YO_2 [151], and CeO_2 doped with Sm, Gd, Y, Yb, Pt [158, 159]. Surface modifications of the ceria NPs can prevent aggregation, minimize protein adsorption and improve biocompatibility [21, 110], and nanoceria synthesized in dextran and PEG was shown to promote cell proliferation and increase cell viability [21].

Recently, NPs of CeO_2 with high Ce^{3+}/Ce^{4+} ratios were shown to mimic super-oxide dismutase activity [20]. In another report polymer coated ceria had intrinsic 'oxidase-like activity' at acidic pH values; this was dependent on the size of the NPs and the thickness of the polymer layer [80, 155, 160]. The regenerative 'enzyme-like activity' of ceria involved a reversible oxidation/reduction cycle, indicating that ceria NPs may have value as therapeutic agents in the treatment of medical diseases related to reactive oxygen species (ROS) [83, 84, 155]. These include utilization in spinal cord repair as neuroprotective agents [82, 142], promotion of cell survival in oxidative conditions [80, 81] and scavenging reactive molecules in the eye to prevent or delay degenerative retinal disorders [81]. When cells were exposed to H_2O_2, a source of reactive oxygen molecules, ceria-based NPs reduced oxidative damage by over 60% [82, 142]. Similar effects were reported with yttria [80]. The antioxidant radical scavenging properties of nanoceria may detoxify reactive oxygen species by one or more of the following mechanisms [21]:

$$Ce^{3+} + OH^\bullet \rightarrow Ce^{4+} + OH^-$$

$$Ce^{4+} + O_2^{\bullet -} \rightleftarrows Ce^{3+} + O_2$$

$$Ce^{4+} + 4OH^- \rightarrow Ce(OH)_4$$

$$Ce^{3+} + 3OH^- \rightarrow Ce(OH)_3$$

These processes take place at the surface of the metal oxide and are affected by conditions such as pH and temperature [21]. It is important to synthesize nanoceria for biomedical applications in "biologically relevant" media that are compatible with cells and tissue, while still conserving the unique redox properties of these particles. There have been very few studies of the redox and antioxidant capacity of doped or mixed oxides, and few studies to elucidate the reaction mechanisms in biological relevant conditions [80, 155]. The next section discusses applications of ceria as scavengers of free radicals and their use in the treatment of oxidative-stress related diseases.

3.6 Ceria Based Therapies for Oxidative Stress-related Diseases

3.6.1 Free Radicals and Biological Consequences of Oxidative Stress

Free radicals are highly reactive molecules that strip electrons from other compounds to compensate for an unpaired electron in an outer orbital shell. They are formed by a variety of both normal and pathological biological process. For example, mitochondria generate free radicals as part of a normal series of steps in which carbon based fuels (glucose, fats and proteins) are oxidized by oxygen. Moreover, because of their intense chemical reactivity, free radicals tend to enter into chemical reactions that generate additional free radicals with lower chemical reactivity; this cascade of free radicals may actually contribute to normal cell signaling mechanisms. However, free radicals are also generated during many pathological processes, such as inflammation, ischemia and reperfusion. In addition, humans are exposed to free radicals in the environment as a result of pollutants and by exposure to ultraviolet light.

Cells and organisms possess intrinsic mechanisms to regulate the levels of free radicals. These detoxifying processes consist of enzymes that convert free radicals to less toxic substances in the presence of appropriate substrates, or chemical compounds that may donate an electron to the free radical to reduce its reactivity. However, the production of free radicals either endogenously or from exogenous sources can easily exceed the capacity of these protective mechanisms to detoxify free radicals and lead to oxidative stress. In living cells, targets of the damaging effects of excess free radicals include molecules indispensable for normal function such as proteins, lipids and DNA. As such, oxidative stress plays a role in aging as well as a variety of disease states.

The hypothesis that free radicals and oxidative stress are part of the normal aging process was initially put forth by Harman in 1956 [161] as the 'free radical theory of aging'. That oxidants contribute to aging is now widely accepted [162], and there is overwhelming support for this idea from both *in vivo* and *in vitro* studies. For example, long term exposure to free radical scavengers/antioxidants is neuroprotective and increases life-span in a variety of model systems [163–166]. The increase in oxidative stress with aging can be attributed to an increase in oxidant load, decrease in antioxidant capacity and decreased ability to repair oxidative damage over time [167–169].

Oxidative stress also plays a prominent role in the pathology of age-related neurological disorders including Alzheimer's disease, Parkinson's Disease, cerebrovascular disease and stroke (recently reviewed) [170–175]. In stroke, reactive oxygen species (ROS) derived from various sources accumulate both during ischemia and during the reperfusion period [176, 177]. This ischemic/oxidative stress can ultimately contribute to cell death via necrotic or apoptotic pathways [173, 178–181]. Large vessel ischemic stroke (due to a clot or atherosclerosis) can lead to the loss of 1.9 million neurons and 13.8 billion synapses per minute, equivalent to what is lost in 3.1 weeks of aging [182]. The dramatic death of brain cells is due to the fast onset of cellular events, termed the ischemic cascade, that contributes to the demise of brain cells; bioenergetic failure leads to loss of ionic homeostasis, excitotoxicity, oxidative stress and eventually cell death via necrotic or apoptotic mechanisms [183]. Reactive oxygen species, including superoxide anion ($O_2^{\bullet-}$), hydrogen peroxide (H_2O_2), hydroxyl radical ($HO^{\bullet-}$), nitric oxide (NO) and peroxynitrite ($ONOO^-$) accumulate during the ischemic period and induce oxidative damage. This damage is worsened when blood flow is restored during the reperfusion period [179, 180, 184–187]. Sources of ROS during ischemia/reperfusion include the mitochondria and calcium-dependent enzymes such as cyclooxygenase, xanthine oxidase and nitric oxide synthase [187]. The involvement of NO in ischemia has received considerable attention because of the myriad deleterious effects associated both with NO and its byproduct, peroxynitrite ($ONOO^-$).

NO normally plays an important physiological role in neurotransmission and the regulation of cerebral blood flow, but elevated concentrations of NO exacerbate tissue damage under pathophysiological conditions such as stroke. Following brain ischemia, NO levels in the ischemic region increase into the micromolar range within 20 min *in vivo* and remain elevated for up to 48 h [188]. Elevated intracellular Ca^{2+} resulting from glutamate excitotoxicity may lead to increased activation of nitric oxide synthase (NOS) and higher concentrations of both NO and peroxynitrite [189–196]. Decreasing the activity of NOS following ischemia using NOS antagonists ameliorates neuronal injury following severe hypoxia or ischemia in animal models, brain slices and cultured neurons [197–201]. NOS knockouts have smaller infarct volumes and fewer neurologic deficits compared to their wild type littermates following acute, focal ischemia [202–205]. In addition, NO donors administered after 12–24 h following ischemic insult exacerbate injury [206]. It has been suggested that nitrosative damage resulting from peroxynitrite (ONOO⁻) formation secondarily activates caspases, poly(ADP-ribose) polymerase (PARP-1), and the translocation of apoptosis inducing factor (AIF) into the nucleus, resulting in programmed cell death [203, 204, 206–218].

Peroxynitrite is both an oxidant and a nitrating agent and is formed by the reaction of NO and superoxide anion. Unlike other strong oxidants, peroxynitrite reacts selectively with biological targets. The major protein modification following the reaction of peroxynitrite with proteins is the nitrosylation of tyrosine residues (3-nitrotyrosine) [219]. Endogenous tyrosine nitration is derived from enzymatically produced NO, although NO itself is not a nitrating agent. Under pathophysiological conditions, it appears that CO_2 is a catalyst for peroxynitrite-mediated nitration of tyrosine residues by nitrosocarbonate. While nitrosocarbonate degrades more quickly than peroxynitrite, it can attack a wider number of substrates. Furthermore the nitrosocarbonate-peroxynitrite interaction appears to enhance protein nitration, one of the most destructive aspects of peroxynitrite [219, 220]. Peroxynitrite can also react with thiol groups, and the nitrosylation of cysteine residues is an important signaling mechanism and is involved in the regulation of enzymatic activity analogous to the process of phosphorylation [219–222]. Given the wealth of data supporting a deleterious role of NO-mediated nitrosative damage following ischemic-reperfusion injury, interventions that reduce nitrosylation are currently being investigated as therapeutic targets.

3.6.2 Use of Ceria Nanoparticles to Mitigate Oxidative Stress

To date, there has been a paucity of antioxidant agents with proven efficacy in the treatment of neurological diseases. For example, vitamin E (alpha tocopherol), vitamin C (ascorbic acid) and carotenoids have been tested in a variety of 'oxidant injury' diseases, but none (with the possible exception of Vitamin E use as a preventative therapy in atherosclerotic heart disease) has proven successful [223, 224] These agents have failed for a variety of reasons including their limited antioxidant capacity and difficulty penetrating the blood brain barrier (they have difficulty gaining access to the site of free radical formation in the brain). Additionally, in many cases the production of free radicals occurs rapidly and early in the disease process, and administration of antioxidant agents after the initial ischemic/hypoxic injury is ineffective.

A potential new therapy for the treatment of oxidative stress diseases is the use of ceria NPs [225]. The best-known mechanism underlying the action of these NPs is thought to originate from their dual oxidation state [21, 22, 83, 84, 149] and their ability to reversibly bind oxygen and shift reversibly between Ce^{4+} and Ce^{3+} states under oxidizing and reducing conditions. The loss of oxygen and the reduction of Ce^{4+} to Ce^{3+} are accompanied by the creation of oxygen vacancies in the NPs' lattice. The unique antioxidant properties of ceria NPs derive from the kinetics and thermodynamics of the redox processes on the NPs surface. These processes can be enhanced if cerium oxide is in contact with noble metals which can facilitate the transfer of oxygen from the bulk material to the surface and vice-versa [147, 148].

3.6.3 Biological Models Used to Study the Antioxidant Effects of Ceria NPs

Because of their potent free radical scavenging properties, ceria NPs have been used to mitigate oxidative stress in various biological systems [80, 81, 163, 226]. *In vitro*, ceria NPs are able to rescue HT22 cells from oxidative stress-induced cell death [142] and protect normal human breast cells from radiation-induced apoptotic cell death [81] and oxidative damage to the retina following intraocular injection of ceria NPs in rats [227]. Treatment of rat neurons with nanoceria increased the lifespan in culture and reduced the cell injury induced by hydrogen peroxide or UV light *in vitro* [228]. In a recent study, Niu et al. showed that ceria NPs decreased both NO and peroxynitrite formation in a murine model of ischemic cardiomyopathy [229]. Other potential biological uses of nanoceria have emerged from studies showing protection of primary cultures from the detrimental effects of radiation therapy [142], prevention of retinal degeneration induced by intracellular peroxides [230] and neuroprotection to spinal cord neurons [163]. Further studies have used either neuronal or mixed neuronal/glial cultures to investigate the scavenging abilities of ceria to a single ROS species [163, 230].

Cell culture systems represent a relatively defined microenvironment where the presence of a vascular compartment is not a confounding variable. And, as outlined above, the use of immortalized cell lines or primary cell cultures has shed considerable light on the protective effects of CeO_2 NPs after a variety of insults that increase ROS production including UV light, H_2O_2, irradiation and glutamate-induced excitotoxicity. One disadvantage of these systems, however, is that the regional organization and neuronal connectivity are lost as are the relationships of these cellular elements to the extracellular matrix. For example, results from recent studies suggest that many extracellular matrix proteins (e.g. integrins, MAP kinases) contribute to cell signaling and play an active role in shaping neuronal-glial output in the brain [231, 232].

In contrast to cell culture systems, brain slices (both acute and organotypic) provide the advantage of maintaining cell stoichiometry and much of the regional connectivity [201, 233, 234], thus preserving the circuitry between neurons and glia and their relationship to the extracellular matrix. The first studies on hippocampal organotypic (cultured) slices assessed their physiological relevance using

Fig. 3.7 Low power, CCD image (1X) of an organotypic brain slice stained with the fluorescent, nucleophilic dye Sytox Green (Molecular Probes). Hippocampus and the dentate gyrus are located at the center of the image

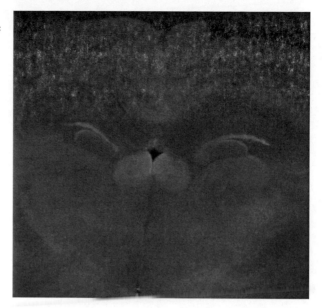

Fig. 3.8 Administration of cerium oxide nanoparticles (1 µg/mL) to organotypic brain slices exposed to ischemia reduces tissue death ~40% (P<0.05; n = 7 paired sections)

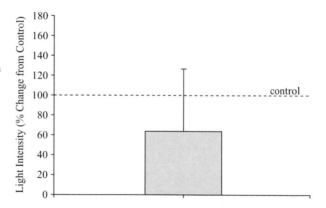

acute slices as comparison [179, 184, 201, 233–235] (Fig. 3.7) and confirmed that synaptic activity as well as chemical signaling is maintained. Cultured slices thus allow the functional analysis of the properties of the nervous tissue and represent a widely accepted experimental model used extensively in molecular biology, electron microscopy, imaging, electrophysiology and immunohistochemical studies. The use of more intact preparations, such as brain slices, is likely a more suitable intermediate model for examining the cellular effects of ceria nanoparticles before they are tested in intact, living animals. Using organotypic brain slices prepared from CD1 mice, we found that treatment with ceria NPs ameliorates ischemic injury in the hippocampus [314] (Fig. 3.8), a region of the brain particularly vulnerable to the effects of oxygen and glucose deprivation associated with stroke.

3.7   Toxicity Considerations of Metal Oxide Nanoparticles for *In Vivo* Biomedical Use

While many engineered nanomaterials are commonly used in commercial products and industrial processed, interactions of NPs with biological processes such as transport mechanisms, kinetic properties, toxicity and accumulation in living organisms as well as their environmental and health effects are largely unknown, although these issues have recently become a significant concern [16, 236–238]. Concerted efforts to understand the behavior of nanomaterials in biological systems and study their possible interactions with living organisms have been promoted at a global scale [239]. In the following sections we discuss the *in vivo* and *in vitro* toxicity of metal oxide NPs relevant to biomedical applications.

3.7.1   Effect of Physico-Chemical Properties

Various types of metal oxide nanostructures (e.g. ZnO, CuO, TiO_2) show varying degrees of cytotoxic effects which are not observed with larger particles of the bulk material [240–242]. Recent studies focus on assessment of health effects of NPs after respiratory, gastrointestinal and dermal exposure using mice [10, 243], cell culture [240, 244, 245], bacteria or crustaceans [246], freshwater alga [241] and zebrafish [247–263]. While research is growing, mechanisms of developmental toxicity remain largely unexplored [248]; understanding the translocation, accumulation and retention pathways of NPs in vital target sites need more systematic studies [264]. Recent research has stressed the need to develop a "predictive toxicological paradigm" for the assessment of nanomaterial toxicity to define which aspects of each material's physicochemical properties "leads to molecular or cellular injury and (the assessment) also has to be valid in terms of disease pathogenesis in whole organisms" [57, 265–267]. The mechanisms toxicity of NPs are complex [268] and may be a result of:

- Nanomaterial size: NPs less than 100 nm can enter cells by receptor mediated endocytosis [97]; NPs less than 40 nm can enter the cell nucleus; and NPs less than 35 nm can pass the blood brain barrier [243, 269–272]
- Material solubility, which varies with particle composition and species [241, 268, 273]
- Aggregation, which results in larger entities ranging from several hundred nm to several μm [241]
- Production of oxidant species: NPs in contact with biological materials (cells or tissues) can trigger production of ROS [274–277], which can further damage the cells through oxidative stress [240, 241, 246]
- Reactivity due to the metabolic alkalosis or intracellular dissolution [10, 268, 276]

- Increased mobility across cell membranes: NPs may be able to cross cell membranes [278]. However, studies have shown that NPs do not necessarily have to enter the cells to produce toxic effects [246]
- Surface coating: coating and physicochemical properties (hydrophobicity and surface charge) [279, 280].

Several studies have shown a direct relationship between the structure of NPs and their impact on biological systems [263, 281]. For example, Au NPs—traditionally an inert material as a bulk compound—show significant biological reactivity in mammalian cells. In some studies, Au NPs have shown reduced toxic effects despite their uptake into the cells by endocytosis [282, 283]. However, when the particles were functionalized with cationic side chains, they were toxic [284]. The aggregation state and surface charge are other important factors that determine cellular uptake and NP's adsorption on cellular membrane. The surface coatings for the 'same' type of material may depend on synthesis and/or processing used [285, 286]. It is also important to correlate the material's biophysical characteristics with *in vivo* and *in vitro* assays and establish a "life cycle" of the behavior and transport of NPs in biological systems starting from the material processing to cellular uptake, tissue response and clearance.

3.7.2   In Vivo *and* In Vitro *Toxicity Studies of Metal Oxides Nanoparticles*

As with any therapeutic agent that is being evaluated for potential use in humans, toxicological studies are important to assess health risks and deleterious side effects. Carbon-based materials, such as fullerenes (buckeyballs) and single-walled nanotubes (SWNTs), have been the most extensively characterized nanostructures in this regard. The major toxicological concern with these nanomaterials is their pro-oxidant effects [287] and the ease with which they are transported across cell membranes and localize to mitochondria. Toxicity of these compounds varies according to tissue/cell types, but generally, cytotoxic effects emerge in a dose- and time-dependent manner for all of the carbon-based nanoparticles tested; cells can only withstand short-term exposure to low concentrations of these particles (< 10 mg/ml). A few direct comparison studies reported that SWNTs have greater toxicity than other carbon nanoparticles [288–290]. Several hypotheses have been put forward to explain this observation. One is related to the manufacturing process; the synthesis of SWNTs requires the use of metal catalysts, which can be toxic themselves. Particle aggregation has also been suggested as a factor in the cytotoxicity of NPs. Studies on multi-wall nanotubes have yielded results similar to those of SWNTs—they cause significant cytotoxicity.

Several cytotoxic studies of metal oxide NPs have been reported recently [242, 291–295]. When comparing the toxicity of nano and micro-meter particles of several metal oxides (Fe_2O_3, Fe_3O_4, TiO_2 and CuO) in A549 human cell lines, it was found that CuO NPs were more toxic than microM sized CuO particles. The CuO NPs caused significant mitochrondrial damage [242]. In the same study, TiO_2 caused more damage to DNA and the iron oxides NPs showed low toxicity and no differences among different size particles. Toxicity of TiO_2, Al_2O_3 and carbon nanotubes toward two strains of bacteria was dependent on their chemical composition, size, surface charge and shape, but also on the bacterial strain [204]. Exposure time, dose and the type of treated cells are other factors that determine *in vitro* cytotoxic effects [293]. Solubility strongly influenced the cytotoxic effect as well [294]. NPs with very low solubility can be persistent within the biological system and induce long-term effects on the organism. In a study of nanosize zinc oxide, investigators tried to establish whether *in vitro* assays can be used to predict lung damage following inhalational exposure *in vivo*. *In vivo* studies in rats showed short-term lung inflammatory or cytotoxic responses while *in vitro* cell culture exposure produced minor responses only at high doses [292]. Thus, cell culture methods did not accurately predict the *in vivo* exposure response [296]. Nonetheless, *in vitro* assays may provide a simple, ready available cytotoxic test [294], but the value of such testing is not fully established.

The toxicological assessment of ceria-based NPs has lagged considerably behind that of the carbon-based structures. The majority of studies of ceria NPs have been performed using *in vitro* preparations. In contrast to the carbon-based forms, ceria NPs are relatively non-toxic and had few deleterious effects on cell cultures prepared from spinal motor neurons [297], retina [298] and immortalized cell lines. These cells withstood high concentrations of ceria NPs without adverse effects. It should be noted, however, that there are some cell types for which this generalization does not apply. For example, immortalized, human bronchial epithelial cells (BEAS-2B) exposed to ceria NPs show reduced viability in some studies [299], though not all studies [300]. Bronchial epithelial cells may be inherently more sensitive to NPs—a matter needing further investigation.

Zebrafish provide a vertebrate model system to test the route of exposure and to identify levels of toxicity of NPs [247–263]. Zebrafish have a full complement of organ systems that are similar in structure and physiology to other vertebrate systems [301]. Recent work has focused on study of toxicity and effects of ceria NPs *in vivo* in zebrafish embryos. In these studies, ceria NPs accumulated to a great extent in the digestive system; yet developmental defects were not observed. By comparison, exposure to different sizes nickel NPs induced developmental defects including thinning of the epithelial barrier, and at higher concentrations, skeletal muscle fiber separation [249]. It is possible that the antioxidant activity of ceria [21, 22] is preventing tissue inflammation while protecting against the release of reactive oxygen species. We have also examined the behavior of iron oxide NPs and their interactions with embryos [302]. Figure 3.9 shows Fe_3O_4 NPs that attach strongly to the chorion. The chorion is an accellular envelope of intercrossing layers protecting the embryos during development [263]. The channels of the chorion are ~0.6 μm diameter; thus the chorion should not prevent the diffusion of the 10 nm particles. The fact that the particles attach to the chorion indicates the presence of specific and non-specific interactions. An effect of surface coating is evident; when

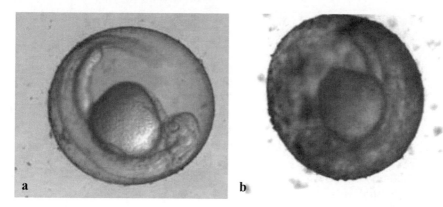

Fig. 3.9 Embryos within the chorion exposed to Au@Fe$_3$O$_4$ **a** and Fe$_3$O$_4$ **b** NPs showing strong attachment of the Fe$_3$O$_4$

Fig. 3.10 Transmission Electron Micrograph of cerium nanoparticles localized to mitochondria (*arrows*) and neurofilaments in the rat cortex overlying the hippocampus

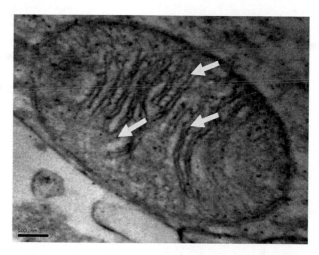

we compared the Fe$_3$O$_4$ with the Au coated Fe$_3$O$_4$ NPs (Fig. 3.9), the coated Fe$_3$O$_4$ did not attach to the chorion, suggesting the occurrence of a different type of surface specific interaction from the uncoated particles.

Few studies have examined the cellular distribution of NPs. Using an organotypic brain slice culture system we found that cerium NPs co-localize extensively with lipid/myelin membranes, mitochondria and neurofilaments (Fig. 3.10). This cellular association with mitochondria is interesting since mitochondria are the principle source of cellular free radical production and neurofilaments are responsible for shuttling mitochondria to regions of the cell that are ATP depleted. Moreover, their concentration in lipid membranes may reduce the extent of lipid peroxidation associated with increased ROS production. Thus, ceria appear to be optimally situated to scavenge free radicals near the source of production. These preliminary studies suggest that although ceria NPs localize to mitochondria, the toxic effects similar to those seen with carbon-based nanomaterials are not observed.

It is important to take into consideration that changes in the size, chemical composition, surface structure, solubility, and shape may influence the biological effects of NPs since these factors can alter protein binding and cellular uptake. NP binding to proteins may generate complexes that are more mobile and can enter tissue sites that are normally inaccessible by free particles. The application of dextran surface coatings to ceria NPs alters the surface charge and can prevent aggregation, minimize protein adsorption, improve biocompatibility and increase cellular uptake and circulation time *in vivo*. Different biocompatible surface coatings, such dextran, PEG, polysorbate and chitosan, can be used to vary the surface charge of the NPs [303, 304]; dextran yields NPs with negative surface charge, whereas the application of chitosan results in a positive surface charge. Nanoceria samples having positive zeta potential were found to adsorb more BSA while the samples with negative zeta potential showed little or no protein adsorption. The cellular uptake studies showed preferential uptake of the negatively charged NPs [110].

Ceria NPs of small sizes (5–10 nm) relative to many proteins, permit NPs to traverse both vascular and cellular compartments and as result, NPs have been investigated as vehicles for drug delivery to the brain [305]. Recently, Yokel et al. 2009 [306] examined the biodistribution of uncoated, 30 nm ceria NPs in rats following i.v. infusion. Infusions were delivered to the rats via tail vein between 0.5–7.5 h, and the doses ranged from 50–750 mg/kg. The doses used in this study were far greater than those used in any previous studies. Administering such large doses makes sense given their interest in determining the biodistribution and toxicology of the ceria NPs. Importantly, few behavioral changes were noted aside from a slight tachypnea and an increase in chewing and licking demonstrated by some animals during infusion. No adverse effects were noted following the administration of the NPs, and none of the animals died prior to the end of the experiment. Up to 20 h following the infusions, the majority of ceria was distributed either in the liver or the spleen; less than 0.1% was located in the brain. TEM analysis revealed few ceria particles in the brain, and these particles appeared to be restricted to astrocytes. The integrity of the blood-brain barrier was intact. This localization of ceria principally in the liver and spleen may have been due to the relative ease with which these particular particles agglomerated *in vivo*. The only marker for oxidative stress in the brain was HNE staining, a product of lipid peroxidation of polyunsaturated omega-6 acyl groups in the hippocampus 20 h post infusion. The highly reactive HNE can bind to proteins, inducing both conformational and functional changes [307–310]. Given the lack of ceria in the brain, it is possible that these markers of lipid peroxidation where induced by the production of pro-inflammatory cytokines that secondarily crossed the blood brain barrier [311–313] The results of this study suggest that short-term, systemic administration of ceria showed little toxicity with very large doses. The lack of cerium entry into the brain noted in this study suggests that the physical and chemical (i.e. size, shape, surface charge and propensity for agglomeration) features of the ceria are critical factors in delivering NPs to the brain. We have found that ceria particles agglomeration approaching 50 nm greatly reduces the mobility of the NPs into the intracellular space in brain tissue. Using biocompatible cell surface coatings (e.g. dextran, PEG) increase dispersion and decrease agglomeration and facilitate cellular uptake of the NPs. As work progresses in this area and the therapeutic window for ceria is established, the benefits associated with the antioxidant capacities of the ceria NPs will need to be balanced against the deleterious pro-oxidant effects. Studies such as Yokel's [306] are an important and necessary first step in delimiting the physiological effects of NPs and determining their potential therapeutic value in treating conditions leading to oxidative tissue injury.

3.8   Conclusions

Metal oxide NPs have already had a major impact in the biomedical field. The biomedically beneficial uses of metal oxide NPs derive from their biophysical properties, which are often unique to the nanoparticulate forms of the metal oxides—larger forms of these metal oxides do not have such interesting biological properties. The important properties of metal oxide NPs range from the surface porosity of TiO_2 nanoparticles, which enhance tissue binding, to the enzyme mimetic properties of CeO_2 nanoparticles either with or without doping by other transitional metals. Although a great deal has already been learned about these NPs, studies of the chemistry, bioavailability and especially toxicology are only in their infancy. It is abundantly clear that metal oxide NPs have tremendous promise as biomedically useful materials, and many interesting uses of these substances remain to be explored. Before the full potential of metal oxide nanoparticles can be realized, there must be more extensive and thorough studies of the toxicity of these NPs. At the present time, metal oxide nanoparticles do not seem to be particularly toxic at low levels. However, this facet of these materials is the least well studied aspect of their chemistry, and this deficiency must be rectified before metal oxide NPs can be used widely in human biomedical applications.

Acknowledgements This material is based upon work supported in part by the National Science Foundation under Grant No. 0954919. Any opinions, findings, and conclusions or recommendations expressed in this material are those of the author(s) and do not necessarily reflect the views of the National Science Foundation.

References

1. Latham, A. H.; Williams, M. E., Controlling transport and chemical functionality of magnetic nanoparticles. *Accounts of Chemical Research* **2008**, 41, (3), 411-420.
2. Lopez-Cruz, A.; Barrera, C.; Calero-DdelC, V. L.; Rinaldi, C., Water dispersible iron oxide nanoparticles coated with covalently linked chitosan. *Journal of Materials Chemistry* **2009**, 19, (37), 6870-6876.
3. Dobrovolskaia, M. A.; Aggarwal, P.; Hall, J. B.; McNeil, S. E., Preclinical studies to understand nanoparticle interaction with the immune system and its potential effects on nanoparticle biodistribution. *Molecular Pharmaceutics* **2008**, 5, (4), 487-495.

4. Liong, M.; Lu, J.; Kovochich, M.; Xia, T.; Ruehm, S. G.; Nel, A. E.; Tamanoi, F.; Zink, J. I., Multifunctional inorganic nanoparticles for imaging, targeting, and drug delivery. *Acs Nano* **2008**, 2, (5), 889-896.
5. Borm, P. J. A., Robbins, D., Haubold, S., Kuhlbusch, T., Fissan, H., Donaldson, K., Schins, R., Stone, V., Kreyling, W., Lamann, J., Krutmann, J., Warheit, D., Oberdorster E., The potential risks of nanomaterials: a review carried out for ECETOC. *Particle and Fibre Toxicology*, **2006**, 3:11.
6. Richman, E. K.; Hutchison, J. E., The Nanomaterial Characterization Bottleneck. *Acs Nano* **2009**, 3, (9), 2441-2446.
7. Won, Y. H.; Jang, H. S.; Kim, S. M.; Stach, E.; Ganesana, M.; Andreescu, S.; Stanciu, L. A., Biomagnetic Glasses: Preparation, Characterization, and Biosensor Applications. *Langmuir* **2009**.
8. Andreescu, S.; Njagi, J.; Ispas, C.; Ravalli, M. T., JEM Spotlight: Applications of advanced nanomaterials for environmental monitoring. *J Environ Monit* **2009**, 11, (1), 27-40.
9. Ispas, C.; Sokolov, I.; Andreescu, S., Enzyme-functionalized mesoporous silica for bioanalytical applications. *Anal Bioanal Chem* **2009**, 393, (2), 543-54.
10. Meng, H.; Chen, Z.; Xing, G. M.; Yuan, H.; Chen, C. Y.; Zhao, F.; Zhang, C. C.; Zhao, Y. L., Ultrahigh reactivity provokes nanotoxicity: Explanation of oral toxicity of nano-copper particles. *Toxicology Letters* **2007**, 175, (1-3), 102-110.
11. Donaldson K., S. A., The Janus faces of nanoparticles. *J Nanosci Nanotechnol* **2007**, 7, (12), 4607-11.
12. Campbell, C. T.; Peden, C. H., Chemistry. Oxygen vacancies and catalysis on ceria surfaces. *Science* **2005**, 309, (5735), 713-4.
13. Silva, G. A., Neuroscience nanotechnology: progress, opportunities and challenges. *Nat Rev Neurosci* **2006**, 7, (1), 65-74.
14. Faraji, A. H.; Wipf, P., Nanoparticles in cellular drug delivery. *Bioorg Med Chem* **2009**, 17, (8), 2950-62.
15. Song, H. T.; Choi, J. S.; Huh, Y. M.; Kim, S.; Jun, Y. W.; Suh, J. S.; Cheon, J., Surface modulation of magnetic nanocrystals in the development of highly efficient magnetic resonance probes for intracellular labeling. *Journal of the American Chemical Society* **2005**, 127, (28), 9992-9993.
16. Suh, W. H.; Suslick, K. S.; Stucky, G. D.; Suh, Y. H., Nanotechnology, nanotoxicology, and neuroscience. *Prog Neurobiol* **2009**, 87, (3), 133-70.
17. Rosi, N. L.; Mirkin, C. A., Nanostructures in biodiagnostics. *Chem Rev* **2005**, 105, (4), 1547-62.
18. Haddow, D. B.; Kelly, J. M.; James, P. F.; Short, R. D.; Scutt, A. M.; Rawsterne, R.; Kothari, S., Cell response to sol-gel derived titania coatings. *Journal of Materials Chemistry* **2000**, 10, (12), 2795-2801.
19. Kim, S. E.; Lim, J. H.; Lee, S. C.; Nam, S. C.; Kang, H. G.; Choi, J., Anodically nanostructured titanium oxides for implant applications. *Electrochimica Acta* **2008**, 53, (14), 4846-4851.
20. Korsvik, C.; Patil, S.; Seal, S.; Self, W. T., Superoxide dismutase mimetic properties exhibited by vacancy engineered ceria nanoparticles. *Chemical Communications* **2007**, (10), 1056-1058.
21. Karakoti, A. S.; Monteiro-Riviere, N. A.; Aggarwal, R.; Davis, J. P.; Narayan, R. J.; Self, W. T.; McGinnis, J.; Seal, S., Nanoceria as antioxidant: Synthesis and biomedical applications. *Jom* **2008**, 60, (3), 33-37.
22. Heckert, E. G.; Karakoti, A. S.; Seal, S.; Self, W. T., The role of cerium redox state in the SOD mimetic activity of nanoceria. *Biomaterials* **2008**, 29, (18), 2705-2709.
23. Schladt, T. D.; Graf, T.; Tremel, W., Synthesis and Characterization of Monodisperse Manganese Oxide Nanoparticles-Evaluation of the Nucleation and Growth Mechanism. *Chemistry of Materials* **2009**, 21, (14), 3183-3190.

24. Nguyen, T. D.; Do, T. O., General Two-Phase Routes to Synthesize Colloidal Metal Oxide Nanocrystals: Simple Synthesis and Ordered Self-Assembly Structures. *Journal of Physical Chemistry C* **2009**, 113, (26), 11204-11214.

25. Gupta, A. K.; Gupta, M., Synthesis and surface engineering of iron oxide nanoparticles for biomedical applications. *Biomaterials* **2005**, 26, (18), 3995-4021.

26. Laurent, S.; Forge, D.; Port, M.; Roch, A.; Robic, C.; Elst, L. V.; Muller, R. N., Magnetic iron oxide nanoparticles: Synthesis, stabilization, vectorization, physicochemical characterizations, and biological applications. *Chemical Reviews* **2008**, 108, (6), 2064-2110.

27. Alivisatos, A. P., Semiconductor clusters, nanocrystals, and quantum dots. *Science* **1996**, 271, (5251), 933-937.

28. Andreescu, D.; Matijevic, E.; Goia, D. V., Formation of uniform colloidal ceria in polyol. *Colloids and Surfaces a-Physicochemical and Engineering Aspects* **2006**, 291, (1-3), 93-100.

29. Koziej, D.; Fischer, F.; Kranzlin, N.; Caseri, W. R.; Niederberger, M., Nonaqueous TiO2 Nanoparticle Synthesis: a Versatile Basis for the Fabrication of Self-Supporting, Transparent, and UV-Absorbing Composite Films. *Acs Applied Materials & Interfaces* **2009**, 1, (5), 1097-1104.

30. Pinna, N.; Niederberger, M., Oxide synthesis as Cornerstone of nanoscience. *European Journal of Inorganic Chemistry* **2008**, (6), 825-825.

31. Niederberger, M.; Garnweitner, G., Organic reaction pathways in the nonaqueous synthesis of metal oxide nanoparticles. *Chemistry-a European Journal* **2006**, 12, (28), 7282-7302.

32. Niederberger, M.; Garnweitner, G.; Pinna, N.; Neri, G., Non-aqueous routes to crystalline metal oxide nanoparticles: Formation mechanisms and applications. *Progress in Solid State Chemistry* **2005**, 33, (2-4), 59-70.

33. Garnweitner, G.; Niederberger, M., Controlled synthesis of metal oxide nanoparticles through organic chemistry. *Abstracts of Papers of the American Chemical Society* **2005**, 230, U2042-U2042.

34. Matijevic, E., Sapieszko, R.S., , Forced hydrolysis in homogeneous solutions. *Fine Particles: synthesis, characterization and mechanisms of growth* **2000**, 92, (Ed. by Tadao Sugimoto), 2-34.

35. Pinna, N.; Hochepied, J. F.; Niederberger, M.; Gregg, M., Chemistry and physics of metal oxide nanostructures. *Physical Chemistry Chemical Physics* **2009**, 11, (19), 3607-3607.

36. Zhang, Q. C.; Yu, Z. H.; Li, G.; Ye, Q. M.; Lin, J. H., Synthesis of quantum-size cerium oxide nanocrystallites by a novel homogeneous precipitation method. *Journal of Alloys and Compounds* **2009**, 477, (1-2), 81-84.

37. Zhang, Q. L.; Yang, Z. M.; Ding, B. J., Synthesis of Cerium Oxide Nanoparticles by the Precipitation Method. *Materials Research, Pts 1 and 2* **2009**, 610-613, 233-238 1430.

38. Tok, A. I. Y.; Du, S. W.; Boey, F. Y. C.; Chong, W. K., Hydrothermal synthesis and characterization of rare earth doped ceria nanoparticles. *Materials Science and Engineering a-Structural Materials Properties Microstructure and Processing* **2007**, 466, (1-2), 223-229.

39. Chen, H. I.; Chang, H. Y., Synthesis of nanocrystalline cerium oxide particles by the precipitation method. *Ceramics International* **2005**, 31, (6), 795-802.

40. Li, J. G.; Ikegami, T.; Lee, J. H.; Mori, T., Characterization and sintering of nanocrystalline CeO2 powders synthesized by a mimic alkoxide method. *Acta Materialia* **2001**, 49, (3), 419-426.

41. Yin, Y.; Alivisatos, A. P., Colloidal nanocrystal synthesis and the organic-inorganic interface. *Nature* **2005**, 437, (7059), 664-670.

42. Amin, N. A., S.; Matijevic, E., Magnetic properties of uniform spherical magnetite particles prepared from ferrous hydroxide gels. *Phys. Status Solidi A* **1987**, 101, 233-238.

43. Ozaki, M. M., E. , Preparation and magnetic properties of monodispersed spindle-type γ-Fe2O3 particles. *Journal of Colloid and Interface Science* **1985**, 107, (1), 199-203

44. Goia, D. V., Preparation and formation mechanisms of uniform metallic particles in homogeneous solutions. *J. Mater Chem.* **2004**, 14, 451-458.

45. Garnweitner, G.; Niederberger, M., Nonaqueous and surfactant-free synthesis routes to metal oxide nanoparticles. *Journal of the American Ceramic Society* **2006**, 89, (6), 1801-1808.

46. Gao, J. H.; Gu, H. W.; Xu, B., Multifunctional Magnetic Nanoparticles: Design, Synthesis, and Biomedical Applications. *Accounts of Chemical Research* **2009**, 42, (8), 1097-1107.
47. Fang, C.; Zhang, M. Q., Multifunctional magnetic nanoparticles for medical imaging applications. *Journal of Materials Chemistry* **2009**, 19, (35), 6258-6266.
48. Pisanic, T. R.; Blackwell, J. D.; Shubayev, V. I.; Finones, R. R.; Jin, S., Nanotoxicity of iron oxide nanoparticle internalization in growing neurons. *Biomaterials* **2007**, 28, (16), 2572-2581.
49. Brown, D. M.; Kinloch, I. A.; Bangert, U.; Windle, A. H.; Walter, D. M.; Walker, G. S.; Scotchford, C. A.; Donaldson, K.; Stone, V., An in vitro study of the potential of carbon nanotubes and nanofibres to induce inflammatory mediators and frustrated phagocytosis. *Carbon* **2007**, 45, (9), 1743-1756.
50. Brown, D. M.; Hutchison, L.; Donaldson, K.; MacKenzie, S. J.; Dick, C. A. J.; Stone, V., The effect of oxidative stress on macrophages and lung epithelial cells: The role of phosphodiesterases 1 and 4. *Toxicology Letters* **2007**, 168, (1), 1-6.
51. Driscoll, K. E.; Deyo, L. C.; Carter, J. M.; Howard, B. W.; Hassenbein, D. G.; Bertram, T. A., Effects of particle exposure and particle-elicited inflammatory cells on mutation in rat alveolar epithelial cells. *Carcinogenesis* **1997**, 18, (2), 423-430.
52. Jiang, W., et al., Nanoparticle-mediated cellular response is size-dependent. Nat Nanotechnol. **2008**, 3, (3), 145-50.
53. Lundqvist, M., et al., Nanoparticle size and surface properties determine the protein corona with possible implications for biological impacts. . *Proc Natl Acad Sci U S A* **2008**, 105, (38), 14265-70.
54. Madl, A. K.; Pinkerton, K. E., Health effects of inhaled engineered and incidental nanoparticles. *Critical Reviews in Toxicology* **2009**, 39, (8), 629-658.
55. Stone, V.; Johnston, H.; Schins, R. P. F., Development of in vitro systems for nanotoxicology: methodological considerations. *Critical Reviews in Toxicology* **2009**, 39, (7), 613-626.
56. Park, M. V. D. Z.; Lankveld, D. P. K.; van Loveren, H.; de Jong, W. H., The status of in vitro toxicity studies in the risk assessment of nanomaterials. *Nanomedicine* **2009**, 4, (6), 669-685.
57. Meng, H.; Xia, T.; George, S.; Nel, A. E., A Predictive Toxicological Paradigm for the Safety Assessment of Nanomaterials. *Acs Nano* **2009**, 3, (7), 1620-1627.
58. Jones, C. F.; Grainger, D. W., In vitro assessments of nanomaterial toxicity. *Advanced Drug Delivery Reviews* **2009**, 61, (6), 438-456.
59. Geiser, M.; Rothen-Rutishauser, B.; Kapp, N.; Schurch, S.; Kreyling, W.; Schulz, H.; Semmler, M.; Hof, V. I.; Heyder, J.; Gehr, P., Ultrafine particles cross cellular membranes by non-phagocytic mechanisms in lungs and in cultured cells. *Environmental Health Perspectives* **2005**, 113, (11), 1555-1560.
60. Hardman, R., A toxicologic review of quantum dots: Toxicity depends on physicochemical and environmental factors. *Environmental Health Perspectives* **2006**, 114, (2), 165-172.
61. Sequeira, R.; Genaidy, A.; Weckman, G.; Shell, R.; Karwowski, W.; Acosta-Leon, A., Health effects of nanomaterials: A critical appraisal approach and research to practice. *Human Factors and Ergonomics in Manufacturing* **2008**, 18, (3), 293-341.
62. Landsiedel, R.; Kapp, M. D.; Schulz, M.; Wiench, K.; Oesch, F., Genotoxicity investigations on nanomaterials: Methods, preparation and characterization of test material, potential artifacts and limitations-Many questions, some answers. *Mutation Research-Reviews in Mutation Research* **2009**, 681, (2-3), 241-258.
63. Kasemo, B.; Lausmaa, J., Material-Tissue Interfaces - the Role of Surface-Properties and Processes. *Environmental Health Perspectives* **1994**, 102, 41-45.
64. Fahmy, B.; Cormier, S. A., Copper oxide nanoparticles induce oxidative stress and cytotoxicity in airway epithelial cells. *Toxicology in Vitro* **2009**, 23, (7), 1365-1371.
65. Limbach, L. K.; Li, Y. C.; Grass, R. N.; Brunner, T. J.; Hintermann, M. A.; Muller, M.; Gunther, D.; Stark, W. J., Oxide nanoparticle uptake in human lung fibroblasts: Effects of particle size, agglomeration, and diffusion at low concentrations. *Environmental Science & Technology* **2005**, 39, (23), 9370-9376.

66. Ge, Y. Q.; Zhang, Y.; Xia, J. G.; Ma, M.; He, S. Y.; Nie, F.; Gu, N., Effect of surface charge and agglomerate degree of magnetic iron oxide nanoparticles on KB cellular uptake in vitro. *Colloids and Surfaces B-Biointerfaces* **2009**, 73, (2), 294-301.
67. Karakoti, A. S.; Hench, L. L.; Seal, S., The potential toxicity of nanomaterials - The role of surfaces. *Jom* **2006**, 58, (7), 77-82.
68. De, M. M., O.R. Rana, S. Rotello,V.M. , Size and geometry dependent protein–nanoparticle self-assembly. *Chem. Commun.* **2009**, 2157-2159.
69. Perrault, S. D.; Walkey, C.; Jennings, T.; Fischer, H. C.; Chan, W. C. W., Mediating Tumor Targeting Efficiency of Nanoparticles Through Design. *Nano Letters* **2009**, 9, (5), 1909-1915.
70. Corti, M.; Lascialfari, A.; Micotti, E.; Castellano, A.; Donativi, M.; Quarta, A.; Cozzoli, P. D.; Manna, L.; Pellegrino, T.; Sangregorio, C., Magnetic properties of novel superparamagnetic MRI contrast agents based on colloidal nanocrystals. *Journal of Magnetism and Magnetic Materials* **2008**, 320, (14), E320-E323.
71. Jun, Y. W.; Huh, Y. M.; Choi, J. S.; Lee, J. H.; Song, H. T.; Kim, S.; Yoon, S.; Kim, K. S.; Shin, J. S.; Suh, J. S.; Cheon, J., Nanoscale size effect of magnetic nanocrystals and their utilization for cancer diagnosis via magnetic resonance imaging. *Journal of the American Chemical Society* **2005**, 127, (16), 5732-5733.
72. Hanessian, S.; Grzyb, J. A.; Cengelli, F.; Juillerat-Jeanneret, L., Synthesis of chemically functionalized superparamagnetic nanoparticles as delivery vectors for chemotherapeutic drugs. *Bioorganic & Medicinal Chemistry* **2008**, 16, (6), 2921-2931.
73. Chen, F. H,; Gao, Q.; Ni, J. Z., The grafting and release behavior of doxorubincin from Fe3O4@SiO2 core-shell structure nanoparticles via an acid cleaving amide bond: the potential for magnetic targeting drug delivery. *Nanotechnology* **2008**, 19, (16), -.
74. Wu, S. H.; Lin, Y. S.; Hung, Y.; Chou, Y. H.; Hsu, Y. H.; Chang, C.; Mou, C. Y., Multifunctional mesoporous silica nanoparticles for intracellular labeling and animal magnetic resonance imaging studies. *Chembiochem* **2008**, 9, (1), 53-57.
75. Koktysh, D. S.; Liang, X. R.; Yun, B. G.; Pastoriza-Santos, I.; Matts, R. L.; Giersig, M.; Serra-Rodriguez, C.; Liz-Marzan, L. M.; Kotov, N. A., Biomaterials by design: Layer-by-layer assembled ion-selective and biocompatible films of TiO2 nanoshells for neurochemical monitoring. *Advanced Functional Materials* **2002**, 12, (4), 255-265.
76. Kern, P.; Schwaller, P.; Michler, J., Electrolytic deposition of titania films as interference coatings on biomedical implants: Microstructure, chemistry and nano-mechanical properties. *Thin Solid Films* **2006**, 494, (1-2), 279-286.
77. Rossi, S.; Tirri, T.; Paldan, H.; Kuntsi-Vaattovaara, H.; Tulamo, R.; Narhi, T., Peri-implant tissue response to TiO2 surface modified implants. *Clinical Oral Implants Research* **2008**, 19, (4), 348-355.
78. Rossi, S.; Moritz, N.; Tirri, T.; Peltola, T.; Areva, S.; Jokinen, M.; Happonen, R. P.; Narhi, T., Comparison between sol-gel-derived anatase- and rutile-structured TiO2 coatings in soft-tissue environment. *Journal of Biomedical Materials Research Part A* **2007**, 82A, (4), 965-974.
79. La Flamme, K. E.; Popat, K. C.; Leoni, L.; Markiewicz, E.; La Tempa, T. J.; Roman, B. B.; Grimes, C. A.; Desai, T. A., Biocompatibility of nanoporous alumina membranes for immunoisolation. *Biomaterials* **2007**, 28, (16), 2638-2645.
80. Schubert, D.; Dargusch, R.; Raitano, J.; Chan, S. W., Cerium and yttrium oxide nanoparticles are neuroprotective. *Biochemical and Biophysical Research Communications* **2006**, 342, (1), 86-91.
81. Silva, G. A., Nanomedicine: seeing the benefits of ceria. *Nat Nanotechnol* **2006**, 1, (2), 92-4.
82. Das, M.; Bhargava, N.; Patil, S.; Riedel, L.; Molnar, P.; Seal, S.; Hickman, J. J., Novel in vitro cell culture model of adult mammalian: Spinal cord cells. *In Vitro Cellular & Developmental Biology-Animal* **2005**, 41, 9A-9A.
83. Rzigalinski, B. A.; Meehan, K.; Davis, R. M.; Xu, Y.; Miles, W. C.; Cohen, C. A., Radical nanomedicine. *Nanomedicine* **2006**, 1, (4), 399-412.

84. Cohen, C. A.; Kurnick, M. D.; Rzigalinski, B. A., Cerium oxide nanoparticles extend lifespan and protect drosophila melanogaster from paraquat (PQ)-induced oxidative stress (OS). *Free Radical Biology and Medicine* **2006**, 41, S20-S20.

85. Park, H. Y.; Schadt, M. J.; Wang, L.; Lim, I. I. S.; Njoki, P. N.; Kim, S. H.; Jang, M. Y.; Luo, J.; Zhong, C. J., Fabrication of magnetic core @ shell Fe oxide @ Au nanoparticles for interfacial bioactivity and bio-separation. *Langmuir* **2007**, 23, (17), 9050-9056.

86. Huang, X. L.; Zhuang, J.; Chen, D.; Liu, H. Y.; Tang, F. Q.; Yan, X. Y.; Meng, X. W.; Zhang, L.; Ren, J., General Strategy for Designing Functionalized Magnetic Microspheres for Different Bioapplications. *Langmuir* **2009**, 25, (19), 11657-11663.

87. Thierry, B.; Al-Ejeh, F.; Brown, M. P.; Majewski, P.; Griesser, H. J., Immunotargeting of Functional Nanoparticles for MRI detection of Apoptotic Tumor Cells. *Advanced Materials* **2009**, 21, (5), 541-+.

88. Kumagai, M.; Imai, Y.; Nakamura, T.; Yamasaki, Y.; Sekino, M.; Ueno, S.; Hanaoka, K.; Kikuchi, K.; Nagano, T.; Kaneko, E.; Shimokado, K.; Kataoka, K., Iron hydroxide nanoparticles coated with poly(ethylene glycol)-poly(aspartic acid) block copolymer as novel magnetic resonance contrast agents for in vivo cancer imaging. *Colloids and Surfaces B-Biointerfaces* **2007**, 56, (1-2), 174-181.

89. Popat, K. C.; Mor, G.; Grimes, C.; Desai, T. A., Poly (ethylene glycol) grafted nanoporous alumina membranes. *Journal of Membrane Science* **2004**, 243, (1-2), 97-106.

90. de Campos, A. M.; Diebold, Y.; Carvalho, E. L.; Sanchez, A.; Alonso, M. J., Chitosan nanoparticles as new ocular drug delivery systems: in vitro stability, in vivo fate, and cellular toxicity. *Pharm Res* **2004**, 21, (5), 803-10.

91. Xia, T. A.; Kovochich, M.; Liong, M.; Meng, H.; Kabehie, S.; George, S.; Zink, J. I.; Nel, A. E., Polyethyleneimine Coating Enhances the Cellular Uptake of Mesoporous Silica Nanoparticles and Allows Safe Delivery of siRNA and DNA Constructs. *Acs Nano* **2009**, 3, (10), 3273-3286.

92. Hafelli, U. O.; Riffle, J. S.; Harris-Shekhawat, L.; Carmichael-Baranauskas, A.; Mark, F.; Dailey, J. P.; Bardenstein, D., Cell Uptake and in Vitro Toxicity of Magnetic Nanoparticles Suitable for Drug Delivery. *Molecular Pharmaceutics* **2009**, 6, (5), 1417-1428.

93. Lundqvist, M.; Stigler, J.; Elia, G.; Lynch, I.; Cedervall, T.; Dawson, K. A., Nanoparticle size and surface properties determine the protein corona with possible implications for biological impacts. *Proceedings of the National Academy of Sciences of the United States of America* **2008**, 105, (38), 14265-14270.

94. Cedervall, T.; Lynch, I.; Lindman, S.; Berggard, T.; Thulin, E.; Nilsson, H.; Dawson, K. A.; Linse, S., Understanding the nanoparticle-protein corona using methods to quantify exchange rates and affinities of proteins for nanoparticles. *Proceedings of the National Academy of Sciences of the United States of America* **2007**, 104, (7), 2050-2055.

95. Aggarwal, P.; Hall, J. B.; McLeland, C. B.; Dobrovolskaia, M. A.; McNeil, S. E., Nanoparticle interaction with plasma proteins as it relates to particle biodistribution, biocompatibility and therapeutic efficacy. *Advanced Drug Delivery Reviews* **2009**, 61, (6), 428-437.

96. Ehrenberg, M. S.; Friedman, A. E.; Finkelstein, J. N.; Oberdorster, G.; McGrath, J. L., The influence of protein adsorption on nanoparticle association with cultured endothelial cells. *Biomaterials* **2009**, 30, (4), 603-610.

97. Lynch, I.; Cedervall, T.; Lundqvist, M.; Cabaleiro-Lago, C.; Linse, S.; Dawson, K. A., The nanoparticle - protein complex as a biological entity; a complex fluids and surface science challenge for the 21st century. *Advances in Colloid and Interface Science* **2007**, 134-35, 167-174.

98. De, M.; Miranda, O. R.; Rana, S.; Rotello, V. M., Size and geometry dependent protein-nanoparticle self-assembly. *Chemical Communications* **2009**, (16), 2157-2159.

99. Deng, Z. J.; Mortimer, G.; Schiller, T.; Musumeci, A.; Martin, D.; Minchin, R. F., Differential plasma protein binding to metal oxide nanoparticles. *Nanotechnology* **2009**, 20, (45), -.

100. Nel, A. E.; Madler, L.; Velegol, D.; Xia, T.; Hoek, E. M. V.; Somasundaran, P.; Klaessig, F.; Castranova, V.; Thompson, M., Understanding biophysicochemical interactions at the nano-bio interface. *Nature Materials* **2009**, 8, (7), 543-557.

101. Teichroeb, J. H.; Forrest, J. A.; Ngai, V.; Jones, L. W., Anomalous thermal denaturing of proteins adsorbed to nanoparticles. *European Physical Journal E* **2006**, 21, (1), 19-24.
102. Teichroeb, J. H.; Forrest, J. A.; Jones, L. W., Size-dependent denaturing kinetics of bovine serum albumin adsorbed onto gold nanospheres. *European Physical Journal E* **2008**, 26, (4), 411-415.
103. Allen, L. T.; Tosetto, M.; Miller, I. S.; O'Connor, D. P.; Penney, S. C.; Lynch, I.; Keenan, A. K.; Pennington, S. R.; Dawson, K. A.; Gallagher, W. M., Surface-induced changes in protein adsorption and implications for cellular phenotypic responses to surface interaction. *Biomaterials* **2006**, 27, (16), 3096-3108.
104. Chithrani, B. D.; Chan, W. C. W., Elucidating the mechanism of cellular uptake and removal of protein-coated gold nanoparticles of different sizes and shapes. *Nano Letters* **2007**, 7, (6), 1542-1550.
105. Markovic, Z.; Trajkovic, V., Biomedical potential of the reactive oxygen species generation and quenching by fullerenes (C60). *Biomaterials* **2008**, 29, (26), 3561-3573.
106. H. Meng, Z. C., G. Xing, H. Yuan, C. Chen, F. Zhao, C. Zhang, Y. Zhao, Ultrahigh reactivity provokes nanotoxicity: Explanation of oral toxicity of nano-copper particles *Toxicology Letters* **2007**, 175, 102-110.
107. Lei, R. H.; Wu, C. Q.; Yang, B. H.; Ma, H. Z.; Shi, C.; Wang, Q. J.; Wang, Q. X.; Yuan, Y.; Liao, M. Y., Integrated metabolomic analysis of the nano-sized copper particle-induced hepatotoxicity and nephrotoxicity in rats: A rapid in vivo screening method for nanotoxicity. *Toxicology and Applied Pharmacology* **2008**, 232, (2), 292-301.
108. Li, S. Q., Zhu R.R., Zhu H., Xue M., Sun X.Y., Yao S.D., Wang S.L., Nanotoxicity of TiO(2) nanoparticles to erythrocyte in vitro. *Food Chem Toxicol* **2008**.
109. Chen, J. M.; Hessler, J. A.; Putchakayala, K.; Panama, B. K.; Khan, D. P.; Hong, S.; Mullen, D. G.; DiMaggio, S. C.; Som, A.; Tew, G. N.; Lopatin, A. N.; Baker, J. R.; Holl, M. M. B.; Orr, B. G., Cationic Nanoparticles Induce Nanoscale Disruption in Living Cell Plasma Membranes. *Journal of Physical Chemistry B* **2009**, 113, (32), 11179-11185.
110. Patil, S.; Sandberg, A.; Heckert, E.; Self, W.; Seal, S., Protein adsorption and cellular uptake of cerium oxide nanoparticles as a function of zeta potential. *Biomaterials* **2007**, 28, (31), 4600-4607.
111. Wilhelm, C.; Billotey, C.; Roger, J.; Pons, J. N.; Bacri, J. C.; Gazeau, F., Intracellular uptake of anionic superparamagnetic nanoparticles as a function of their surface coating. *Biomaterials* **2003**, 24, (6), 1001-1011.
112. Corot, C.; Robert, P.; Idee, J. M.; Port, M., Recent advances in iron oxide nanocrystal technology for medical imaging. *Advanced Drug Delivery Reviews* **2006**, 58, (14), 1471-1504.
113. Qiao, R. R.; Yang, C. H.; Gao, M. Y., Superparamagnetic iron oxide nanoparticles: from preparations to in vivo MRI applications. *Journal of Materials Chemistry* **2009**, 19, (35), 6274-6293.
114. Jun, Y. W.; Lee, J. H.; Cheon, J., Chemical design of nanoparticle probes for high-performance magnetic resonance imaging. *Angewandte Chemie-International Edition* **2008**, 47, (28), 5122-5135.
115. Burda, C.; Chen, X. B.; Narayanan, R.; El-Sayed, M. A., Chemistry and properties of nanocrystals of different shapes. *Chemical Reviews* **2005**, 105, (4), 1025-1102.
116. Muench, G. A., S.; Matijevic, E., Effects of magnetic field, strain, and size on the Morin temperature of spherical α-Fe2O3 particles. *IEEE Transactions on Magnetics* **1982**, 18, (6), 1583-1585.
117. Amin, N. A., S.; Matijevic, E. , Magnetic properties of submicronic α-Fe2O3 particles of uniform size distribution at 300 K. *Physica Status Solidi A* **1987**, 104, (1), K65-K68.
118. Ozaki, M., Egami, T., Sugiyama, N., Matijevis, E.,, Agglomeration in colloidal hematite dispersions due to weak magnetic interactions: II. The effects of particle size and shape *J. Colloid Interface Sci.* **1988**, 126, 212-219.
119. Ozaki, O. O., N.; Matijevic, E., Preparation and magnetic properties of uniform hematite platelets. *J. Colloid Interface Sci.* **1990**, 137, 546-549.

120. Kim, J.; Piao, Y.; Hyeon, T., Multifunctional nanostructured materials for multimodal imaging, and simultaneous imaging and therapy. *Chemical Society Reviews* **2009**, 38, (2), 372-390.

121. Debbage, P.; Jaschke, W., Molecular imaging with nanoparticles: giant roles for dwarf actors. *Histochemistry and Cell Biology* **2008**, 130, (5), 845-875.

122. Wang, G. P.; Song, E. Q.; Xie, H. Y.; Zhang, Z. L.; Tian, Z. Q.; Zuo, C.; Pang, D. W.; Wu, D. C.; Shi, Y. B., Biofunctionalization of fluorescent-magnetic-bifunctional nanospheres and their applications. *Chemical Communications* **2005**, (34), 4276-4278.

123. Koo, Y. E. L.; Reddy, G. R.; Bhojani, M.; Schneider, R.; Philbert, M. A.; Rehemtulla, A.; Ross, B. D.; Kopelman, R., Brain cancer diagnosis and therapy with nanoplatforms. *Advanced Drug Delivery Reviews* **2006**, 58, (14), 1556-1577.

124. Jain, T. K.; Reddy, M. K.; Morales, M. A.; Leslie-Pelecky, D. L.; Labhasetwar, V., Biodistribution, clearance, and biocompatibility of iron oxide magnetic nanoparticles in rats. *Mol Pharm* **2008**, 5, (2), 316-27.

125. Berry, C. C.; Wells, S.; Charles, S.; Aitchison, G.; Curtis, A. S. G., Cell response to dextran-derivatised iron oxide nanoparticles post internalisation. *Biomaterials* **2004**, 25, (23), 5405-5413.

126. Horak, D.; Babic, M.; Jendelova, P.; Herynek, V.; Trchova, M.; Likavcanova, K.; Kapcalova, M.; Hajek, M.; Sykova, E., Effect of different magnetic nanoparticle coatings on the efficiency of stem cell labeling. *Journal of Magnetism and Magnetic Materials* **2009**, 321, (10), 1539-1547.

127. Bulte, J. W. M.; Kraitchman, D. L., Iron oxide MR contrast agents for molecular and cellular imaging. *Nmr in Biomedicine* **2004**, 17, (7), 484-499.

128. Diebold, U., The surface science of titanium dioxide. *Surface Science Reports* **2003**, 48, (5-8), 53-229.

129. Niederberger, M.; Garnweitner, G.; Krumeich, F.; Nesper, R.; Colfen, H.; Antonietti, M., Tailoring the surface and solubility properties of nanocrystalline titania by a nonaqueous in situ functionalization process. *Chemistry of Materials* **2004**, 16, (7), 1202-1208.

130. Santos, E.; Kuromoto, N. K.; Soares, G. A., Mechanical properties of titania films used as biomaterials. *Materials Chemistry and Physics* **2007**, 102, (1), 92-97.

131. Petersson, I. U.; Loberg, J. E. L.; Fredriksson, A. S.; Ahlberg, E. K., Semi-conducting properties of titanium dioxide surfaces on titanium implants. *Biomaterials* **2009**, 30, (27), 4471-4479.

132. Liu, X. Y.; Chu, P. K.; Ding, C. X., Surface modification of titanium, titanium alloys, and related materials for biomedical applications. *Materials Science & Engineering R-Reports* **2004**, 47, (3-4), 49-121.

133. Nair, A. S.; Suryanarayanan, V.; Pradeep, T.; Thomas, J.; Anija, M.; Philip, R., AuxAgy @ ZrO2 core-shell nanoparticles: synthesis, characterization, reactivity and optical limiting. *Materials Science and Engineering B-Solid State Materials for Advanced Technology* **2005**, 117, (2), 173-182.

134. Njagi, J.; Ispas, C.; Andreescu, S., Mixed ceria-based metal oxides biosensor for operation in oxygen restrictive environments. *Analytical Chemistry* **2008**, 80, (19), 7266-7274.

135. Ispas, C.; Njagi, J.; Cates, M.; Andreescu, S., Electrochemical studies of ceria as electrode material for sensing and biosensing applications. *Journal of the Electrochemical Society* **2008**, 155, (8), F169-F176.

136. Njagi, J., Chernov, M.M., Leiter, J.C., Andreescu S., Amperometric detection of dopamine in vivo with an enzyme based carbon fiber microbiosensor. *Analytical Chemistry* **2010**, in press.

137. Popat, K. C.; Eltgroth, M.; La Tempa, T. J.; Grimes, C. A.; Desai, T. A., Titania nanotubes: A novel platform for drug-eluting coatings for medical implants? *Small* **2007**, 3, (11), 1878-1881.

138. Ainslie, K. M.; Tao, S. L.; Popat, K. C.; Daniels, H.; Hardev, V.; Grimes, C. A.; Desai, T. A., In vitro inflammatory response of nanostructured titania, silicon oxide, and polycaprolactone. *Journal of Biomedical Materials Research Part A* **2009**, 91A, (3), 647-655.

139. Ma, L. L.; Liu, J.; Li, N.; Wang, J.; Duan, Y. M.; Yan, J. Y.; Liu, H. T.; Wang, H.; Hong, F. S., Oxidative stress in the brain of mice caused by translocated nanoparticulate TiO2 delivered to the abdominal cavity. *Biomaterials* **2010**, 31, (1), 99-105.

140. Duan, Y., Liu, J., Ma, L., Li, H., Wang, J., Zheng, L., Liu, C., Wang, X., Zhao, X., Yan, J., Wang, S., Wang, H., Zhang, X., Hong, F., Toxicological characteristics of nanoparticulate anatase titanium dioxide in mice. *Biomaterials* **2010**, (31), 894-899.

141. T.J. Jonaitis, J. W. C., B. Magnuson, Concerns regarding nano-sized titanium dioxide dermal penetration and toxicity study. *Toxicology Letters* **2009**.

142. Tarnuzzer, R. W.; Colon, J.; Patil, S.; Seal, S., Vacancy engineered ceria nanostructures for protection from radiation-induced cellular damage. *Nano Lett* **2005**, 5, (12), 2573-7.

143. Manorama, S. V.; Izu, N.; Shin, W.; Matsubara, I.; Murayama, N., On the platinum sensitization of nanosized cerium dioxide oxygen sensors. *Sensors and Actuators B-Chemical* **2003**, 89, (3), 299-304.

144. Bamwenda, G. R.; Arakawa, H., Cerium dioxide as a photocatalyst for water decomposition to O-2 in the presence of Ce-aq(4+) and Fe-aq(3+) species. *Journal of Molecular Catalysis a-Chemical* **2000**, 161, (1-2), 105-113.

145. Campbell, C. T.; Peden, C. H. F., Chemistry - Oxygen vacancies and catalysis on ceria surfaces. *Science* **2005**, 309, (5735), 713-714.

146. Li, R.; Yin, S.; Yabe, S.; Yamashita, M.; Momose, S.; Yoshida, S.; Sato, T., Preparation and photochemical properties of nanoparticles of ceria doped with zinc oxide. *British Ceramic Transactions* **2002**, 101, (1), 9-13.

147. Trovarelli, A , Structural and oxygen storage/release properties of CeO2-based solid solutions. *Comments on Inorganic Chemistry* **1999**, 20, (4-6), 263-284.

148. Schalow, T.; Laurin, M.; Brandt, B.; Schauermann, S.; Guimond, S.; Kuhlenbeck, H.; Starr, D. E.; Shaikhutdinov, S. K.; Libuda, J.; Freund, H. J., Oxygen storage at the metal/oxide interface of catalyst nanoparticles. *Angewandte Chemie-International Edition* **2005**, 44, (46), 7601-7605.

149. Dutta, P.; Pal, S.; Seehra, M. S.; Shi, Y.; Eyring, E. M.; Ernst, R. D., Concentration of Ce3+ and oxygen vacancies in cerium oxide nanoparticles. *Chemistry of Materials* **2006**, 18, (21), 5144-5146.

150. Sinha, A. K.; Suzuki, K., Preparation and characterization of novel mesoporous ceria-titania. *Journal of Physical Chemistry B* **2005**, 109, (5), 1708-1714.

151. Boaro, M.; Trovarelli, A.; Hwang, J. H.; Mason, T. O., Electrical and oxygen storage/release properties of nanocrystalline ceria-zirconia solid solutions. *Solid State Ionics* **2002**, 147, (1-2), 85-95.

152. Reddy, B. M.; Saikia, P.; Bharali, P.; Yamada, Y.; Kobayashi, T.; Muhler, M.; Grunert, W., Structural Characterization and Catalytic Activity of Nanosized Ceria-Terbia Solid Solutions. *Journal of Physical Chemistry C* **2008**, 112, (42), 16393-16399.

153. Izu, N.; Oh-hori, N.; Itou, M.; Shin, W.; Matsubara, I.; Murayama, N., Resistive oxygen gas sensors based on Ce1-xZrxO2 nano powder prepared using new precipitation method. *Sensors and Actuators B-Chemical* **2005**, 108, (1-2), 238-243.

154. Rajabbeigi, N.; Elyassi, B.; Khodadadi, A. A.; Mohajerzadeh, S.; Mortazavi, Y.; Sahimi, M., Oxygen sensor with solid-state CeO2-ZrO2-TiO2 reference. *Sensors and Actuators B-Chemical* **2005**, 108, (1-2), 341-345.

155. Tsai, Y. Y.; Oca-Cossio, J.; Lin, S. M.; Woan, K.; Yu, P. C.; Sigmund, W., Reactive oxygen species scavenging properties of ZrO2-CeO2 solid solution nanoparticles. *Nanomedicine* **2008**, 3, (5), 637-645.

156. Wang, Z. L.; Feng, X. D., Polyhedral shapes of CeO2 nanoparticies. *Journal of Physical Chemistry B* **2003**, 107, (49), 13563-13566.

157. Zhang, F.; Jin, Q.; Chan, S. W., Ceria nanoparticles: Size, size distribution, and shape. *Journal of Applied Physics* **2004**, 95, (8), 4319-4326.

158. Babu, S.; Thanneeru, R.; Inerbaev, T.; Day, R.; Masunov, A. E.; Schulte, A.; Seal, S., Dopant-mediated oxygen vacancy tuning in ceria nanoparticles. *Nanotechnology* **2009**, 20, (8), -.

159. Yeung, C. M. Y.; Yu, K. M. K.; Fu, Q. J.; Thompsett, D.; Petch, M. I.; Tsang, S. C., Engineering Pt in ceria for a maximum metal-support interaction in catalysis. *Journal of the American Chemical Society* **2005**, 127, (51), 18010-18011.

160. Asati, A.; Santra, S.; Kaittanis, C.; Nath, S.; Perez, J. M., Oxidase-Like Activity of Polymer-Coated Cerium Oxide Nanoparticles. *Angewandte Chemie-International Edition* **2009**, 48, (13), 2308-2312.

161. Harman, D., Aging: a theory based on free radical and radiation chemistry. *J Gerontol* **1956**, 11, (3), 298-300.

162. Beckman, K. B.; Ames, B. N., The free radical theory of aging matures. *Physiol Rev* **1998**, 78, (2), 547-81.

163. Das, M.; Patil, S.; Bhargava, N.; Kang, J. F.; Riedel, L. M.; Seal, S.; Hickman, J. J., Autocatalytic ceria nanoparticles offer neuroprotection to adult rat spinal cord neurons. *Biomaterials* **2007**, 28, (10), 1918-1925.

164. Kim, J.; Takahashi, M.; Shimizu, T.; Shirasawa, T.; Kajita, M.; Kanayama, A.; Miyamoto, Y., Effects of a potent antioxidant, platinum nanoparticle, on the lifespan of Caenorhabditis elegans. *Mech Ageing Dev* **2008**, 129, (6), 322-31.

165. Carney, J. M.; Starke-Reed, P. E.; Oliver, C. N.; Landum, R. W.; Cheng, M. S.; Wu, J. F.; Floyd, R. A., Reversal of age-related increase in brain protein oxidation, decrease in enzyme activity, and loss in temporal and spatial memory by chronic administration of the spin-trapping compound N-tert-butyl-alpha-phenylnitrone. *Proc Natl Acad Sci U S A* **1991**, 88, (9), 3633-6.

166. Head, E., Oxidative damage and cognitive dysfunction: antioxidant treatments to promote healthy brain aging. *Neurochem Res* **2009**, 34, (4), 670-8.

167. Calabrese, V.; Scapagnini, G.; Ravagna, A.; Colombrita, C.; Spadaro, F.; Butterfield, D. A.; Giuffrida Stella, A. M., Increased expression of heat shock proteins in rat brain during aging: relationship with mitochondrial function and glutathione redox state. *Mech Ageing Dev* **2004**, 125, (4), 325-35.

168. Rodrigues Siqueira, I.; Fochesatto, C.; da Silva Torres, I. L.; Dalmaz, C.; Alexandre Netto, C., Aging affects oxidative state in hippocampus, hypothalamus and adrenal glands of Wistar rats. *Life Sci* **2005**, 78, (3), 271-8.

169. Sofic, E.; Sapcanin, A.; Tahirovic, I.; Gavrankapetanovic, I.; Jellinger, K.; Reynolds, G. P.; Tatschner, T.; Riederer, P., Antioxidant capacity in postmortem brain tissues of Parkinson's and Alzheimer's diseases. *J Neural Transm Suppl* **2006**, (71), 39-43.

170. Uttara, B.; Singh, A. V.; Zamboni, P.; Mahajan, R. T., Oxidative stress and neurodegenerative diseases: a review of upstream and downstream antioxidant therapeutic options. *Curr Neuropharmacol* **2009**, 7, (1), 65-74.

171. Pratico, D., Evidence of oxidative stress in Alzheimer's disease brain and antioxidant therapy: lights and shadows. *Ann N Y Acad Sci* **2008**, 1147, 70-8.

172. Miller, R. L.; James-Kracke, M.; Sun, G. Y.; Sun, A. Y., Oxidative and inflammatory pathways in Parkinson's disease. *Neurochem Res* **2009**, 34, (1), 55-65.

173. Niizuma, K.; Endo, H.; Chan, P. H., Oxidative stress and mitochondrial dysfunction as determinants of ischemic neuronal death and survival. *Journal of Neurochemistry* **2009**, 109, 133-138.

174. Danielson, S. R.; Andersen, J. K., Oxidative and nitrative protein modifications in Parkinson's disease. *Free Radical Biology and Medicine* **2008**, 44, (10), 1787-1794.

175. Chrissobolis, S.; Faraci, F. M., The role of oxidative stress and NADPH oxidase in cerebrovascular disease. *Trends in Molecular Medicine* **2008**, 14, (11), 495-502.

176. Yamato, M.; Egashira, T.; Utsumi, H., Application of in vivo ESR spectroscopy to measurement of cerebrovascular ROS generation in stroke. *Free Radical Biology and Medicine* **2003**, 35, (12), 1619-1631.

177. Matsuda, S.; Umeda, M.; Uchida, H.; Kato, H.; Araki, T., Alterations of oxidative stress markers and apoptosis markers in the striatum after transient focal cerebral ischemia in rats. *Journal of Neural Transmission* **2009**, 116, (4), 395-404.

178.   Floyd, R. A.; Carney, J. M., Free-Radical Damage to Protein and DNA - Mechanisms In-
       volved and Relevant Observations on Brain Undergoing Oxidative Stress. *Annals of Neu-
       rology* **1992**, 32, S22-S27.
179.   Oliver, C. N.; Starke-Reed, P. E.; Stadtman, E. R.; Liu, G. J.; Carney, J. M.; Floyd, R. A.,
       Oxidative damage to brain proteins, loss of glutamine synthetase activity, and production
       of free radicals during ischemia/reperfusion-induced injury to gerbil brain. *Proc Natl Acad
       Sci U S A* **1990**, 87, (13), 5144-7.
180.   Chrissobolis, S.; Faraci, F. M., The role of oxidative stress and NADPH oxidase in cerebro-
       vascular disease. *Trends Mol Med* **2008**, 14, (11), 495-502.
181.   Chong, Z. Z.; Li, F. Q.; Maiese, K., Oxidative stress in the brain: Novel cellular targets that
       govern survival during neurodegenerative disease. *Progress in Neurobiology* **2005**, 75, (3),
       207-246.
182.   Saver, J. L., Time is brain–quantified. *Stroke* **2006**, 37, (1), 263-6.
183.   Brouns, R.; De Deyn, P. P., The complexity of neurobiological processes in acute ischemic
       stroke. *Clin Neurol Neurosurg* **2009**, 111, (6), 483-95.
184.   Floyd, R. A.; Carney, J. M., Free radical damage to protein and DNA: mechanisms involved
       and relevant observations on brain undergoing oxidative stress. *Ann Neurol* **1992**, 32 Suppl,
       S22-7.
185.   Yamato, M.; Egashira, T.; Utsumi, H., Application of in vivo ESR spectroscopy to measure-
       ment of cerebrovascular ROS generation in stroke. *Free Radic Biol Med* **2003**, 35, (12),
       1619-31.
186.   Matsuda, S.; Umeda, M.; Uchida, H.; Kato, H.; Araki, T., Alterations of oxidative stress
       markers and apoptosis markers in the striatum after transient focal cerebral ischemia in rats.
       J Neural Transm **2009**, 116, (4), 395-404.
187.   Chan, P. H., Reactive oxygen radicals in signaling and damage in the ischemic brain. *J
       Cereb Blood Flow Metab* **2001**, 21, (1), 2-14.
188.   Eliasson, M. J.; Huang, Z.; Ferrante, R. J.; Sasamata, M.; Molliver, M. E.; Snyder, S. H.;
       Moskowitz, M. A., Neuronal nitric oxide synthase activation and peroxynitrite formation in
       ischemic stroke linked to neural damage. *J Neurosci* **1999**, 19, (14), 5910-8.
189.   Uttenthal, L. O.; Alonso, D.; Fernandez, A. P.; Campbell, R. O.; Moro, M. A.; Leza, J. C.;
       Lizasoain, I.; Esteban, F. J.; Barroso, J. B.; Valderrama, R.; Pedrosa, J. A.; Peinado, M. A.;
       Serrano, J.; Richart, A.; Bentura, M. L.; Santacana, M.; Martinez-Murillo, R.; Rodrigo, J.,
       Neuronal and inducible nitric oxide synthase and nitrotyrosine immunoreactivities in the
       cerebral cortex of the aging rat. *Microscopy Research and Technique* **1998**, 43, (1), 75-88.
190.   Alonso, D.; Serrano, J.; Rodriguez, I.; Ruiz-Cabello, J.; Fernandez, A. P.; Encinas, J. M.;
       Castro-Blanco, S.; Bentura, M. L.; Santacana, M.; Richart, A.; Fernandez-Vizarra, P.; Ut-
       tenthal, L. O.; Rodrigo, J., Effects of oxygen and glucose deprivation on the expression and
       distribution of neuronal and inducible nitric oxide synthases and on protein nitration in rat
       cerebral cortex. *Journal of Comparative Neurology* **2002**, 443, (2), 183-200.
191.   Merrill, J. E.; Murphy, S. P., Inflammatory events at the blood brain barrier: Regulation of
       adhesion molecules, cytokines, and chemokines by reactive nitrogen and oxygen species.
       Brain Behavior and Immunity **1997**, 11, (4), 245-263.
192.   Ginsberg, M. D., Neuroprotection for ischemic stroke: Past, present and future. *Neurophar-
       macology* **2008**, 55, (3), 363-389.
193.   Endres, M.; Scott, G.; Namura, S.; Salzman, A. L.; Huang, P. L.; Moskowitz, M. A.; Szabo,
       C., Role of peroxynitrite and neuronal nitric oxide synthase in the activation of poly(ADP-
       ribose) synthetase in a murine model of cerebral ischemia-reperfusion. *Neuroscience Let-
       ters* **1998**, 248, (1), 41-44.
194.   Bal-Price, A.; Brown, G. C., Inflammatory neurodegeneration mediated by nitric oxide
       from activated glia-inhibiting neuronal respiration, causing glutamate release and excito-
       toxicity. *Journal of Neuroscience* **2001**, 21, (17), 6480-6491.
195.   Romanos, E.; Planas, A. M.; Amaro, S.; Chamorro, A., Uric acid reduces brain damage and
       improves the benefits of rt-PA in a rat model of thromboembolic stroke. *Journal of Cerebral
       Blood Flow and Metabolism* **2007**, 27, (1), 14-20.

196. Suzuki, M.; Tabuchi, M.; Ikeda, M.; Tomita, T., Concurrent formation of peroxynitrite nitric oxide synthase in the brain with the expression of inducible during middle cerebral artery occlusion and reperfusion in rats. *Brain Research* **2002**, 951, (1), 113-120.

197. Ashwal, S.; Cole, D. J.; Osborne, T. N.; Pearce, W. J., Low dose L-NAME reduces infarct volume in the rat MCAO/reperfusion model. *J Neurosurg Anesthesiol* **1993**, 5, (4), 241-9.

198. Dawson, V. L.; Dawson, T. M., Nitric oxide in neurodegeneration. *Prog Brain Res* **1998**, 118, 215-29.

199. Ginsberg, M. D., Neuroprotection for ischemic stroke: past, present and future. *Neuropharmacology* **2008**, 55, (3), 363-89.

200. Moncada, S.; Bolanos, J. P., Nitric oxide, cell bioenergetics and neurodegeneration. *J Neurochem* **2006**, 97, (6), 1676-89.

201. Martinez-Lazcano, J. C.; Perez-Severiano, F.; Escalante, B.; Ramirez-Emiliano, J.; Vergara, P.; Gonzalez, R. O.; Segovia, J., Selective protection against oxidative damage in brain of mice with a targeted disruption of the neuronal nitric oxide synthase gene. *J Neurosci Res* **2007**, 85, (7), 1391-402.

202. Harukuni, I.; Bhardwaj, A., Mechanisms of brain injury after global cerebral ischemia. *Neurol Clin* **2006**, 24, (1), 1-21.

203. Li, X.; Nemoto, M.; Xu, Z.; Yu, S. W.; Shimoji, M.; Andrabi, S. A.; Haince, J. F.; Poirier, G. G.; Dawson, T. M.; Dawson, V. L.; Koehler, R. C., Influence of duration of focal cerebral ischemia and neuronal nitric oxide synthase on translocation of apoptosis-inducing factor to the nucleus. *Neuroscience* **2007**, 144, (1), 56-65.

204. Park, E. M.; Cho, S.; Frys, K.; Racchumi, G.; Zhou, P.; Anrather, J.; Iadecola, C., Interaction between inducible nitric oxide synthase and poly(ADP-ribose) polymerase in focal ischemic brain injury. *Stroke* **2004**, 35, (12), 2896-901.

205. Iadecola, C.; Zhang, F.; Casey, R.; Nagayama, M.; Ross, M. E., Delayed reduction of ischemic brain injury and neurological deficits in mice lacking the inducible nitric oxide synthase gene. *J Neurosci* **1997**, 17, (23), 9157-64.

206. Szabo, C.; Dawson, V. L., Role of poly(ADP-ribose) synthetase in inflammation and ischaemia-reperfusion. *Trends Pharmacol Sci* **1998**, 19, (7), 287-98.

207. Ha, H. C.; Snyder, S. H., Poly(ADP-ribose) polymerase-1 in the nervous system. *Neurobiol Dis* **2000**, 7, (4), 225-39.

208. Endres, M.; Scott, G. S.; Salzman, A. L.; Kun, E.; Moskowitz, M. A.; Szabo, C., Protective effects of 5-iodo-6-amino-1,2-benzopyrone, an inhibitor of poly(ADP-ribose) synthetase against peroxynitrite-induced glial damage and stroke development. *Eur J Pharmacol* **1998**, 351, (3), 377-82.

209. Goretski, J.; Zafiriou, O. C.; Hollocher, T. C., Steady-state nitric oxide concentrations during denitrification. *J Biol Chem* **1990**, 265, (20), 11535-8.

210. Susin, S. A.; Lorenzo, H. K.; Zamzami, N.; Marzo, I.; Snow, B. E.; Brothers, G. M.; Mangion, J.; Jacotot, E.; Costantini, P.; Loeffler, M.; Larochette, N.; Goodlett, D. R.; Aebersold, R.; Siderovski, D. P.; Penninger, J. M.; Kroemer, G., Molecular characterization of mitochondrial apoptosis-inducing factor. *Nature* **1999**, 397, (6718), 441-6.

211. Daugas, E.; Nochy, D.; Ravagnan, L.; Loeffler, M.; Susin, S. A.; Zamzami, N.; Kroemer, G., Apoptosis-inducing factor (AIF): a ubiquitous mitochondrial oxidoreductase involved in apoptosis. *FEBS Lett* **2000**, 476, (3), 118-23.

212. Cao, G.; Clark, R. S.; Pei, W.; Yin, W.; Zhang, F.; Sun, F. Y.; Graham, S. H.; Chen, J., Translocation of apoptosis-inducing factor in vulnerable neurons after transient cerebral ischemia and in neuronal cultures after oxygen-glucose deprivation. *J Cereb Blood Flow Metab* **2003**, 23, (10), 1137-50.

213. Yu, S. W.; Wang, H.; Poitras, M. F.; Coombs, C.; Bowers, W. J.; Federoff, H. J.; Poirier, G. G.; Dawson, T. M.; Dawson, V. L., Mediation of poly(ADP-ribose) polymerase-1-dependent cell death by apoptosis-inducing factor. *Science* **2002**, 297, (5579), 259-63.

214. Yu, S. W.; Wang, H.; Dawson, T. M.; Dawson, V. L., Poly(ADP-ribose) polymerase-1 and apoptosis inducing factor in neurotoxicity. *Neurobiol Dis* **2003**, 14, (3), 303-17.

215. Du, L.; Zhang, X.; Han, Y. Y.; Burke, N. A.; Kochanek, P. M.; Watkins, S. C.; Graham, S. H.; Carcillo, J. A.; Szabo, C.; Clark, R. S., Intra-mitochondrial poly(ADP-ribosylation) contributes to NAD+ depletion and cell death induced by oxidative stress. *J Biol Chem* **2003**, 278, (20), 18426-33.

216. Hong, S. J.; Dawson, T. M.; Dawson, V. L., Nuclear and mitochondrial conversations in cell death: PARP-1 and AIF signaling. *Trends Pharmacol Sci* **2004**, 25, (5), 259-64.

217. El Kossi, M. M.; Zakhary, M. M., Oxidative stress in the context of acute cerebrovascular stroke. *Stroke* **2000**, 31, (8), 1889-92.

218. Yu, S. W.; Andrabi, S. A.; Wang, H.; Kim, N. S.; Poirier, G. G.; Dawson, T. M.; Dawson, V. L., Apoptosis-inducing factor mediates poly(ADP-ribose) (PAR) polymer-induced cell death. *Proc Natl Acad Sci U S A* **2006**, 103, (48), 18314-9.

219. Hess, D. T.; Matsumoto, A.; Kim, S. O.; Marshall, H. E.; Stamler, J. S., Protein S-nitrosylation: purview and parameters. *Nat Rev Mol Cell Biol* **2005**, 6, (2), 150-66.

220. Brown, G. C.; Borutaite, V., Inhibition of mitochondrial respiratory complex I by nitric oxide, peroxynitrite and S-nitrosothiols. *Biochim Biophys Acta* **2004**, 1658, (1-2), 44-9.

221. Beckman, J. S.; Beckman, T. W.; Chen, J.; Marshall, P. A.; Freeman, B. A., Apparent hydroxyl radical production by peroxynitrite: implications for endothelial injury from nitric oxide and superoxide. *Proc Natl Acad Sci U S A* **1990**, 87, (4), 1620-4.

222. Reiter, C. D.; Teng, R. J.; Beckman, J. S., Superoxide reacts with nitric oxide to nitrate tyrosine at physiological pH via peroxynitrite. *Journal of Biological Chemistry* **2000**, 275, (42), 32460-32466.

223. Kamat, C. D.; Gadal, S.; Mhatre, M.; Williamson, K. S.; Pye, Q. N.; Hensley, K., Antioxidants in Central Nervous System Diseases: Preclinical Promise and Translational Challenges. *Journal of Alzheimers Disease* **2008**, 15, (3), 473-493.

224. Cherubini, A.; Ruggiero, C.; Morand, C.; Lattanzio, F.; Dell'Aquila, G.; Zuliani, G.; Di Iorio, A.; Andres-Lacueva, C., Dietary antioxidants as potential pharmacological agents for ischemic stroke. *Current Medicinal Chemistry* **2008**, 15, (12), 1236-1248.

225. Rzigalinski, B. A., Nanoparticles and cell longevity. *Technol Cancer Res Treat* **2005**, 4, (6), 651-9.

226. Rzigalinski, B. A., Danelisen, I., Strawn, E.T, Cohen, A.A. and Liang, C. , Nanoparticles for Cell Engineering- A Radical Concept. *Nanotechnologies for the Life Sciences* **2006**, 9, Tissue, Cell and Organ Engineering.

227. Chen, J.; Patil, S.; Seal, S.; McGinnis, J. F., Nanoceria particles prevent ROI-induced blindness. *Recent Advances in Retinal Degeneration* **2008**, 613, 53-59.

228. Rzigalinski, B. A.; Bailey, D.; Chow, L.; Kuiry, S. C.; Patil, S.; Merchant, S.; Seal, S., Cerium oxide nanoparticles increase the lifespan of cultured brain cells and protect against free radical and mechanical trauma. *Faseb Journal* **2003**, 17, (4), A606-A606.

229. Niu, J. L.; Azfer, A.; Rogers, L. M.; Wang, X. H.; Kolattukudy, P. E., Cardioprotective effects of cerium oxide nanoparticles in a transgenic murine model of cardiomyopathy. *Cardiovascular Research* **2007**, 73, (3), 549-559.

230. Chen, J. P.; Patil, S.; Seal, S.; McGinnis, J. F., Rare earth nanoparticles prevent retinal degeneration induced by intracellular peroxides. *Nature Nanotechnology* **2006**, 1, (2), 142-150.

231. Sawe, N.; Steinberg, G.; Zhao, H., Dual roles of the MAPK/ERK1/2 cell signaling pathway after stroke. *Journal of Neuroscience Research* **2008**, 86, (8), 1659-1669.

232. Denes, A.; Thornton, P.; Rothwell, N. J.; Allan, S. M., Inflammation and brain injury: Acute cerebral ischaemia, peripheral and central inflammation. *Brain Behav Immun* **2009**.

233. Yu, S. W.; Wang, H. M.; Poitras, M. F.; Coombs, C.; Bowers, W. J.; Federoff, H. J.; Poirier, G. G.; Dawson, T. M.; Dawson, V. L., Mediation of poly(ADP-ribose) polymerase-1-dependent cell death by apoptosis-inducing factor. *Science* **2002**, 297, (5579), 259-263.

234. Yu, S. W.; Wang, H. M.; Dawson, T. A.; Dawson, V. L., Poly(ADP-ribose) polymerase-1 and apoptosis inducing factor in neurotoxicity. *Neurobiology of Disease* **2003**, 14, (3), 303-317.

235. Rashid, P. A.; Whitehurst, A.; Lawson, N.; Bath, P. M., Plasma nitric oxide (nitrate/nitrite)
      levels in acute stroke and their relationship with severity and outcome. *J Stroke Cerebro-
      vasc Dis* **2003**, 12, (2), 82-7.
236. Englert, B. C., Nanomaterials and the environment: uses, methods and measurement. *Jour-
      nal of Environmental Monitoring* **2007**, 9, (11), 1154-1161.
237. Hood, E., Nanotechnology: looking as we leap. *Environ Health Perspect* **2004**, 112, A740-
      A749.
238. Nel, A.; Xia, T.; Madler, L.; Li, N., Toxic potential of materials at the nanolevel. *Science*
      2006, 311, (5761), 622-627.
239. Thomas, K., Aguar, P., Kawasaki, H., Morris, J., Nakanishi, Savage, N., *Toxicological Sci-
      ences* **2006**, 92, (1), 23-32.
240. A. Elder, H. Y., R. Gwiazda, X. Teng, S. Thurston, H. He, G. Oberdörster, Testing Nano-
      materials of Unknown Toxicity: An Example Based on Platinum Nanoparticles of Different
      Shapes. *Adv. Mater.* **2007**, 19, 3124-3129.
241. N. M. Franklin, N. J. R., S. C. Apte, G. E. Batley, G. E. Gadd, P. S. Casey, Comparative
      Toxicity of Nanoparticulate ZnO, Bulk ZnO, andZnCl2 to a Freshwater Microalga (Pseu-
      dokirchneriella subcapitata): The Importance of Particle Solubility. *Environ. Sci. Technol.*
      2007, 41, 8484-8490.
242. Karlsson, H. L.; Gustafsson, J.; Cronholm, P.; Moller, L., Size-dependent toxicity of metal
      oxide particles-A comparison between nano- and micrometer size. *Toxicology Letters* **2009**,
      188, (2), 112-118.
243. Z. Chen, H. M., G. Xing, C. Chen, Y. Zhao, G. Jia, T. Wang, H. Yuan, C. Ye, F. Zhao, Z.
      Chai, C. Zhu, X. Fang, B. Ma, L. Wan, Acute toxicological effects of copper nanoparticles
      in vivo. *Toxicology Letters* **2006**, 163, 109-120.
244. J. A. Khan, B. P., T. K. Das, Y. Singh, S. Maiti, Molecular effects of uptake of gold nanopar-
      ticles in HeLa cells. *Chem. Bio. Chem.* **2007**, 8, 1237-1240.
245. Brunner, T. J. W., P.; Manser, P.; Spohn, P.; Grass, R. N.; Limbach, L. K.; Bruinink, A.;
      Stark, W. J, In vitro cytotoxicity of oxide nanoparticles: Comparison to asbestos, silica, and
      the effect of particle solubility. . *Environ. Sci. Technol.* **2006**, 40, 4374-4381.
246. M. Heinlaan, A. I., I. Blinova, H.-C. Dubourguier, A. Kahru, Toxicity of nanosized and
      bulk ZnO, CuO and TiO2 to bacteria Vibrio fischeri and crustaceanus Daphnia magna and
      Thamnocephalus platyurus. *Chemosphere* **2008**, 71, 1308-1316.
247. Park, M. V. D. Z.; Annema, W.; Salvati, A.; Lesniak, A.; Elsaesser, A.; Barnes, C.; McKerr,
      G.; Howard, C. V.; Lynch, I.; Dawson, K. A.; Piersma, A. H.; de Jong, W. H., In vitro devel-
      opmental toxicity test detects inhibition of stem cell differentiation by silica nanoparticles.
      Toxicology and Applied Pharmacology **2009**, 240, (1), 108-116.
248. Bar-Ilan, O.; Albrecht, R. M.; Fako, V. E.; Furgeson, D. Y., Toxicity Assessments of Multi-
      sized Gold and Silver Nanoparticles in Zebrafish Embryos. *Small* **2009**, 5, (16), 1897-1910.
249. Ispas, C.; Andreescu, D.; Patel, A.; Goia, D. V.; Andreescu, S.; Wallace, K. N., Toxicity and
      Developmental Defects of Different Sizes and Shape Nickel Nanoparticles in Zebrafish.
      Environmental Science & Technology **2009**, 43, (16), 6349-6356.
250. Kovochich, M.; Espinasse, B.; Auffan, M.; Hotze, E. M.; Wessel, L.; Xia, T.; Nel, A. E.;
      Wiesner, M. R., Comparative Toxicity of C-60 Aggregates toward Mammalian Cells: Role
      of Tetrahydrofuran (THF) Decomposition. *Environmental Science & Technology* **2009**, 43,
      (16), 6378-6384.
251. Koeneman, B. A.; Zhang, Y.; Hristovski, K.; Westerhoff, P.; Chen, Y. S.; Crittenden, J. C.;
      Capco, D. G., Experimental approach for an in vitro toxicity assay with non-aggregated
      quantum dots. *Toxicology in Vitro* **2009**, 23, (5), 955-962.
252. Cheng, J. P.; Chan, C. M.; Veca, L. M.; Poon, W. L.; Chan, P. K.; Qu, L. W.; Sun, Y. P.;
      Cheng, S. H., Acute and long-term effects after single loading of functionalized multi-
      walled carbon nanotubes into zebrafish (Danio rerio). *Toxicology and Applied Pharmacol-
      ogy* **2009**, 235, (2), 216-225.
253. King-Heiden, T. C.; Wiecinski, P. N.; Mangham, A. N.; Metz, K. M.; Nesbit, D.; Pedersen,
      J. A.; Hamers, R. J.; Heideman, W.; Peterson, R. E., Quantum Dot Nanotoxicity Assess-

ment Using the Zebrafish Embryo. *Environmental Science & Technology* **2009**, 43, (5), 1605-1611.

254. Yeo, M. K.; Pak, S. W., Exposing Zebrafish to Silver Nanoparticles during Caudal Fin Regeneration Disrupts Caudal Fin Growth and p53 Signaling. *Molecular & Cellular Toxicology* **2008**, 4, (4), 311-317.

255. Harper, S.; Usenko, C.; Hutchison, J. E.; Maddux, B. L. S.; Tanguay, R. L., In vivo biodistribution and toxicity depends on nanomaterial composition, size, surface functionalisation and route of exposure. *Journal of Experimental Nanoscience* **2008**, 3, (3), 195-206.

256. Zhu, X. S.; Zhu, L.; Lang, Y. P.; Chen, Y. S., Oxidative stress and growth inhibition in the freshwater fish Carassius auratus induced by chronic exposure to sublethal fullerene aggregates. *Environmental Toxicology and Chemistry* **2008**, 27, (9), 1979-1985.

257. Asharani, P. V.; Wu, Y. L.; Gong, Z. Y.; Valiyaveettil, S., Toxicity of silver nanoparticles in zebrafish models. *Nanotechnology* **2008**, 19, (25), -.

258. Handy, R. D.; Henry, T. B.; Scown, T. M.; Johnston, B. D.; Tyler, C. R., Manufactured nanoparticles: their uptake and effects on fish-a mechanistic analysis. *Ecotoxicology* **2008**, 17, (5), 396-409.

259. Lee, K. J.; Nallathamby, P. D.; Browning, L. M.; Osgood, C. J.; Xu, X. H. N., In vivo imaging of transport and biocompatibility of single silver nanoparticles in early development of zebrafish embryos. *Acs Nano* **2007**, 1, (2), 133-143.

260. Griffitt, R. J.; Weil, R.; Hyndman, K. A.; Denslow, N. D.; Powers, K.; Taylor, D.; Barber, D. S., Exposure to copper nanoparticles causes gill injury and acute lethality in zebrafish (Danio rerio), *Environmental Science & Technology* **2007**, 41, (23), 8178-8186.

261. Usenko, C. Y.; Harper, S. L.; Tanguay, R. L., In vivo evaluation of carbon fullerene toxicity using embryonic zebrafish. *Carbon* **2007**, 45, (9), 1891-1898.

262. Zhu, X. S.; Zhu, L.; Li, Y.; Duan, Z. H.; Chen, W.; Alvarez, P. J. J., Developmental toxicity in zebrafish (Danio rerio) embryos after exposure to manufactured nanomaterials: Buckminsterfullerene aggregates (nC(60)) and fullerol. *Environmental Toxicology and Chemistry* **2007**, 26, (5), 976-979.

263. Fako, V. E.; Furgeson, D. Y., Zebrafish as a correlative and predictive model for assessing biomaterial nanotoxicity. *Advanced Drug Delivery Reviews* **2009**, 61, (6), 478-486.

264. Oberdorster, G.; Oberdorster, E.; Oberdorster, J., Nanotoxicology: An emerging discipline evolving from studies of ultrafine particles. *Environmental Health Perspectives* **2005**, 113, (7), 823-839.

265. Weiss, P. S., The Big Picture. *Acs Nano* **2009**, 3, (7), 1603-1604.

266. Godwin, H. A.; Chopra, K.; Bradley, K. A.; Cohen, Y.; Harthorn, B. H.; Hoek, E. M. V.; Holden, P.; Keller, A. A.; Lenihan, H. S.; Nisbet, R. M.; Nel, A. E., The University of California Center for the Environmental Implications of Nanotechnology. *Environmental Science & Technology* **2009**, 43, (17), 6453-6457.

267. Wiesner, M. R.; Lowry, G. V.; Jones, K. L.; Hochella, M. F.; Di Giulio, R. T.; Casman, E.; Bernhardt, E. S., Decreasing Uncertainties in Assessing Environmental Exposure, Risk, and Ecological Implications of Nanomaterials. *Environmental Science & Technology* **2009**, 43, (17), 6458-6462.

268. R.J. Griffitt, J. L., J. Gao, J-C. Bonzongo, D. S. Barber, *Environ. Toxicol. Chem.* **2008**, 27, 1972-1978.

269. Arnold, M.; Cavalcanti-Adam, E. A.; Glass, R.; Blummel, J.; Eck, W.; Kantlehner, M.; Kessler, H.; Spatz, J. P., Activation of integrin function by nanopatterned adhesive interfaces. *Chemphyschem* **2004**, 5, (3), 383-388.

270. Cassee FR, M. H., Duistermaat E, Freijer JJ, Geerse KB, Marijnissen; JC, e. a., Particle size-dependent total mass deposition in lungs determines inhalation toxicity of cadmium chloride aerosols in rats. Application of a multiple path dosimetry model. *Arch Toxicol* **2002**, 76, 277-286.

271. Huang M, K. E., Lim LY, Uptake and cytotoxicity of chitosan molecules and nanoparticles: effects of molecular weight and degree of deacetylation. *Pharm Res* **2004**, 21, 344-353.

272. Warheit, D. B., Nanoparticles: Health impacts? . *Materials Today* **2004**, 7, 32-35.

273. R. J . Griffitt, R. W., K. H yndman, N. D. Denslow, K. Powers, D. Taylor, D. S . Barber, Exposure to Copper Nanoparticles Causes Gill Injury and Acute Lethality in Zebrafish (Danio rerio). *Environ. Sci. Technol.* **2007**, 41, 8178–8186.
274. A.R. Badireddy, E. M. H., S. Chellam, P. Alvarez, M. R. Wiesner, *Environ. Sci. Technol.* **2007**, 41, 6627-6632.
275. K. D. Pickering, M. R. W., *Environ. Sci. Technol.* **2005**, 39, 1359-1365.
276. X. Zhu, L. Z., Y. Li, Z. Duan, W. Chen, P. J. J. Alvarez, Developmental toxicity in zebrafish (Danio Rerio) embryos after exposure to manufactured nanomaterials: buckminsterfullerene aggregates (nC_{60}) and fullerol. *Environ. Toxicol. Chem.* **2007**, 26, (5), 976-979.
277. Zhu, X. S.; Wang, J. X.; Zhang, X. Z.; Chang, Y.; Chen, Y. S., The impact of ZnO nanoparticle aggregates on the embryonic development of zebrafish (Danio rerio). *Nanotechnology* **2009**, 20, (19), -.
278. Z. Chen, G. X., H. Yuan, C. Chen, F. Zhao, C. Zhang, Y. Zhao, *Toxicology Letters* **2007**, 175, 102-110.
279. Heiden, T. C. K.; Dengler, E.; Kao, W. J.; Heideman, W.; Peterson, R. E., Developmental toxicity of low generation PAMAM dendrimers in zebrafish. *Toxicology and Applied Pharmacology* **2007**, 225, (1), 70-79.
280. Ginzburg, V. V.; Balijepailli, S., Modeling the thermodynamics of the interaction of nanoparticles with cell membranes. *Nano Letters* **2007**, 7, (12), 3716-3722.
281. J. M. Balbus, A. D. M., V. L. Colvin, V. Castranova, G. P. Daston, R. A. Denison, K. L. Dreher, P. L. Goering, A. M. Goldberg, K. M. Kulinowski, N. A. Monteiro-Riviere, G. Oberdörster, G. S. Omenn, K. E. Pinkerton, K. S. Ramos, K. M. Rest, J. B. Sass, E. K. Silbergeld, B. A. Wong, Meeting Report: Hazard Assessment for Nanoparticles—Report from an Interdisciplinary Workshop. *Environmental Health Perspectives* **2007**, 115, (11), 1654-1659.
282. R. Shukla, V. B., M. Chaudhary, A. Basu, R. R. Bhonde, M. Sastry, *Langmuir* **2005**, 21, 10644-10654.
283. Connor, E. E., Mwamuka, J., Gole,A., Murphy, C. J., Wyatt, M. D., *Small* **2005**, 1, 325-327.
284. Goodman, C. M. M., C. D., Yilmaz, T., Rotello, V. M., *Bioconjugate Chem.* **2004**, 15, 897-900.
285. Ogle, R., Application of Industrial Hygiene Tools and Tenets to Controlling Nanomaterials in R&D Operations. *Presentation at the Commercialization of NanoMaterials 2007, Pittsburgh, PA, 12 November 2007* **2007**.
286. Rickabaugh, K., Laboratory Workplace Safety Practices and Sampling and Analysis Considerations". *(Presentation at the Commercialization of NanoMaterials 2007, Pittsburgh, PA, 12 November 2007). ***2007**.
287. Colvin, V. L., The potential environmental impact of engineered nanomaterials. *Nat Biotechnol* **2003**, 21, (10), 1166-70.
288. Brown, G. C.; Borutaite, V., Inhibition of mitochondrial respiratory complex I by nitric oxide, peroxynitrite and S-nitrosothiols. *Biochimica Et Biophysica Acta-Bioenergetics* **2004**, 1658, (1-2), 44-49.
289. Gahwiler, B. H.; Capogna, M.; Debanne, D.; McKinney, R. A.; Thompson, S. M., Organotypic slice cultures: a technique has come of age. *Trends in Neurosciences* **1997**, 20, (10), 471-477.
290. Park, E. M.; Cho, S.; Frys, K.; Racchumi, G.; Zhou, P.; Anrather, J.; Iadecola, C., Interaction between inducible nitric oxide synthase and poly(ADP-ribose) polymerase in focal ischemic brain injury. *Stroke* **2004**, 35, (12), 2896-2901.
291. Simon-Deckers, A.; Loo, S.; Mayne-L'Hermite, M.; Herlin-Boime, N.; Menguy, N.; Reynaud, C.; Gouget, B.; Carriere, M., Size-, Composition- and Shape-Dependent Toxicological Impact of Metal Oxide Nanoparticles and Carbon Nanotubes toward Bacteria. *Environmental Science & Technology* **2009**, 43, (21), 8423-8429.
292. Warheit, D. B.; Sayes, C. M.; Reed, K. L., Nanoscale and Fine Zinc Oxide Particles: Can in Vitro Assays Accurately Forecast Lung Hazards following Inhalation Exposures? *Environmental Science & Technology* **2009**, 43, (20), 7939-7945.

293.   Di Virgilio, A. L.; Reigosa, M.; de Mele, M. F., Response of UMR 106 cells exposed to titanium oxide and aluminum oxide nanoparticles. *J Biomed Mater Res A* 92, (1), 80-6.

294.   Brunner, T. J.; Wick, P.; Manser, P.; Spohn, P.; Grass, R. N.; Limbach, L. K.; Bruinink, A.; Stark, W. J., In vitro cytotoxicity of oxide nanoparticles: Comparison to asbestos, silica, and the effect of particle solubility. *Environmental Science & Technology* **2006**, 40, (14), 4374-4381.

295.   Wiench, K.; Wohlleben, W.; Hisgen, V.; Radke, K.; Salinas, E.; Zok, S.; Landsiedel, R., Acute and chronic effects of nano- and non-nano-scale TiO2 and ZnO particles on mobility and reproduction of the freshwater invertebrate Daphnia magna. *Chemosphere* **2009**, 76, (10), 1356-1365.

296.   Sayes, C. M.; Reed, K. L.; Subramoney, S.; Abrams, L.; Warheit, D. B., Can in vitro assays substitute for in vivo studies in assessing the pulmonary hazards of fine and nanoscale materials? *Journal of Nanoparticle Research* **2009**, 11, (2), 421-431.

297.   S.; Sarita S. H.; Allan B., R. A. Y. R. L. F. J. M. U. M. T. T. U. M. G. P. W. E. A. G. R., Biodistribution and oxidative stress effects of a systemically-introduced commercial ceria engineered nanomaterial *Nanotoxicology* **2009**, 3, (3), 234 - 248

298.   Chen, J.; Patil, S.; Seal, S.; McGinnis, J. F., Nanoceria particles prevent ROI-induced blindness. *Adv Exp Med Biol* **2008**, 613, 53-9.

299.   Eom, H. J.; Choi, J., Oxidative stress of CeO2 nanoparticles via p38-Nrf-2 signaling pathway in human bronchial epithelial cell, Beas-2B. *Toxicol Lett* **2009**, 187, (2), 77-83.

300.   Xia, T.; Kovochich, M.; Liong, M.; Madler, L.; Gilbert, B.; Shi, H.; Yeh, J. I.; Zink, J. I.; Nel, A. E., Comparison of the mechanism of toxicity of zinc oxide and cerium oxide nanoparticles based on dissolution and oxidative stress properties. *ACS Nano* **2008**, 2, (10), 2121-34.

301.   Wilming, L. G.; Gilbert, J. G.; Howe, K.; Trevanion, S.; Hubbard, T.; Harrow, J. L., The vertebrate genome annotation (Vega) database. *Nucleic Acids Res* **2008**, 36, (Database issue), D753-60.

302.   Ispas, C.; Wallace, K. N.; Andreescu, S., Cytotoxicty studies of metal and metal oxide nanoparticles using zebrafish embryos as model toxicological target. *ACS meeting, Philadelphia, Aug 17-22, 2008*. **2008**.

303.   Perez, J. M.; Asati, A.; Nath, S.; Kaittanis, C., Synthesis of biocompatible dextran-coated nanoceria with pH-dependent antioxidant properties. *Small* **2008**, 4, (5), 552-556.

304.   Zhang, J.; Han, H. W.; Wu, S. J.; Xu, S.; Zhou, C. H.; Yang, Y.; Zhao, X. Z., Ultrasonic irradiation to modify the PEO/P(VDF-HFP)/TiO2 nanoparticle composite polymer electrolyte for dye sensitized solar cells. *Nanotechnology* **2007**, 18, (29), -.

305.   Koziara, J. M.; Lockman, P. R.; Allen, D. D.; Mumper, R. J., The blood-brain barrier and brain drug delivery. *Journal of Nanoscience and Nanotechnology* **2006**, 6, (9-10), 2712-2735.

306.   Yokel, R. A., Manganese flux across the blood-brain barrier. *Neuromolecular Med* **2009**, 11, (4), 297-310.

307.   Aksenov, M. Y.; Aksenova, M. V.; Carney, J. M.; Butterfield, D. A., Oxidative modification of glutamine synthetase by amyloid beta peptide. *Free Radical Research* **1997**, 27, (3), 267-281.

308.   Subramaniam, R.; Roediger, F.; Jordan, B.; Mattson, M. P.; Keller, J. N.; Waeg, G.; Butterfield, D. A., The lipid peroxidation product, 4-hydroxy-2-trans-nonenal, alters the conformation of cortical synaptosomal membrane proteins. *J Neurochem* **1997**, 69, (3), 1161-9.

309.   Reed, T.; Perluigi, M.; Sultana, R.; Pierce, W. M.; Klein, J. B.; Turner, D. M.; Coccia, R.; Markesbery, W. R.; Butterfield, D. A., Redox proteomic identification of 4-Hydroxy-2-nonenal-modified brain proteins in amnestic mild cognitive impairment: Insight into the role of lipid peroxidation in the progression and pathogenesis of Alzheimer's disease. *Neurobiology of Disease* **2008**, 30, (1), 107-120.

310.   Perluigi, M.; Sultana, R.; Cenini, G.; Di Domenico, F.; Memo, M.; Pierce, W. M.; Coccia, R.; Butterfield, D. A., Redox proteomics identification of 4-hydroxynonenal-modified

brain proteins in Alzheimer's disease: Role of lipid peroxidation in Alzheimer's disease pathogenesis. *Proteomics Clinical Applications* **2009**, 3, (6), 682-693.

311. Tangpong, J.; Sompol, P.; Vore, M.; Clair, W.; Ratanachaiyavong, S.; Butterfield, D. A.; Clair, D. K., TNF-mediated NO production enhance MnSOD nitration and mitochondrial dysfunction. *Free Radical Biology and Medicine* **2006**, 41, S169-S169.

312. Tangpong, J.; Cole, M. P.; Sultana, R.; Estus, S.; Vore, M.; St Clair, W.; Ratanachaiyavong, S.; St Clair, D. K.; Butterfield, D. A., Adriamycin-mediated nitration of manganese superoxide dismutase in the central nervous system: insight into the mechanism of chemobrain. *Journal of Neurochemistry* **2007**, 100, (1), 191-201.

313. Joshi, G.; Hardas, S.; Sultana, R.; St Clair, D. K.; Vore, M.; Butterfield, D. A., Glutathione elevation by gamma-glutamyl cysteine ethyl ester as a potential therapeutic strategy for preventing oxidative stress in brain mediated by in vivo administration of adriamycin: Implication for chemobrain. *J Neurosci Res* **2007**, 85, (3), 497-503.

314. Estevez, A. Y.; Lynch, A.; Lucky, J.; Ludington, J.; Mosenthal, W.; Pritchard, S.; Leiter, J. C.; Andreescu, S.; Erlichman, J. S., Neuroprotective mechanisms of cerium oxide nanoparticles in a mouse hippocampal brain slice model of ischemia, *Free Radic Biol Med* **2011**. doi:10.1016/j.freeradbiomed.2011.06.006.

Chapter 4
Biomedical Applications of Gold Nanoparticles

Tapan K. Sau and Dan V. Goia

Abstract Due to their unique physical and chemical properties, gold nanoparticles are poised to play an important role in this exciting and dynamic field. This article discusses the properties of colloidal gold nanoparticles and how they are exploited in the most notable applications in bioimaging, biosensing, diagnostics, drug delivery, and other therapies. The most important methods used for preparing colloidal gold and some of the strategies used in tailoring the properties of the nanoparticles to specific biomedical applications are also reviewed.

4.1   Introduction

Applications of nanoparticles in diagnosis, treatment, and monitoring of biological systems are slowly coalescing into a new field, often referred to as 'nanomedicine' [1]. Materials with nanoscale dimensions are of great interest in biomedical applications because their size is comparable to, or smaller than, that of many important biological entities such as genes (2 nm wide and 10–100 nm long), proteins (5–50 nm), viruses (20–450 nm), or cells (10–100 μm) [2]. These tiny particles can access otherwise unreachable regions of the organism and engage in interactions at molecular level or deliver a therapeutic load. For these reasons, it is widely accepted that systems incorporating either inorganic or organic nanoparticles have the potential to change dramatically the landscape of the biomedical field. A confirmation of this prediction is the large body of research focused on the development of nanoparticles and nano-structures for bio-sensing, diagnostics, drug delivery, and therapeutic purposes. Indeed, hundreds of nanomaterials-based imaging and diagnostic devices and a similar number of drug delivery systems are already in pre-clinical, clinical, or commercial development stage [3] and their number increases rapidly each year.

Due to their unique physical and chemical properties, gold nanoparticles are poised to play an important role in this exciting and dynamic field. This article discusses the properties of colloidal gold nanoparticles and how they are exploited in

D. V. Goia (✉)
Center for Advanced Materials Processing, Department of Chemistry & Biomolecular Science, Clarkson University, 8 Clarkson Avenue, CAMP, Box 5814, Potsdam, NY 13699, USA
e-mail: goiadanv@clarkson.edu

the most notable applications in bioimaging, biosensing, diagnostics, drug delivery, and other therapies. The most important methods used for preparing colloidal gold and some of the strategies used in tailoring the properties of the nanoparticles to specific biomedical applications are also reviewed.

4.2   Historical Background

Due to its esthetically-pleasing qualities, resistance to corrosion, and rarity, gold has been viewed since ancient times as the most valuable metal. These characteristics led to its extensive use in jewelry, art, decorations, and especially as money. In different forms, gold has also a long history of uses for therapeutic purposes. People in ancient India, Egypt, and China believed in the curative powers of gold for treating various diseases such as smallpox, skin ulcers, syphilis, and measles [4–7]. Chrysotherapy, or the use of gold complexes for treating arthritis, is also known for long time [8, 9]. We do not know for sure if gold colloids were used during those times to treat diseases and, if they were, we do not have any information about how they might have been prepared. We do know, however, that the optical properties of colloidal gold were widely used already in antiquity for decorative applications, which included stains and cosmetic ointments. Given the good skin tolerance, which was for sure noticed in the later case, it is reasonable to assume that old societies have attempted to use dispersions of colloidal gold for various external and, possibly, internal therapies. The work of Faraday in the mid-nineteen century has dramatically improved the understanding of the scientific aspects related to the preparation and properties of gold sols. In 1857, he was the first to suggest that the colored stable gold colloids contain the metal in finely divided state [10]. He further concluded that the burgundy red color of gold sols was the result of the interaction of light with the dispersed gold particles. Considering that the particulate nature of colloidal gold was experimentally substantiated only a century later (1942) by Thiessen [11], these insights are a testimony of Faraday's genius. His work laid the foundations for the theoretical and practical discoveries of the early twentieth century in the field of colloid science and paved the way for the use of gold colloids in bio-medical applications. Indeed, color reactions between colloidal gold particles and proteins present in cerebrospinal fluid or blood serum were documented already in 1920s and 1930s [12, 13] and in 1939 Kausche and Ruska reported the effect of colloidal gold adsorption on tobacco mosaic viruses [14]. Harford et al. in 1957 and Feldherr and Marschall 5 years later reported the use of colloidal gold as an electron-dense tracer in cellular uptake and microinjection experiments [15]. In the same decade it was discovered that attaching gold particles to proteins hardly affects their activities [16]. Later, the invention of immunogold staining procedure by Faulk and Taylor in 1971 confirmed the potential of colloidal gold nanoparticles in biological applications [17, 18]. Gold particles containing radioactive isotope ^{198}Au have been

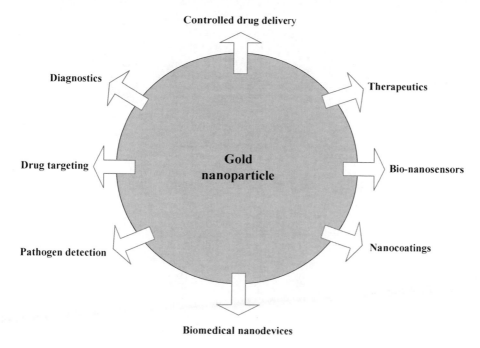

Fig. 4.1 Schematic illustrating various applications of gold nanoparticles in the bio-medical field

used recently for the treatment of liver cancer and sarcoma [16, 19]. Gold bulk or complexes have been widely used in the biomedical field as well. Gold metal is incorporated in medical devices like pacemakers, gold plated stents [20, 21], middle ear gold implants [22, 23], and gold alloys in dental restoration [24, 25]. Finally, several organo-gold complexes with promising antimicrobial, antitumor, antimalarial, and anti-HIV activities have been also reported [6, 8, 9].

As this short overview clearly indicates, the medical and biological use of gold nanoparticles is widespread and their applications are multiplying rapidly. The following section surveys the properties of colloidal gold which makes them suitable for applications in the bio-medical field.

4.3 Properties of Colloidal Gold Relevant for Biomedical Applications

The unique properties of gold nanoparticles exploited in the bio-medical field depend on the size, shape, morphology, surface chemistry, and electrical charge. The ability to tailor these features as well as the biocompatibility of colloidal gold is central to all biomedical applications (Fig. 4.1).

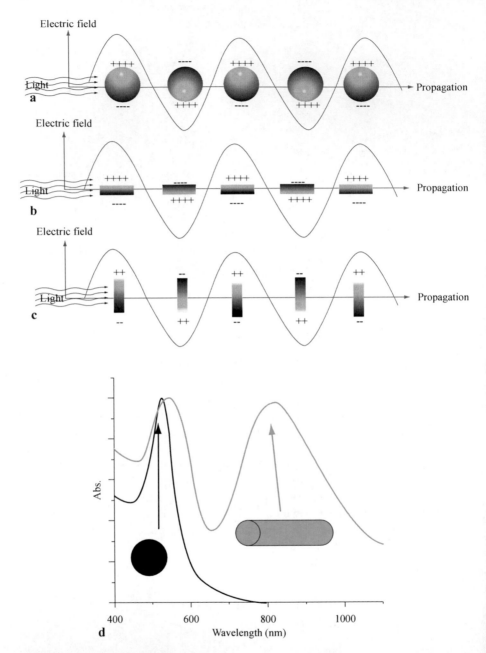

Fig. 4.2 When light impinges on a metal nanoparticle with a size much smaller than the wavelength of the incident radiation ($d \ll \lambda$), it induces a dipole (indicated by '+' and '−' signs) due to the oscillation of the conduction electrons across the metallic nanoparticle. The dipole oscillates in phase with the electric field of the incoming electromagnetic wave. The resonant coherent oscillation of the conduction electrons in the metal nanoparticle causes absorption of radiation energy generating the so-called localized surface plasmon resonance (LSPR). Top schematic shows coherent dipolar oscillation of the conduction electrons for spherical/isometric nanopar

4.3.1 Tunable Localized Surface Plasmon Resonance (LSPR)

Both bulk and colloidal gold show fascinating optical properties. The pleasing yellow color of bulk gold is caused by the electronic transition from the filled $5d$ band to the Fermi level. Relativistic/kinematical effects in gold result in a smaller gap between the full $5d$ band and the Fermi level of the half-filled $6s$ band than the corresponding gap for non-relativistic set of orbitals. Spin-orbit couplings also play an important role in the optical properties observed [26]. Similar transitions occur in other metals as well. In silver, for example, the gap between $4d$ and $5s$ is wider and the transition occurs in the ultraviolet region [27, 28]. Nanometer sized gold particles exhibit characteristic colors when interact with light [29]. UV–visible–NIR light impinging on an interface between a metal and a dielectric can excite collective oscillations of the electrons in the conduction band. These oscillations, called surface plasmons, are localized in the case of nanoparticles (Fig. 4.2). The optical properties of colloidal gold in the UV–visible–NIR spectral range are mainly determined by the localized surface plasmon resonances (LSPR) observable in their extinction/absorption spectra [30]. Associated with the LSPR are strongly enhanced and highly localized electromagnetic fields surrounding the nanoparticles that give rise to the 'local field enhancement' [31]. The peak wavelength, width, and intensity of the LSPR and the enhanced local field in the vicinity of the gold nanoparticles depend on the particle size, shape, composition, surface charge, surface-adsorbed species, refractive index of surrounding medium, and interparticle interactions [29, 32]. The typical wavelengths of LSPR for spherical gold nanoparticles lie in the visible spectrum around 530 nm [33]. In the case of isometric/spherical nanoparticles the LSPR can be tuned by simply changing their size. However, at larger sizes the plasmon band is significantly broadened due to radiation damping [34, 35]. A more elegant tuning of the LSPR, which avoids significant peak broadening, can be achieved by changing the geometry of the nanoparticle. For example, the LSPR is red-shifted all the way in the near infrared region (800–1200 nm) for nanorods, nanoshells, nanobox/nanocages, nanoprisms, or nanostars. The extinction in the NIR region is an extremely useful property for *in vivo* optical applications as this wavelength domain is an effective 'biological window' ('tissue transparency window' or 'water window'). Since NIR radiation can penetrate body tissues with minimum loss due to hemoglobin and water absorption, it provides an opportunity for light-guided effects in deep tissues [31, 36–38]. The LSPRs of non-spherical gold nanoparticles such as rods, disks, triangular prisms, etc. typically split into distinctive dipole and quadrupole plasmon modes due to their anisotropy [32, 39, 40]. Because of the changes in the optical properties, the gold nanoparticles can be used for labeling with sources of different wavelengths [41–43].

---

ticles (**a**) and rod-shaped metal nanoparticle oriented transversely (**b**) and longitudinally (**c**) to the incident radiation. (**d**) Depicts the resulting absorption spectra for gold nanospheres and nanorods. (Eustis and El-Sayed, *Chem. Soc. Rev.*, **2006**, *35*, 209–217. Copyright © 2006, Royal Society of Chemistry)

Table 4.1 SPR wavelengths and refractive index sensitivities (in RIU) of gold nanoparticles of different sizes and shapes. (Adopted from Chen et al., *Langmuir* **2008**, *24*, 5233. Copyright © 2008 American Chemical Society)

Au nanoparticles	Length (nm)[a]	Diameter (nm)[b]	Aspect ratio[c]	Plasmon wavelength (nm)[d]	Index sensitivity (nm/RIU)[e]	Figure of merit
Nanospheres		15 (1)		527	44 (3)	0.6
Nnanocubes	44 (2)			538	83 (2)	1.5
Branched nanoparticles	80 (14)			1141	703 (19)	0.8
Nanorods	40 (6)	17 (2)	2.4 (0.3)	653	195 (7)	2.6
Nanorods	55 (7)	16 (2)	3.4 (0.5)	728	224 (4)	2.1
Nanorods	74 (6)	17 (2)	4.6 (0.8)	846	288 (8)	1.7
Nanobipyramids	27 (4)	19 (7)	1.5 (0.3)	645	150 (5)	1.7
Nanobipyramids	50 (6)	18 (1)	2.7 (0.2)	735	212 (6)	2.8
Nanobipyramids	103 (7)	26 (2)	3.9 (0.2)	886	392 (7)	4.2
Nanobipyramids	189 (9)	40 (2)	4.7 (0.2)	1096	540 (6)	4.5

The numbers in the parentheses are standard deviations.
[a] The length for nanocubes is the edge length; for branched nanoparticle it is the distance from the center of the particle to the tip of the branch.
[b] The diameter for nanobipyramid is the central width.
[c] The ratio between the length and the diameter.
[d] Values for Au nanoparticles dispersed in aqueous solutions.
[e] For nanobranches it is the value for the longer-wavelength plasmon peak; for nanorods and nanobipyramids it is value for the longitudinal plasmon resonance peaks.

The LSPR dependence on the external refractive index and the plasmon coupling due to interparticle interactions are very important in sensing applications [44, 45]. The sensitivity of LSPR to changes in the refractive index depends on particle morphology [46–48]. Between the longitudinal and transversal resonance modes, the former responds more strongly to the change in the refractive index of the dispersion medium. For this reason, in the case of the nanorods the sensitivity increases with the aspect ratio of the particles [49–51]. The refractive index sensitivity becomes also more pronounced as the wavelength of LSPR increases [49]. Single particle spectroscopy investigations clearly show the interrelationship between particle morphology and the sensitivity to the change in local refractive index of the surrounding medium [52, 53]. The LSPR generally undergoes a red shift as the local refractive index increases [52]. The magnitude of the red shift per unit of refractive index varies with the particle geometry (Table 4.1) [48]. According to Chen et al., the sensitivities increase when gold nanoparticles become elongated or their apexes become sharper. Thus, spherical gold nanoparticles exhibit the smallest refractive index sensitivity [44 nm/refractive index unit (RIU)] whereas anisotropic gold nanoparticles exhibit the largest index sensitivity (703 nm/RIU). LSPR sensitivity to the refractive index makes gold nanoparticles promising materials for constructing biomolecular sensors. Typically, for these applications gold nanoparticles are first modified with selective binding agents such as receptor proteins, DNA, or immunoreagents. Adsorption or attachment of biomolecular recognition agents to the nanoparticle surface results in a change in the local refractive index with a corresponding shift in the LSPR signal. Such LSPR shifts can be monitored easily by spectroscopic techniques. This principle has been utilized to detect small quantities of bio-molecules [54–56].

Interparticle coupling of individual particle plasmons results in the red-shift of the LSPR band [57, 58]. Aggregation of the nanoparticles can be induced by specific molecular recognition events, where the bio-molecules act as bridges connecting the nanoparticles [57]. Self- and induced-aggregation of biofunctionalized gold nanoparticles have been used to study DNA hybridization and other selective biomolecular recognitions and to develop colorimetric assays in which the particles are used as detection probes [59–61]. Gold nanoparticle-based colorimetric methods can detect ~ 10 fmol of an oligonucleotide, which is approximately 50 times more sensitive than the conventional fluorescence based sandwich hybridization detection method. By monitoring the LSPR one can extract information regarding the local environment and interparticle distance, while also providing avenues for fast and sensitive detection of target analytes. This enables construction of sensitive colorimetric sensors for various applications.

4.3.2 Large Absorption and Scattering Cross-Sections

The plasmon resonant particles (e.g., Au, Ag nanoparticles) can replace or complement the colorimetric, fluorescent, or radioactive labels that are routinely used in immunoassays and cellular imaging [62–66]. These particles have very high scattering cross-sections compared to the commonly used labels. For example, the elastic light scattering cross-section of a ~ 80 nm plasmon resonant particle is equivalent to that of 500,000 individual fluorescein molecules and is > 10^5-fold larger than that of typical semiconductor quantum dots [64, 67]. The non-blinking and high photostability characteristics of these ultra-bright plasmon resonant particles allows for them to be individually identified and counted. The result is a dramatic reduction in the detection limit of the method, which allows for the development of ultrasensitive assays based on single-target molecule detection.

4.3.3 Localized Enhanced Electromagnetic Field

Plasmon resonant particles are surrounded by strongly enhanced and highly localized electromagnetic fields when excited at their LSPRs. These fields are the direct result of the polarization associated with the collective electron oscillation in the metal particles [29, 32]. For example, a ~ 20-fold enhancement in the electric field has been estimated near a 50 nm gold nanoparticle [68]. The local field enhancement factor depends on the particle shape. The field enhancement at the tip of a spheroid is more than the one created by a sphere of similar dimensions [69–72]. Other factors, such as surface-induced relaxation (surface scattering) and radiative decay (radiation damping) effects, can decrease the enhancement and cause a significant broadening of the plasmon band of metal nanoparticles [69]. A narrow and long lasting plasmon band is beneficial for all applications that rely on a local field enhancement [73]. For example, the effective SERS cross section [74] and third order susceptibility of metal nanoparticles [75] are proportional to the fourth power of the inverse of the homogeneous linewidth of the plasmon band. Interparticle interactions and the coupling of individual particle plasmon bands give rise to strongly localized and enhanced electromagnetic fields (also called 'hot-spots') at the junctions between aggregated plasmon resonant particles [76]. As Raman scattering intensity is approximately proportional to the fourth power of the local electric field enhancement, even a modest increase in the latter near the metal particle leads to a substantial increase of Raman scattering intensities. As a result, signals from single molecules located in these 'hot-spots' can be detected [77–79]. However, it is difficult to create such hot-spots in a complex and dynamic environment such as a cell or a cellular membrane and placing the analyte molecules in those hot-spots present a challenge. In order to take advantage of Raman imaging in such complex systems, the use of gold nanostars has been proposed. Gold nanostars consist of a spherical core and a few sharp-tipped arms pointing outward. The local field enhancements have been predicted to be on the order of 10^2 in the vicinity of their tips [80, 81]. The potential of using SERS effects associated with nanostars, either isolated or connected to other metal surfaces, has been already reported [82, 83].

Due to the enhanced local fields, a molecule residing on or near a plasmon resonant nanoparticle experiences a light intensity far stronger than the intensity of the incident light [84]. Like SERS, the optical local field enhancement provides the basis for other applications in molecular detection, sensing, nanophotonic devices via surface-enhanced fluorescence (SEF), two photon photoluminescence (TPPL), and other nonlinear optical processes. Surface enhanced spectroscopic techniques (like SERS and SEF), which rely on the local field enhancement, can strongly enhance the spectral intensity of cellular chemical constituents near the particles and serve as tools for ultra-sensitive monitoring of the intracellular distribution of various species and events. Colloidal nanoparticles of gold (and silver) are particularly attractive as SERS and SEF substrates, as the localized surface plasmon resonance wavelengths of these particles can be tuned over a broad range in the UV–visible–NIR region [85–87].

Fluorescence has been widely used as signal transduction tool for imaging and single molecule studies of biological systems as well as for detection of trace levels of analytes. Metal nanoparticles have been found both to quench or enhance fluorescence intensity of a fluorophore [76, 84, 88–92]. The particle-mediated quenching or enhancement is determined by several factors including the type of interactions between fluorophores and metal surfaces, the location of the fluorophore around the particle, the orientation of its dipole moment relative to the particle surface, and the particle-fluorophore distance [76, 89–94]. Förster resonant energy transfer (FRET) at the surface plasmon absorption of the metal particle results in quenching. FRET has been widely used to measure short distances between sites or bio-molecules due to its $1/r^6$ distance dependence [63–65]. When energetically excited donors and acceptors are present in proximity of subwavelength metal particles, the rates of energy transfer increase at distances up to 70 nm, which is ~ 10-fold larger than typical Förster distance [63, 64]. Fluorescence enhancement occurs through the increase of the strength of the incident light field, which increases the rate of excitation and the radiative decay rate of the fluorophore [90]. An appropriate prox-imity and orientation of fluorophores to metal particles modifies the photonic mode density around the fluorophore and leads, under favorable circumstances, to a ra-diative rate outranging the rate of losses [90, 93]. The quenching or enhancement of the fluorescence, or changing the lifetime of a fluorophore is also dependent on the geometry of the metal particles [85, 95, 96]. Fluorescence is either enhanced or quenched depending if the fluorophore molecular transition dipole is enhanced or canceled by the corresponding image dipole induced in the metal [95]. When a fluo-rophore transition moment is oriented perpendicular to the surface of an ellipsoid, depending on the aspect ratio, a radiative rate enhancement by a factor of thousand or greater can be obtained [95]. Li et al. have demonstrated that sufficiently long gold nanorods (aspect ratio > 13) exhibit intense fluorescence [87]. The emission observed in nanorods of different aspect ratios is a function of local field enhance-ment factors and is greater in the spectral vicinity of the longitudinally polarized surface plasmon resonance [97–99]. Gold nanorods, under pulsed laser excitation, emit a two photon photoluminescence (TPPL) many times brighter than the TPPL from single dye molecules [100, 101]. The strong TPPL is thought to arise from the plasmon-enhanced two photon absorption cross-section due to coupling of weak electronic transitions in the metal [101, 102]. Two photon photoluminescence has been recently used as a nonlinear optical imaging mode. It has also been reported that small gold nanoparticles emit fluorescence upon photoexcitation, thus enabling their visualization with fluorescence microscopy [103, 104].

4.3.4 Surface Functionalization and Biocompatibility

Despite the high oxidation potential and the 'noble' character of gold, the nanopar-ticles of this metal can bind readily to a wide range of molecules with functional groups that have high affinity for the gold surface. Sulfur containing compounds (e.g., thiols, disulfides), organic phosphates, amines, PEG, etc. are some of the most well known surface modifiers. In order to interact specifically with a biological target, a bioactive and selective interface is created by coating the nanoparticles with a broad range of biomolecules such as amino acids, proteins, DNA oligonucle-otides, peptides, etc [105–112]. The applied coating makes the nanoparticles bio-compatible and imparts colloidal stability in both water and physiological media. In addition, modification of the particle surface by suitable (bio)molecules provides desired characteristics for the intended applications. For example, probability of the interaction between nanoparticles and the immune system can be substantially

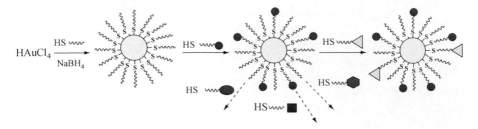

Fig. 4.3 Schematic of formation of thiol-capped gold nanoparticles by the Brust-Schiffrin reaction and a mixed monolayer of gold nanoparticles using the Murray place-exchange reaction. (Kim et al., *Nanoscale* **2009**, *1*, 61–67. Copyright © 2009 Royal Society of Chemistry)

reduced by coating the nanoparticle with a layer of a PEG (poly-ethyleneglycol) or liposomes [1, 113, 114]. The adsorbed layer also offers a possibility for introducing further functional diversity and attaching a range of other molecules to the particle surface by using bioconjugate techniques [115–117] or electrostatic interactions (e.g., layer-by-layer deposition of polyelectrolytes) [118–120]. The surface molecules can also be replaced by other molecules of interest [e.g., polyethylene glycol (PEG)] via a ligand exchange reaction [121]. Place-exchange reactions of thiols on gold nanoparticle surfaces with glutathione have been used for the release of drugs in the intracellular space (Fig. 4.3) [122, 123]. Surface of gold nanoparticles has also been modified by encapsulating gold nanoparticles into a silica shell [124]. The wide variety of functionalized gold nanoparticles that can be produced by different methods widens the breadth of their applications in developing bioassays, studies of biomolecular interactions, and other biomedical applications. Surface modified gold nanoparticles sometimes have dimensions similar to biomolecules and therefore these bio-conjugated nanoparticles can be used as tools for handling and manipulating biological components and for probing mechanisms of biological processes [125, 126]. In addition to binding ability to a range of biomolecules, another important property of gold nanoparticles is their biocompatibility. In biological applications, cells and tissues are exposed to the gold nanoparticles for extended periods of time. Well dispersed gold nanoparticles are nontoxic or at least well tolerated by the cellular environment [112, 127–132].

The surface of gold nanoparticles has been modified with antibodies and other molecules to achieve tumor specific delivery and cell membrane penetration (Fig. 4.4) [133–135]. Subsequently, these nanoparticles have been used for therapeutic purposes. At their LSPRs, gold nanoparticles of certain shapes show photon capture cross-sections four to five orders of magnitude greater than those of photothermal dyes. Aided by large absorption cross-section and poor light emitting ability, gold nanoparticles can efficiently convert absorbed light into heat via non-

Fig. 4.4 Examples of detection and delivery strategies. **a** Schematic diagram of the two main approaches to label-based biomolecules detection: (*a*) direct method; (*b*) indirect method. **b** Schematic diagram showing the use of a gold nanoparticle as a label and possible detection tools. **c** Gold nanoparticles as a multimodal drug delivery system. Hydrophobic drug molecules are attached to gold nanoparticles via cleavable linkers. The payloads can be controllably released by either

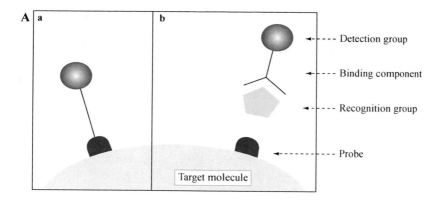

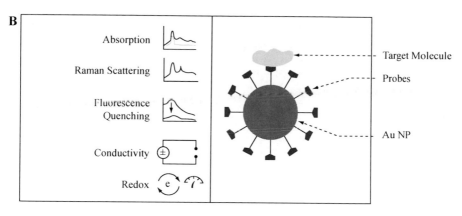

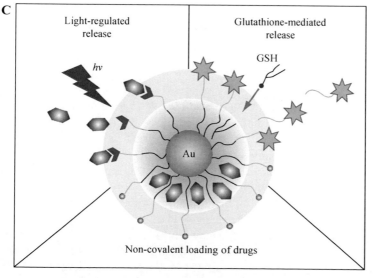

internal (e.g., glutathione, GSH) or external (e.g., light) stimuli. (Schematics (**A**) and (**B**) are adopted from Kim et al., *Microfluid. Nanofluid.* **2009**, *7*, 149–167; Schematic (**C**) is adopted from Kim et al., *Nanoscale* **2009**, *1*, 61–67. Copyright © 2009 Springer and Royal Society of Chemistry) radiative electron relaxation dynamics, known as photothermal effect [136–141]. The magnitude of heat generated will depend on the particle size and geometry. Therefore, gold nanoparticles can be used as nanometric heat sources and probes for changing the local temperature [136, 142–144]. Heat induced changes in the surrounding medium can be useful for photothermal imaging [145–147] and drug release from polymer capsules inside living cells [148–151]. Photothermal therapy is another important application of the photothermal effect of gold nanoparticles [152, 153]. In photothermal therapy, suitably functionalized gold nanoparticles are first incorporated into the target tissue/cells (i.e. tumors) and then irradiated by short laser pulses. The energy absorbed is transferred to the target resulting in its destruction by thermal denaturation/coagulation, mechanical stress caused by microbubble formation, or acoustic shock wave generation [154, 155]. Theoretical calculations have shown that silica core-gold nanoshells (core diameter 50–100 nm, shell thickness 3–8 nm) and gold nanorods [aspect ratio (15–20 nm)/(50–70 nm)] are more efficient photothermal labels and sensitizers than single gold spheres with equivalent volume [156]. Gold nanoparticles of other morphologies such as nanocages and nanoshells have also been tested for their photothermal therapeutic properties [38, 157]. Alternatively, gold-coated magnetic nanoparticles can be heated up by applying an external alternating magnetic field. In this case, the transfer of energy occurs from the external field to the nanoparticles through hysteresis [158–161]. Thus, gold nanoparticles and gold-coated magnetic nanoparticles can generate sufficient heat to cause either irreversible cellular destruction of tumors or just a moderate degree of tissue warming resulting in enhanced chemo- and radio-therapeutic effects [139].

4.4   Biological and Medical Applications of Colloidal Gold

4.4.1   Bioimaging/Labeling

High electron density, characteristic high absorption/scattering in the visible–NIR region of the electromagnetic spectrum, and the ability to scatter and emit secondary electrons are some attributes which enable facile visualization/detection of gold nanoparticles by a variety of techniques like microscopy, photometry, and flow cytometry. As a result, gold nanoparticles can be used in biosensing, cellular and molecular imaging, and diagnostics [67, 162, 163]. The various types of sensors exploit the optical properties (light absorption, scattering, fluorescence, reflectance, Raman scattering, and refractive index) of either bulk gold or individual nanoparticles.

Gold nanoparticles have been used for long time as contrast agents in cellular or molecular imaging [67, 162]. The technique permits the characterization and measurement of biological processes at the cellular and/or molecular level [164]. The contrast agents directed and accumulated at the target/organ provide signals based on which different cells and sub-cellular organelles can be visualized through colorimetric methods. Exogenous contrast agents used to differentiate and visualize cellular and tissue components include lanthanide chelates [165], organic fluorophores [165], Q-dots [166–169], magnetic nanoparticles [170–172], and gold nanoparticles [145, 173–178]. Lanthanide chelates show nonselective localization in the extravascular space [165]. Organic fluorophores are prone to photobleaching and are characterized by low quantum yields and a broad emission window [166, 167]. Q-dots are often associated with cytotoxicity [179–181]. Gold nanoparticles seem to be ideal contrast agents for bioimaging as they are stable under observation, bind strongly to the target without affecting the target activity appreciably, and can be distinguished easily from the surroundings. Large nanoparticles will not enter the cells by endocytosis, because a size of ~150 nm represents the upper limit for passage through caveolae [182]. For this reason, nanoparticles with a size between 30 and 150 nm are the most suitable for use in molecular imaging [182]. In this range the scattered light from gold metallic nanoparticles is so intense that it enables imaging the location of individual particles with conventional optical microscopy like dark field microscopy (DFM) [175, 183–187]. Other optical imaging techniques, such as photothermal imaging [145–147], optical coherence tomography (OCT) [37, 188–192], photoacoustic imaging [193–199], and two-photon photoluminescence (TPPL) microscopy [101, 200–204], use gold nanoparticles as contrasting agents as well. Gold nanoparticles have also been used in electron microscopy (immunostaining) [10, 18], magnetic resonance imaging (MRI) [205, 206], and X-ray imaging (X-ray computed tomography) [170, 207–209]. Techniques such as immunostaining and single particle tracking are used for visualizing sub-cellular components/structures, whereas X-ray computed tomography (X-ray CT) and MRI can be employed for imaging whole cells/organs.

4.4.1.1 Optical Imaging

Dark-field microscopy (DFM) is an important technique for imaging of plasmon resonant nanoparticles and gold nanoparticles are one of the most promising contrast agents for such studies [175, 183–187]. The light scattering of gold nanoparticles after appropriate bioconjugation can be used to distinguish between noncancerous and cancerous cells [183]. In the case of intracellular imaging, detection of low concentrations of plasmon resonant particles by this technique is difficult due to a strong background signal. Two-photon luminescence (TPPL) microscopy is an alternative imaging option, which can deal with the problem of tissue auto-fluorescence and can provide three dimensional spatial resolution. Since gold nanoparticles can sustain surface plasmon resonance with little damping after the photon excitation, they can efficiently emit two-photon luminescence [73]. TPPL of gold nanoparticles has been used to image the location of these particles both *in vitro* and *in vivo* [101, 200–204]. As TPPL excitation falls in the 'biological window' (NIR), the low power densities required for imaging do not damage biological tissues. The imaging performance of other noninvasive imaging techniques, such as OCT [37, 188–192, 210] and photoacoustic tomography (PAT) [193–198], have also been improved by using gold nanoparticles as exogenous contrast agents. OCT, which detects the depth of reflections of a low coherence light source directed into a tissue, can produce real-time high resolution (typically 1–15 μm) images of biological tissues [202]. In photoacoustic imaging, heat generated by the absorption of light (via nonradiative decay) in a sample produces pressure waves (and hence sound) in the surrounding gas, which are detected by ultrasonic transducers [203]. As sound can easily penetrate tissues, this technique offers deep tissue 3D imaging. An additional feature of photoacoustic imaging is that it detects only signals from the regions where antibody-modified gold nanoparticles accumulate on the receptor-dense outer cell surface and not from isolated colloidal gold nanoparticles [211]. This is based on the fact that the LSPR wavelength of aggregated gold nanoparticles shifts more towards red than that for dispersed colloidal gold nanoparticles. When a wavelength higher than that of the plasmon resonance of individual gold particles is used for excitation, gold nanoparticles randomly distributed on the cell surface due to nonspecific adsorption will not produce a photoacoustic signal. In contrast, the aggregates of the nanoparticles would have a pronounced one.

4.4.1.2 Immunostaining

Gold nanoparticles provide high contrast in transmission electron microscopy (TEM) due to the high atomic weight of the element [10]. Faulk and Taylor reported the 'immunocolloid method' for the electron microscope as early as 1971 [18, 212]. Since then, colloidal gold nanoparticles have been the labels of choice for the visualization of cellular and extra-cellular components via in-situ hybridization, immunogold, lectin-gold, and enzyme-gold labeling. In immunostaining, cells are typically fixed and permeabilized first. Immunostaining is also possible without immobilization and permeabilization, in which case only certain domains on the surface of the cell can be labeled. The labeling of specific molecules or cell compartments is done by guided molecular recognition after modification of the nanoparticles with specific antibodies. Antibody-modified gold nanoparticles bind to the target containing the antigen. These molecules or cell compartments cannot be well visualized without labeling due to the lack of contrast among each other. The molecules or cell compartments are labeled with an excess of gold nanoparticles in order to provide high contrast. Due to higher lateral resolution of electron microscopy, individual receptors on the labeled target can be located by visualizing the attached gold nanoparticles with TEM or scanning electron microscope (SEM) [213–215]. Nie et al. have shown that an excellent alternative to rolling circle amplification (RCA) labeling of the cleaved DNA products (that involves fluorescence microscopy) is tagging with gold nanoparticles followed by imaging with a SEM [213]. In addition to the high contrast and lateral resolution provided by the gold nanoparticles, the latter do not undergo photobleaching upon irradiation like fluorescent labels. Recently, Lin et al. have reported synthesis of ultra-small water-soluble fluorescent gold nanoclusters that have a decent quantum yield, high colloidal stability, and can be readily conjugated with biological molecules for specific staining of cells [216].

4.4.1.3 X-ray Computed Tomography (X-ray CT)

X-ray computed tomography is a noninvasive diagnostic method which visualizes differences in sample density due to X-ray attenuation and generates three-dimensional images of the tissue [207]. It can be used for investigations deep inside the body due to the great penetration power of X-rays. As tissues absorb X-rays, radiopaque contrast agents are required to enhance the degree of contrast between normal and diseased tissues. Often organic dyes like Ultravist (iopromide) containing iodine are used as contrast agent in X-ray CT [170, 207, 208]. A number of limitations is associated with the use of iodine-based contrast agents, the most significant being the limited imaging period due to short residence in the bloodstream, rapid renal excretion, and renal toxicity [207–209, 217, 218]. Also, these agents cannot provide sufficient image contrast between smaller vessels and the surrounding tissue because of the fast plasma clearance and rapid diffusion of the contrast agents into the extravascular space. This makes visualization of intraparenchymal vessels quite challenging [208]. The efficacy of the colloidal gold nanoparticles as X-ray contrast materials has been investigated in several studies. The higher atomic number of gold (Au: 79 vs. I: 53) and higher absorption coefficient (Au: $5.16 \, cm^2 \, g^{-1}$ vs. I: $1.94 \, cm^2 \, g^{-1}$ at 100 keV) provide greater contrast per unit weight [207, 208]. Furthermore, the higher k-edge of gold (80.7 vs. 33 keV for iodine) is better matched with the peak intensity of the energy spectrum produced during current CT scanning methods (80–140 keV). Thus, gold can absorb a greater fraction of the energy spectrum than iodine and reduces interference from bone and soft tissue absorptions and radiation dose. Hainfeld et al. reported that, when used as a CT contrast agent, gold nanoparticles (1.9 nm in diameter, ~50 kDa) showed enhanced CT contrast of the vasculature, kidneys, and tumor in mice [207]. Cai et al. showed enhanced efficacy of colloidal gold nanoparticles as a blood-pool imaging agent for X-ray CT [208]. Blood-pool agents should have limited or no leak across normal capillaries, which would allow their diffusion in the intravascular space.

Prior to being introduced into the blood circulation, gold nanoparticles are first conjugated with suitable ligands or antibodies to provide specific binding capabilities at the designated organ. Suitably surface-modified gold nanoparticles not only can be delivered to desired sites at detectable concentrations (unlike iodine) but also stay longer in the blood permitting extended imaging times and providing contrast for resolving the structure of the organ. Kim et al. tested *in vivo* application of polyethylene gycol (PEG)-coated gold nanoparticles as CT contrast agents for angiography and hepatoma detection [209]. The blood circulation time of PEG-coated gold nanoparticles in mice (without apparent loss of contrast) was ~4 h compared to only ~10 min for Ultravist. *In vitro* X-ray absorption coefficient measurements indicated that, at equal concentration, the attenuation of PEG-coated gold nanoparticles was 5.7 times more than Ultravist. Recently, Jackson et al. evaluated iopromide and gold nanoparticles for contrast enhancement at various X-ray tube potentials in an imaging phantom [219]. According to these authors, at the highest energies typically available in computed tomography, significant improvement in contrast enhancement was obtained when gold nanoparticles were used.

4.4.1.4 Magnetic Resonance Imaging

Magnetic resonance imaging (MRI) is another noninvasive diagnostic tool used to obtain functional and anatomic information with high spatial and temporal resolution. MRI records the changes in magnetization of nuclear spins, such as hydrogen protons (^{1}H), in the presence of a large external magnetic field and radiowaves [205]. Protons under different environments in human body relax at different rates. Differences in proton relaxation rates in the presence of a large external magnetic field and radiowaves translate into contrasting anatomical images [205]. The signal intensity from a particular tissue arises from the local values of spin-lattice (longitudinal) relaxation rate ($1/T_1$) and spin-spin (transverse) relaxation rate ($1/T_2$) of the proton spins (T_1 being the longitudinal relaxation time and T_2 the transverse relaxation time). Protons having shorter T_1 will relax rapidly to their equilibrium state. The result is a higher net electromagnetic signal giving rise to positive contrast enhancement in MRI (known as T_1-weighted MRI). On the other hand, a faster transverse relaxation (shorter T_2) results in rapid dephasing of individual spins, which leads to reduced signal intensity or negative contrast enhancement (known as T_2-weighted MRI). The MRI image contrast can be enhanced by decreasing the longitudinal and transverse relaxation times of the proton [220, 221] by the use of (contrast) agents that have the ability to shorten relaxation times. Gadolinium-chelates [165,221,222] or superparamagnetic iron oxides [206, 220] are often used as contrast enhancing agents. Paramagnetic Gd^{3+} ions in gadolinium-chelates (positive contrast agents) affect T_1, whereas superparamagnetic iron oxides (negative contrast agents) due to their large magnetization shorten T_2 relaxation times. However, rapid renal clearance of Gd-chelates impairs its imaging application [205]. In order to improve the retention, Gd-chelates have been conjugated to gold nanoparticles [205]. Gold nanoparticles have been used as delivery vehicles to convey multiple Gd-complexes into selective cellular targets. Gold nanoparticle/Gd-chelate conjugates have been found not only to retain the intrinsic contrasting property of Gd(III) but also to provide enhanced contrast compared to Gd-chelate alone [205]. Gold-shell superparamagnetic nanoparticles have unique optical and magnetic properties which are promising for diagnostics and therapies [206]. Gold-shell magnetic nanoparticles consist of a magnetic core (composed of magnetic materials such as magnetite, Fe_3O_4, maghemite, γ-Fe_2O_3, Co, Ni) coated with gold. The gold shell offers enhanced stability, unique optical signature, and desirable surface chemistry while reducing the toxicity of the core materials. The magnetic core provides a high T_2 relaxation time. The *relaxivity* value in MRI is the measure of how much the magnetic particle can change the relaxation time of the water proton. The higher the relaxation time, the better the contrast obtained in MRI. The magnetic core in core–shell structures also provides an opportunity for directing the nanoparticles to the target site using an external magnetic field and for the therapeutic heating in the presence of an alternating magnetic field.

4.4.1.5   Phagokinetic Tracks Imaging

Gold nanoparticles have also been used in 'phagokinetic tracks imaging' [223–226]. In 'phagokinetic tracking' the cell movement on a substrate covered with a layer of colloidal gold nanoparticles is imaged. It works like a cloud chamber. While migrating along the substrate, cells pick up gold nanoparticles from the deposited layer leaving behind a nanoparticle-free trail on the substrate. The 'phagokinetic tracks' of mother cells, as well as daughter cells after cell division, can be viewed by dark field light microscopy, SEM, or TEM.

4.4.2   Biosensing

Gold nanoparticles are being extensively used in many biosensing applications [46, 50, 227–231]. Their use has provided excellent sensitivity, in some cases even the detection of single molecular binding events. Studies of biomolecular interactions are essential for the understanding of cell physiology and disease progression. Biosensing is extremely useful in biological and medical research for determining blood glucose levels, detection of bacteria, viruses, biological toxins, biowarfare agents, pollutants, and pesticides in the environment, monitoring pathogens in the food, and so on [232–236]. A biosensor provides selective quantitative or semi-quantitative analytical information using a biological recognition element [237]. Biological molecules such as antibodies, carbohydrates, nucleic acids, enzymes, etc. are employed as biological recognition elements [232, 238].

4.4.2.1   Optical

It has been discussed earlier that the opto-electronic properties (such as LSPR, enhanced local field, etc.) of gold nanoparticles, dispersed in liquids, surface-confined, or singly isolated, are affected in various ways upon receptor/analyte binding [239]. Thus, gold nanoparticles can function as efficient opto-electronic transducers to quantify biospecific interactions. When various colorimetric and spectroscopic methods are employed, changes in the optical properties of gold nanoparticles can provide information regarding the receptor/analyte binding, the quantity of analyte molecules present in the sample, and/or the course of a biological event [232]. For example, Mirkin and co-workers employed mercaptoalkyloligonucleotide-modified gold nanoparticles as probes for a colorimetric determination of polynucle-

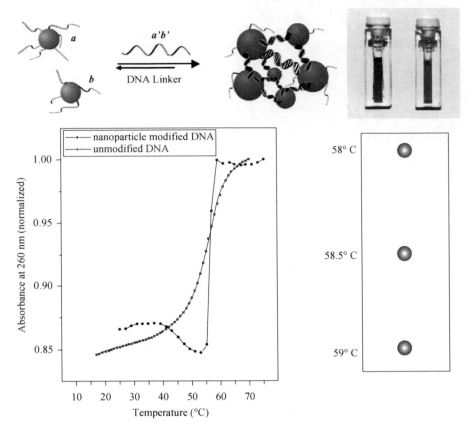

Fig. 4.5 Gold nanoparticle probes for the detection of nucleic acid targets. A mixture of gold nanoparticles with surface-immobilized non-complementary DNA sequences (*a*, *b*) appears *red* in color and has a strong absorbance at 520 nm. When a complementary DNA sequence (*a'b'*) is added to the solution, the particles are reversibly aggregated causing a red shift in the surface plasmon absorbance to 574 nm, thus, appearing *purple* or *blue* in color (*top right*). Heating the aggregates above the melting temperature of the DNA strands reverses the aggregation and the color returns to *red*. The transition is very sharp as demonstrated by the plot (*bottom left*). A small sample aliquot can be placed on the surface of a reverse-phase TLC plate to monitor the degree of hybridization as a function of temperature (*bottom right*). (Adopted from Thaxton et al., *Clin. Chim. Acta* **2006**, *363*, 120–126. Copyright © 2005 Elsevier B.V.)

otides [59]. As each gold nanoparticle carries several oligonucleotides, oligonucleotide-modified gold nanoparticle probes generate cross-linked aggregates upon hybridization with a complementary oligonucleotide target, leading to color change (Fig. 4.5). The difference in the transition temperatures of the nanoparticle aggregates determined colorimetrically were used to differentiate between matched and mismatched DNA duplex. This method is capable of quantitative detection of DNA sequences at very low concentrations [240]. Later, noncross-linking target DNA hybridization with DNA-functionalized gold nanoparticles has been developed for rapid colorimetric sensing of DNA hybridization events. This approach is capable of detecting terminal single-base-pair mismatches and can be eventually employed for reliable genetic diagnosis [233, 241, 242].

Protein–protein interactions and protein folding–unfolding play key roles in structural and functional organization of living cells and are very important in biology. Gold nanoparticles have been used in sensing events like protein aggregation, protein folding, and protein–protein interaction [243–247]. For example, Tsai et al. reported a competitive colorimetric assay based on gold nanoparticles for sensitive detection of protein–protein interactions in solution. The method allows direct visualization without the need for protein labeling or the aid of instruments [248]. Chah et al. have reported an inexpensive method for studying reversible conformational changes of proteins in solution by using gold nanoparticles [243]. In this study it was shown that the protein cytochrome c bound to gold nanoparticles unfolds at low pH and refolds at high pH. This is accompanied by a concomitant reversible color change in solution. The red-to-blue color change results from particle aggregation that occurs when the protein unfolds. The color reverts from blue to red as the protein refolds and the particles peptize. In a similar approach, Ghoshmoulick et al. employed the pH dependent shifts in the gold nanoparticle plasmon resonance to track protein structural changes induced by glycation [245]. Bhattacharya et al. reported a simple colorimetric method for studying protein conformation and demonstrated its effectiveness for both nonheme and heme proteins [247]. Protein variants with defects in folding (caused by subunit misassembly or mutation) could also be classified using this method, which has possible application in hemoglobinopathy (e.g., thalassemia carrier detection). Other researchers have developed sensitive colorimetric biosensors based on the enzymatic cleavage of nucleic acids/proteins on well-dispersed gold nanoparticles [249, 250]. Zhao et al. claim that this biosensing system can be adapted to other enzymes or substrates for detection of analytes such as small molecules, proteases, and protease inhibitors [250]. Leuvering et al. reported a new immunoassay method, called the sol particle immunoassay (SPIA) [251, 252]. In the SPIA, solutions of a gold-containing marker are mixed with the sample of interest in a reaction vessel, and the color change (from red to blue or gray) caused by the aggregation of gold particles due to the biospecific reaction on the particle surface is monitored visually or spectrophotometrically. This method has been used to detect human chorionic gonadotropin in the urine in pregnancy [253, 254], cystatin C (an endogenous marker for the filtration ability of kidney glomeruli [255]), and anti-protein A [256, 257].

Proteins possessing carbohydrate-binding domains are known as lectins. Carbohydrate–protein interactions are worth-studying because they play important roles in many cellular processes, including cell–cell communication, proliferation, differentiation, and inflammation caused by bacteria and viruses. Carbohydrate-encapsulated gold nanoparticles have been employed in the detection of the interactions between the lectin and the gold surface and for the evaluation of the ligand-density effects [238, 258, 259]. Chen et al. developed a method where carbohydrate-encapsulated gold nanoparticles could be used as efficient affinity probes for separation and enrichment of target proteins from a mixture at the femtomole level [260]. Aslan et al. developed a plasmonic-type glucose sensing platform that could potentially be used to monitor μM to mM glucose levels in physiological fluids, such as tears, blood, and urine [261–263]. Monitoring of glucose concentrations is of paramount interest in the management of diabetes. Glucose oxidase immobilized-gold nanoparticles have been used as glucose sensors with enhanced sensitivity and stability [264]. Using the light scattering properties of gold nanoparticles Liu et al. developed a one-step homogeneous immunoassay for the detection of a prostate cancer biomarker, free-PSA (prostate specific antigen), in a concentration range from 0.1 to 10 ng/mL [265].

Molecular recognition of streptavidin and biotin has been routinely used for various proof-of-concept studies of the bio-recognition phenomenon with respect to protein–DNA, protein–ligand, protein–peptide, and protein–protein interactions. A clear understanding of these interactions is essential for the design of new drugs and ligands for proteins and nucleic acids. Biotin, a water-soluble B-complex vitamin, is commonly found in many cosmetic and health products for hair and skin. It binds very strongly to streptavidin, a tetrameric protein. The high affinity of streptavidin–biotin interactions has been exploited for several applications in immunology and histochemistry [266–272]. Change in LSPR properties of biotin-conjugated gold nanoparticles have been extensively used to study biotin–streptavidin interactions [54, 273–279]. Spectral changes induced by changes in the refractive index in the vicinity of individual gold nanparticles due to target binding have been utilized for label-free (multiplex) biosensing [276, 280–282]. Recently, very sensitive single gold nanoparticle-based biosensors have been developed based on this concept. For example, Nusz et al. reported a single-gold nanorod LSPR biosensor for the binding of streptavidin to biotin where the lowest streptavidin concentration that was experimentally measured was 1 nM [282]. This concentration level was substantially lower than that detected with biotinylated single gold nanospheres [54, 282, 283].

Along the same lines, it is worthwhile mentioning other interesting applications of gold nanoparticles. For example, Sönnichsen et al. demonstrated that in addition to the detection of analytes, the color changes arising from gold particles in close proximity can also be used to measure lengths on the nanometer scale and detect conformation changes in molecules (Fig. 4.6) [58]. The same group reported that gold nanorods could be exploited for the study of rotational motion in biomolecules [284]. Using image correlation techniques, Murphy and co-workers demonstrated the use of optical patterns produced by resonant Rayleigh scattering gold nanorods as markers for studying local deformations of stretchable polymers [285]. They proposed that modified gold nanorods could be used for examining mechanical effects in biological tissues. Recently, Selhuber-Unkel et al. have shown that gold nanorods can be used as efficient optical handles in nanoscale experiments [286]. These authors anticipated that gold nanorods could be employed as force sensors in future in-vitro and in-vivo studies as well as force transducers in single-molecule experiments.

Fig. 4.6 Molecular plasmonic ruler. **a** Schematic illustration of principle of transmission dark-field microscopy (*top left corner*) and nanoparticle functionalization and immobilization. **b** *Left*: dark-field images of single gold nanoparticles (*green*); middle: gold nanoparticle pairs (optical and TEM images); *right*: scattering spectra of single and pairs of gold nanoparticles. **c** Representative spectral shift between a gold nanoparticle pair connected with single-stranded DNA (*red*) and double-stranded DNA (*blue*). **d** Spectral position as a function of time after the addition of complementary DNA. The scattered intensity, I_{sca}, is shown color-coded on the *bottom* and the peak position obtained by fitting each spectrum is traced on the *top*. Discrete states are observed, indicated by horizontal dashed lines. (Adopted from Sönnichsen et al., *Nat. Biotechnol.* **2005**, *23*, 741–745. Copyright © 2005 Nature Publishing Group)

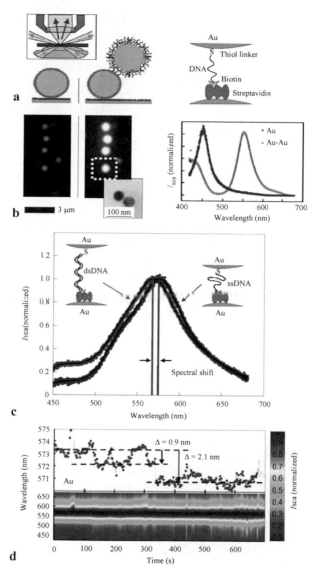

4.4.2.2   Gold Stains

Gold nanoparticles have been used as replacement for fluorophores or dyes in ELISA-type assays. In this case, the analyte is immobilized on an immunofunctionalized surface via simple adsorption or specific binding (e.g., capture antibody). The surface of gold nanoparticles is conjugated with analyte-specific antibodies that bind the metal particles to the surface [287, 288]. Sometimes a 'silver enhancement' procedure has been employed to amplify the signal [289]. This approach has been used for the detection of target DNA sequences with complementary DNA [290–292], detection of proteins, etc [293–296].

4.4.2.3   SERS and Fluorescence

As discussed earlier, due to enhanced local field at LSPR, Raman scattering of some analyte molecules is dramatically enhanced when the molecules are located near a gold nanoparticle. As a result, surface-enhanced (resonance) Raman scattering, or SE(R)RS, can be used under some conditions as a powerful tool for ultrasensitive chemical analysis down to the single-molecule or single particle level [82, 297]. This method can simultaneously detect a single molecule and examine its chemical structure. The nanoparticles are modified with ligands that can specifically bind to the analyte, thereby bringing the analyte near the gold surface and increasing its Raman signal [298]. Though the use of Raman probes is a relatively new approach for the detection and examination of complex biomolecules, investigations of the potential of SERS for DNA/RNA detection, identification of proteins, and detection of protein–ligand interactions are already well underway [293, 299–305].

Fluorescence has been widely used as signal transduction tool for the detection of trace levels of analytes and for single-molecule imaging in biological systems [306]. Metal nanoparticles can quench or enhance fluorescence intensity of a fluorophore [84, 90, 307]. Factors such as the position dependent interactions between fluorophores and metallic surfaces, orientation of its dipole moment relative to the particle surface, and particle-fluorophore distance determine whether there will be quenching or enhancement of fluorescence [84, 90, 93, 94]. Quenching may occur through nanoparticle surface energy transfer (NSET) or fluorescence resonance energy transfer (FRET). FRET is a dipole–dipole interaction mechanism in which a donor fluorophore in its excited state transfers (radiationless) its energy to an acceptor. Due to the $1/r^6$ distance dependence, FRET has been widely used to measure distances between sites or biomolecules [63–65]. The quenching based on NSET has some advantages over FRET. For example, in addition to an extremely low signal to background ratio, it can handle simultaneous fluorescence quenching of multiple fluorescent sources with different energies. NSET-based gold nanoprobes have been used to detect single mismatch of DNA sequences, measure the activity of proteases with high sensitivity, estimate the level of protein glycosylation, etc [308, 309]. Recently, Lee et al. used fluorescein-hyaluronic acid conjugated gold nanoparticles as probes for real-time monitoring of the generation of intracellular reactive oxygen species (ROS) in live cells via NSET [310, 311]. Excess ROS generation in various pathogenic processes has been known as an early indicator of cellular disorders and cytotoxic events [312–314]. The gold nanoparticle-induced quenching property has also been utilized by Wang et al. for biomolecular sensing [57]. Similarly, Selvaraj and Alagar used the gold nanoparticle-induced quenching property for the detection of antileukemic fluorescent drug 5-fluorouracil (5FU) [315]. The colloidal gold-5FU complex was also reported to show antibacterial and antifungal activity against

Micrococcus luteus, Staphylococcus aureus, Pseudomonas aeruginosa, Escherichia coli, Aspergillus fumigatus, and *Aspergillus niger*.

4.4.2.4   Electrochemical Biosensors

Electrochemical biosensors play an important role in biosensor research [316–318]. An electrochemical transducer is used in all biosensors. The gold nanoparticle-based electrochemical biosensors are made by coupling biological recognition elements with gold nanoparticle-modified electrochemical transducers. Gold nanoparticles can be deposited onto the transducer surface using electrochemical methods or by physical attachment procedures. More frequently, nanostructure composites consisting of gold nanoparticles embedded in conductive polymers and silica gels are used as sensing materials for such applications. These composites often contain biological reagents to provide biomolecular recognition sites. The introduction of the gold nanoparticles to an electrode offers increased roughness and a larger surface area that leads to higher currents and increased binding for biomolecules. Furthermore, the curvature of small gold particles allows closer contact of the particle with the redox sites of the enzyme facilitating electron transport. Gold nanoparticle-modified electrodes have been often used for sensing molecules like NADH, hydrogen peroxide, oxygen, etc. that are involved in redox processes in biochemical systems [319–328]. In addition to the development of (redox) enzyme electrodes, the gold nanoparticle-based electrochemical approach has also been used for the design of DNA and immunological sensors [329–333].

4.4.3   Drug Delivery and Therapeutics

When administered orally or intravenously, therapeutic drugs disperse in the entire body. The primary aim of the targeted drug delivery is to deliver the therapeutics more effectively to the site of interest. Other aims include enhancing bioavailability and retention, controlled release of drugs, extending the period of drug circulation in the body, and minimizing side effects [334]. In order to achieve site-specific drug delivery while avoiding drug interaction with surrounding normal cells, targeting utilizes molecular and cellular changes within diseased tissues under pathophysiological conditions [335, 336]. As nanoparticles can be engineered in various ways to achieve their enhanced availability at the targeted site, nanoparticle platforms are being extensively used to control the accumulation of drug molecules or therapeutic agents into the targeted site. Nanoparticles with sizes ($<\sim$100 nm) and shapes that allow them to pass through capillaries and penetrate cells have been designed and prepared [335, 337–344]. Upon suitable surface modification, these particles can circulate for long times in the body undetected by the immune system before reaching the targeted site [345–347]. The unique optical, magnetic, and electrical

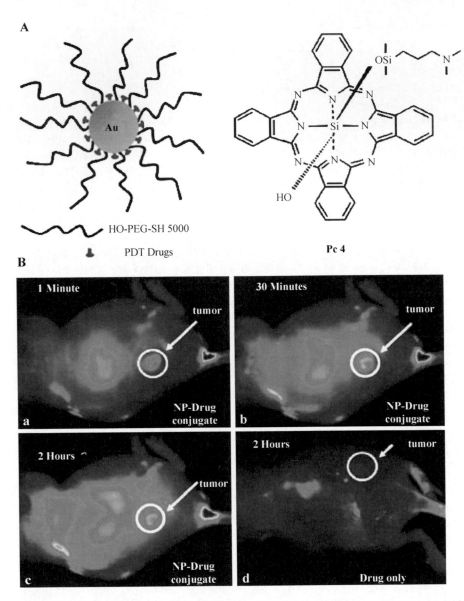

Fig. 4.7 Gold nanoparticles as drug delivery agents. **A** Schematic structures of a water-soluble gold nanoparticle and Pc 4. **B** Fluorescence images of a tumor-bearing mouse after being injected with a solution containing gold nanoparticle-Pc 4 conjugates (normal saline medium; 0.9% NaCl, pH 7.2) at different times after intravenous tail injection (*a–c*). The bright signal is due to Pc 4 fluorescence. (*d*) Image of the mouse tissue receiving a Pc 4 formulation without the gold nanoparticle vector added. Without the gold nanoparticles as drug vector no circulation of the drug in the body or into the tumor was detected 2 h after injection. (Adopted from Cheng et al., *J. Am. Chem. Soc.* **2008**, *130*, 10643–10647. Copyright © 2008 American Chemical Society)

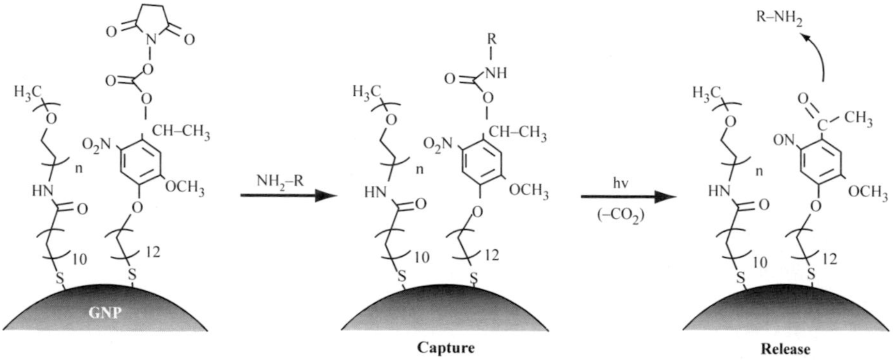

Fig. 4.8 An example of controlled drug delivery. Capture and release of amines on gold nanoparticles functionalized with a photocleavable succinimidyl ester. Histamine shows no biological activity when attached to gold nanoparticles but becomes active when is released from the particles upon photo-irradiation. (Adopted from Nakanishi et al, *J. Am. Chem. Soc.* **2009**, *131*, 3822–3823. Copyright © 2009 American Chemical Society).

properties of nanoparticles allows to track their intracellular transport and localization [183, 348].

4.4.3.1 Drug Delivery

Due to their facile surface conjugation chemistry, size, and shape-tunable optical properties, gold nanoparticles constitute excellent vectors for targeted drug delivery (Fig. 4.7) [349, 350]. They can be used not only for carrying drug molecules, but also for controlled site- and time-specific release of drugs triggered by specific external stimuli (Fig. 4.8). Molecules with high target binding affinities are grafted onto the gold nanoparticles. Targeting molecules could be proteins, peptides, nucleic acid aptamers, or other organic molecules (e.g., folic acid, paclitaxel) [175, 179, 344, 350–352]. Drug molecules and secondary coating molecules (such as PEG and BSA) ensure the binding to the surface of specific cells and their cellular uptake while minimizing nonspecific adsorption to tissues or to each other [345–347, 353]. Drug molecules can be either attached to nanoparticle or entrapped in surrounding coating layers [335, 354].

Nanoparticle-based targeting techniques can be 'passive' or 'active'. 'Passive targeting' takes advantage of the difference in the permeability, size-specific filtration, and retention of drug-loaded nanoparticles between the desired target and the surrounding normal organs/tissues. In diseased tissues, the enhanced permeation and retention effect is facilitated by the highly leaky nature of tumor vasculatures and poor lymphatic drainage of the interstitial fluid surrounding the tumor [355–358]. In 'active targeting', surfaces of gold nanoparticles are modified with a suitable ligand, peptide or antibody which can specifically bind to the receptors on the tissue of interest [1, 38, 351].

Gold nanoparticles modified with transferrin [359], anti-EGFR (epidermal growth factor receptor) antibodies or tumor necrosis factor-α (TNF-α) [16, 344] have been employed to target cancer cells. TNF-α, which overexpresses in solid tumors [16] and mediates hemorrhagic necrosis in tumors [344, 352, 360], are employed for cancer therapy. Linking TNF-α to the surfaces of gold nanoparticles of sizes ranging between 20 and 100 nm have brought preferential accumulation of TNF-gold nanoparticle conjugates in the tumor vasculature and reduced systemic toxicity in comparison with native unconjugated TNF [16, 361, 362]. Aurimune™, a chemotherapy agent which is undergoing clinical trials for the treatment of melanoma, colorectal cancer, and urinary tract cancer, contains TNF-α grafted 33-nm colloidal gold nanoparticles [344, 352]. Like TNF-α, epidermal growth factor receptor (EGFR) is overexpressed in several types of cancers [363–369]. EGFR-enriched cancer cells can be targeted by anti-EGFR antibody and anti-EGFR antibody grafted gold nanoparticles have been used successfully to target cancer cells [178, 179, 370, 371]. Patra et al. reported that the delivery of cetuximab as a targeting agent and gemcitabine as an anticancer drug using a gold nanoparticle-based transport system results in significant inhibition of pancreatic tumor cell proliferation [371].

Since it stores genetic information, the cell nucleus is an important target in therapeutics. Recent research has demonstrated that it is possible to bypass the impermeable nature of the plasma and nuclear membranes by utilizing suitably modified gold nanoparticles. A number of nuclear membrane-penetrating peptides, such as Simian virus nuclear localization peptides [127, 183, 372], HIV 1 Tat-protein-derived peptides [351], etc. [127, 133, 347, 353, 373, 374] have been grafted onto surfaces of gold nanoparticles to be delivered into the cell nuclei. Joshi et al. reported an insulin delivery method by using amino acid-modified gold nanoparticles that can bind insulin via hydrogen bonds. This appears to be a promising approach for the development of a nonparenteral delivery system for insulin to replace subcutaneous administration [375].

A number of small molecules, such as folic acid, methotrexate, paclitaxel, coumarin isocyanate, mercaptopurine, etc., have been attached to gold nanoparticles to be delivered to tumor tissues. In these cases, often 'passive targeting' occurs [335]. Folate receptor is overexpressed in a number of human cancers such as the prostate, kidney, lung, ovary, breast cancers, etc [376–378]. Folic acid grafted gold nanoparticles can selectively target folate receptor-positive tumors [376, 377, 379–381]. Conjugation of a number of chemotherapeutic agents (e.g., methotrexate, paclitaxel) to gold nanoparticles has been reported to improve the drug efficacy in comparison with free forms of these agents [344, 382–384]. Similar results have been observed upon conjugation of antileukemic and antiinflammatory drugs 6-mercaptopurine and its riboside derivative to gold nanoparticles [385]. Furthermore, 6-mercaptopurine-gold nanoparticle conjugates have also shown enhanced antibacterial and antifungal activities [315]. Li et al. reported similar enhancement effect of gold nanoparticles in drug delivery and as biomarkers of drug-resistant cancer cells [386]. Shenoy et al. have reported that a significant enhancement of emission intensity can be obtained by linking coumarin (a small fluorescent molecule) to

PEG-functionalized gold nanoparticles through a carbamate bond. The resulting conjugates have been employed as cellular probes as well as delivery agents [387].

4.4.3.2   Hyperthermal Therapy

Localized heating by using infrared lamps, lasers, or ultrasound has been used in therapeutic medicine for the treatment of cancer and other conditions [388]. Hyperthermic properties of gold or gold-coated magnetic nanoparticles are being explored for highly localized and controlled destruction/treatment of targets varying in size from a few nanometer to tens of microns (from bacteria to cancer cells) [38, 139, 152, 153, 175, 179, 389–392]. Hyperthermal therapy is a promising, relatively simple to perform, and less invasive therapeutic procedure. In this therapeutic mode, it is not only possible to access intracellular targets but also to achieve without surgical intervention the destruction of tumors that are embedded in vital regions. Gold nanoparticles have been conjugated to antibodies or viral vectors [393] for selective and efficient targeting. For example, Huang et al. have demonstrated that anti-EGFR conjugated gold-nanorods can destroy EGFR-positive cells at half the energy required to kill EGFR-negative cells [175]. Similarly, gold nanospheres and nanoshells have been used for the treatment of breast cancer and other cell lines [153, 174, 370, 394–398]. Compared to gold nanospheres, gold nanoshells (gold-on-silica), nanorods, and nanocages have been shown as promising candidates for simultaneous imaging and photothermal therapy [31, 153, 175, 399–405]. The hyperthermal property of gold nanoparticles has been also used for the treatment of parasite infections. For example, Pissuwan et al. reported the destruction of more than 80% T. gondii tachyzoites by photothermal effects of gold nanorods conjugated with an antibody selective for Toxoplasma gondii [406]. Sometimes X-ray radiation has been used for localized heat generation. Enhanced absorption of X-ray radiation at the interface between materials of high and low atomic numbers leads to a localized generation of heat. Hainfeld et al. used this concept to destroy murine tumors in mice [128]. Gold nanoparticles have been used to enhance radiotherapy in prostate cancer as well [407].

Photothermal properties of gold nanoparticles have been used for controlled drug delivery. Controlled drug delivery systems represent advanced systems that can release the payload at a predetermined time or in a predetermined sequence of pulses in response to external controlled stimuli [408]. Controlled drug delivery systems can overcome many of the hurdles of conventional drug delivery systems thereby increasing drug efficacy and decreasing toxicity. Gold nanoparticles have been used for photo-activated drug-release when the drug payload is encapsulated in a temperature-sensitive polymer shell surrounding the gold nanoparticles [118, 119, 409–412]. Takahashi et al. have demonstrated that gold nanorods can be used as carriers for a photo-triggered gene delivery [413]. They have used 1064 nm-laser irradiation to trigger the release of plasmid DNA from PC-nanorod-DNA complexes via morphological transformation of gold nanorods into spherical nanoparticles. Similarly, Kitagawa et al. have demonstrated the controlled release of myoglobin protein from myoglobin–gold nanorod complexes by the near IR (pulsed) laser ir-radiation [414].

Some of the recently published examples of therapies include applications of gold nanoparticles to detection and control of microorganisms, enhanced activity of antibiotics, and so on [415–419]. Dykman et al. studied the capacity of colloidal gold for enhancing immune response in laboratory animals [420]. Immunization of the animals with colloidal gold conjugates with haptens or complete antigens induced the production of highly active antibodies. Nie et al. demonstrated that the antioxidant-functionalized gold nanoparticles can enhance radical scavenging ac-tivity of the vitamin E-derived antioxidant [421]. This can be a step forward in the direction of developing new strategy for design of artificial antioxidants.

4.5   Synthesis of Colloidal Gold Nanoparticles

Although colloidal gold nanoparticles have been used for many centuries, the first scientific approach for their preparation was documented only in the nineteenth century by Faraday [422]. Later, many other methods yielding gold nanoparticles with varying degree of monodispersity have been reported. More recently, advances in the preparation of colloidal gold nanoparticles with well controlled size, shape, and structure have fuelled a rush for innovative applications in many fields of tech-nology and medicine.

The most frequently used route for the preparation of colloidal gold involve the reduction in solutions of gold compounds (particularly tetrachloroauric acid) via dedicated reducing agents, electrodeposition, or radiation assisted techniques (e.g., pulse or laser radiolysis, UV, and ultrasonic irradiation,) [4, 423–430]. Organic and inorganic compounds, such as formaldehyde, sodium citrate, ascorbic acid, (mono- and poly-hydroxy) alcohols, sodium borohydride, hydrazine, hydroxyl amine, etc. have been used for supplying the electrons needed in the reduction process. Recent-ly, methods relying on microorganisms, fungi, and plant extracts as reductants have also been reported [431–439]. The biological routes, which may have been likely used by early societies, have great potential today as well because they are simple, cost-efficient, and environment-friendly.

Despite significant recent progresses in the synthesis of metallic particles in gen-eral and gold particles in particular, there are still relatively few preparative methods available to generate gold colloids suitable for biomedical applications. One reason is that very few methods capable to synthesize gold nanoparticles with well con-trolled size and shape are available. Also, many reported systems involve reactants and additives which make the surface of the gold nanoparticles unsuitable for medi-cal and biological applications. Molecules with functional groups containing sulfur and phosphorous, for example, may be attached too strongly to the metal surface and cannot be easily displaced. This may prevent the particles to interact with the biological substrates/targets of interest or limit the ability to functionalize their sur-face so that they can fulfill their intended role. The unique needs of the biomedical applications and their rapid multiplication will likely give new impetus to the efforts to develop gold nanoparticles with improved properties.

4.5.1 Synthesis of Spherical Gold Nanoparticles

Most preparation methods reported in the literature tend to yield gold nanoparticles with a broad particle size distribution. However, a few synthesis strategies have been developed that allow for the preparation of nearly monodispersed gold nanoparticles [440–442]. One of the most widely used route is the Turkevich–Frens method [440, 441]. The Turkevich–Frens method involves the reduction of tetrachloroauric acid with sodium citrate. An aqueous solution of $HAuCl_4$ (typically ~ 1.0 mM) is first heated to reflux under stirring followed by the addition of a freshly prepared aqueous trisodium citrate solution containing up to 40 M stoichiometric excess. The gold ions are reduced in ~ 25 min as indicated by the change in the color of the solution from yellow to deep purple and then dark red. The excess citrate molecules act as a stabilizer. The negatively-charged citrate ions adsorb on the gold nanoparticles and provide a sufficiently high surface charge (-45 to -50 mV) to prevent the aggregation of gold nanoparticles due to electrostatic repulsion. This method produces nearly monodispersed spherical gold nanoparticles in the 10–20 nm diameter range. By varying the gold-to-citrate ratio, larger colloidal gold nanoparticles (diameters between 40 and 120 nm) can be also produced. However, the larger gold particles tend to be more polydisperse. Chitosan-capped gold nanoparticles can be prepared by a modified Turkevich process in which trisodium citrate is replaced by monosodium glutamate [443]. Polysaccharides, such as heparin, chitosan, dextran, etc., have been used both as reducing and stabilizing agents for preparing gold glyco-nanoparticles in aqueous solution [126, 261, 444]. Goia and Matijevic have developed a method for the preparation of larger sized nearly monodispersed gold sols [442]. This method involves the reaction of a concentrated $HAuCl_4$ solution by isoascorbic acid in the presence of gum Arabic as colloid stabilizer. By this method, uniform gold particles ranging in modal diameter from 60 nm to 5 μm can be produced. The same authors have prepared concentrated dispersions of modispersed gold particles with diameters ranging from 15 to 40 nm by reducing aqueous solutions of $HAuCl_4$ with diethylaminodextran (Fig. 4.9) [445].

Another approach to prepare gold nanoparticles with a narrow particle size distribution is the so-called 'seed-mediated' synthesis [424, 446]. Small gold particles are first produced by reducing gold ions with a strong reducing agent (e.g., sodium borohydride). These particles are then added to 'growth solutions' where they act as seeds for further gold deposition. The 'growth solutions' contain Au^{3+} ions, a reducing agent, and often a stabilizing agent. The reducing agent used in the 'growth solution' is a relatively weak reducing agent (e.g., ascorbic acid, hydroxylamine, etc). The reducing agent is chosen such that the reduction of Au^{3+} ions occurs only after the addition of seed particles. A weak reducing agent in this step allows only the diffusional growth of pre-existing seed par-

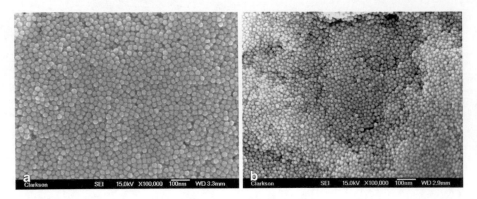

Fig. 4.9 Highly dispersed uniform spherical gold nanoparticles: **a** 35 nm and **b** 18 nm

ticles [447–451]. The final size of the particles depends mainly on the size of the 'seeds' and the amount of the precursor ions subsequently reduced [452]. Gold nanoparticles in the 30–150 nm range with narrow size distributions (relative standard deviation < 10%) have been prepared by this method [452, 453]. The 'seed-mediated' synthesis protocol has been also used to prepare isomorphous particles [424].

Several less known routes have been used to prepare size-controlled gold nanoparticles. For example, reduction in the restricted space provided by the droplets of the dispersed phase in microemulsions has been reported to produce nearly monodispersed gold nanoparticles [455]. Refluxing a polydispersed colloidal suspension (i.e., digestive ripening) in a solvent containing ligands, such as alkanethiols, amines, phosphines, silanes, and halides, was used to convert a polydispersed colloidal suspension of gold into a monodispersed sol [456].

Brust et al. have developed an important synthesis protocol that results in alkylthiol-stabilized gold nanoparticles of controlled size and diameters ranging from 1 to 3 nm [457]. In this method, a gold salt (typically $AuCl_4^-$) is first transferred into an organic solvent (toluene) using tetraoctylammonium bromide as the phase-transfer reagent and is reduced by $NaBH_4$ in the presence of alkylthiol. A rapid addition of the reducing agent or larger thiol/gold molar ratio gives a smaller core (gold particle) sizes. Later, Brust and coworkers extended this synthesis to a single phase system [458]. The original method of Brust et al. has been later modified to provide water-soluble surface functionality [121, 238, 459, 460].

4.5.2 Synthesis of Non-Spherical Gold Nanoparticles

There are several other particles of intriguing shapes (nanoshells, nanorods, nanocages, etc.), which have shown great potentials in the biomedical field [187, 404, 461]. The optical properties of these particles can be tuned over a wide range. Nanorods can also be used in two-photon luminescence and second harmonic generation [46]. Nanocages have emerged as a novel class of biocompatible vectors with potential applications in drug delivery, tissue imaging, and photothermal therapy [188, 404].

4.5.2.1 Synthesis of Gold Nanoshells

The number of applications of gold nanoshells in biomedical applications has increased rapidly in recent years [153, 174, 461–463]. Gold nanoshells consist usually of a thin gold shell encapsulating a dielectric core. Silica (or sometimes polystyrene or gold sulfide) is the most frequently used dielectric core. The spectral properties of these particles can be tuned by varying the relative dimensions of the core and the shell. By using this protocol, gold nanoshells with optical resonances extending from the visible to approximately 3 µm in the infrared region have been prepared [461]. In a typical synthesis [464–467], silica nanoparticles are first synthesized by using the Stöber method. The silica surface is next functionalized with a bifunctional molecule (e.g., 3-aminopropyltrimethoxysilane) that can bind to preformed gold nanoparticles through one of the functional groups (typically the amino group). The small gold particles (1–2 nm) are usually formed by reducing $HAuCl_4$ with alkaline tetrakis(hydroxymethyl)phosphonium chloride (THPC). These gold particles attach to the surface functionalized silica and act as seeds that facilitate further nucleation of clusters/particles, which ultimately form a continuous metal shell upon further addition of gold. Kah et al. reported a single-step process to form the precursor seed particles [468]. They first precipitated gold oxide on the silica surface. The gold nanoparticles subsequently generated on the surface in the reduction step provided the seeds for the formation of the shell.

Gold can also be deposited by various methods onto magnetic core nanoparticles [469–471]. Gold nanoshells can be formed either directly on the surface of the magnetic nanoparticles or on a silica layer previously deposited onto their surface. In the synthesis of gold nanoshells on silica-embedded magnetic nanoparticles, the latter are obtained by the Stöber method. The deposition of the gold nanoshells follows the usual functionalization of silica surface and subsequent attachment and growth of small gold seed particles [472].

4.5.2.2 Synthesis of Gold Nanorods

Gold nanorods have been used in various bio-analytical and biomedical applications. They offer superior NIR absorption and scattering at much smaller particle sizes compared to nanospheres and nanoshells [31]. Gold nanorods of a higher aspect ratio and a smaller effective radius are the best photoabsorbing nanoparticles, whereas the highest scattering contrast (for imaging applications) can be obtained from nanorods of high aspect ratio and a larger effective radius [31]. Existing methods for synthesizing gold nanorods with high aspect ratio are characterized typically by poor yields [423, 424]. Short gold nanorods (20–100 nm long), which are very interesting from biomedical applications point of view as well, can be prepared with higher yields by using a small quantity of silver via the seed-mediated or non-seed-mediated approach [473–475]. In a typical synthesis, 4.75 mL of 0.10 M CTAB, 0.200 mL of 0.01 M $HAuCl_4 \cdot 3H_2O$, and 0.030 mL of 0.01 M $AgNO_3$ solutions are added in this order to a test tube and gently mixed by inversion. The resulting solution becomes colorless following the addition of 0.032 mL of 0.10 M ascorbic acid solution. Next, a small quantity of gold seed solution (e.g., 0.010 mL) is added and the reaction mixture is gently mixed for 10 s and then left undisturbed. The rod development takes about one and half hour and the particles can be used after ~3 h. Variable quantities of seed solution can be used to change the aspect ratio of the rods. Short gold nanorods of different aspect ratio can also be prepared by varying the amount of silver nitrate for a given amount of gold [473].

4.5.2.3 Synthesis of Gold Nanocages

Nanocages are nanostructures that possess hollow interiors and porous walls [210]. Synthesis of gold nanocages uses the principle of galvanic displacement. Typically, these particles are obtained by adding $HAuCl_4$ to a dispersion of preformed silver nanocubes. The electrochemical potential difference between the gold and silver drives the displacement reaction. Gold ions are reduced and deposited on the surface of the silver nanoparticles. At the same time, metallic silver from the cores is oxidized and dissolved in the external solution leading to the formation of hollow and porous gold structures. By changing the amount of $HAuCl_4$ added to the suspension of silver nanocubes, the LSPR of nanocages can be tuned into the biological near-infrared window. Xia and co-workers later reported that the use of a wet etchant other than $HAuCl_4$ (e.g., based on $Fe(NO_3)_3$ or NH_4OH) allows better control of the wall thickness and porosity of the resultant nanocages [476].

References

1. Moghimi, S. M.; Hunter, A. C.; Murray, J. C. *FASEB J.* **2005**, *19*, 311–330.
2. Pankhurst, Q. A.; Connoly, J.; Dobson, J. *J. Phys. D: Appl. Phys.* **2003**, *36*.
3. Lux Research, The Nanotech ReportTM: Investment Overview and Market Research for Nanotechnology **2006**.
4. Daniel, M.; Astruc, D. *Chem. Rev.* **2004**, *104*, 293–346.
5. Huaizhi, Z.; Yuantao, N. *Gold Bull.* **2001**, *34*, 24–29.
6. Gielen, M.; Tiekink, E. *Metallotherapeutic Drugs and Metal-Based Diagnostic Agents: The Use of Metals in Medicine*; John Wiley and Sons: Hoboken, NJ, 2005.
7. Kumar, C. *Nanomaterials for Cancer Diagnosis*; Wiley-VCH: Weinheim, Germany, 2007.
8. Sun, R. W.; Ma, D.; Wong, E. L.; Che, C. *Dalton Trans.* **2007**, 4884–4892.
9. Shaw III, F. *Chem. Rev.* **1999**, *99*, 2589–2600.
10. Roth, J. *Histochem. Cell Biol.* **1996**, *106*, 1–8.
11. Thiessen, P. A. *Kolloid Z.* **1942**, *101*, 241–248.

12. Green, F. *The Colloidal Gold Reaction of the Cerebrospinal Fluid*; Medizin Fritz-Dieter Söhn: Berlin, 1925.
13. Maclagan, N. F. *Br. J. Exp. Pathol.* **1944**, *25*, 15.
14. Kausche, G.; Ruska, H. *Kolloid Z.* **1939**, *89*, 21–26.
15. Feldherr, C. M.; Marshall, J. M. *J. Cell Biol.* **1962**, 640.
16. Paciotti, G. F.; Myer, L.; Weinreich, D.; Goia, D.; Pavel, N.; McLaughlin, R. E.; Tamarkin, L. *Drug Deliv.* **2004**, *11*, 169–183.
17. Polak, J. M.; Varndell, I. M. *Immunolabeling for Electron Microscopy*; Elsevier: Amsterdam, 1984.
18. Robinson, J. M.; Takizawa, T.; Vandre, D. D. *J. Histochem. Cytochem.* **2000**, *48*, 487–492.
19. Rogoff, E. E.; Romano, R.; Hahn, E. W. *Radiology* **1975**, *114*, 225–226.
20. Edelman, E. R.; Seifert, P.; Groothuis, A.; Morss, A.; Bornstein, D.; Rogers, C. *Circulation* **2001**, *103*, 429–434.
21. Svedman, C.; Tillman, C.; Gustavsson, C. G.; Möller, H.; Frennby, B.; Bruze, M. *Contact Dermatitis* **2005**, *52*, 192–196.
22. Thelen, A.; Bauknecht, H.; Asbach, P.; Schrom, T. *Eur. Arch. Oto-Rhino-L.* **2006**, *263*, 900–905.
23. Danscher, G. *Histochem. Cell Biol.* **2002**, *117*, 447–452.
24. Demann, E. T. K.; Stein, P. S.; Haubenreich, J. E. *J. Long-Term Eff. Med.* **2005**, *15*, 687–698.
25. Svedman, C.; Dunér, K.; Kehler, M.; Möller, H.; Gruvberger, B.; Bruze, M. *Clin. Res. Cardiol.* **2006**, *95*, 689–691.
26. Sihelníková, I.; Tvaročka, I. *Chem. Pap.* **2007**, *61*, 237–255.
27. Pyykko, P. *Chem. Rev.* **1988**, *88*, 563–594.
28. Schmidbaur, H.; Cronje, S.; Djordjevic, B.; Schuster, O. *Chem. Phys.* **2005**, *311*, 151–161.
29. Kreibig, U.; Vollmer, M. *Optical properties of metal clusters*; Springer: Berlin, 1995.
30. Kreibig, U.; Genzel, L. *Surf. Sci.* **1985**, *156*, 678–700.
31. Jain, P. K.; Lee, K. S.; El-Sayed, I. H.; El-Sayed, M. A. *J. Phys. Chem. B* **2006**, *110*, 7238–7248.
32. Bohren, C.; Huffmann, D. *Absorption and scattering of light by small particle*; John–Wiley: New York, 1983.
33. Andreescu, D.; Sau, T. K.; Goia, D. V. *J. Colloid Interface Sci.* **2006**, *298*, 742–751.
34. Yguerabide, J.; Yguerabide, E. E. *Anal. Biochem.* **1998**, *262*, 137–156.
35. Yguerabide, J.; Yguerabide, E. E. *Anal. Biochem.* **1998**, *262*, 157–176.
36. Mahmood, U.; Weissleder, R. *Mol. Cancer Ther.* **2003**, *2*, 489–496.
37. Loo, C.; Lin, A.; Hirsch, L.; Lee, M.; Barton, J.; Halas, N.; West, J.; Drezek, R. *Technol. Cancer Res. Treat.* **2004**, *3*, 33–40.
38. O'Neal, D. P.; Hirsch, L. R.; Halas, N. J.; Payne, J. D.; West, J. L. *Cancer Lett.* **2004**, *209*, 171–176.
39. Nelayah, J.; Kociak, M.; Stephan, O.; Abajo, F. J. G. D.; Tence, M.; Henrard, L.; Taverna, D.; Pastoriza-Santos, I.; Liz-Marzan, L. M.; Colliex, C. *Nat. Phys.* **2007**, *3*, 348–353.
40. Hao, E.; Schatz, G. C. *J. Chem. Phys.* **2004**, *120*, 357–366.
41. Orendorff, C. J.; Sau, T. K.; Murphy, C. J. *Small* **2006**, *2*, 636–639.
42. Sönnichsen, C.; Franzl, T.; Wilk, T.; Plessen, G. V.; Feldmann, J. *New J. Phys.* **2002**, *4*, 93–93.
43. Jin, R. C.; Cao, Y. W.; Mirkin, C. A.; Kelly, K. L.; Schatz, G. C.; Zheng, J. G. *Science* **2001**, *294*, 1901–1903.
44. Jain, P. K.; Huang, W.; El-Sayed, M. A. *Nano Lett.* **2007**, *7*, 2080–2088.
45. Jain, P. K.; El-Sayed, M. A. *Nano Lett.* **2007**, *7*, 2854–2858.
46. Murphy, C. J.; Gole, A. M.; Hunyadi, S. E.; Stone, J. W.; Sisco, P. N.; Alkilany, A.; Kinard, B. E.; Hankins, P. *Chem. Commun.* **2008**, 544–557.
47. Lal, S.; Link, S.; Halas, N. J. *Nat. Photonics* **2007**, *1*, 641–648.
48. Chen, H.; Kou, X.; Yang, Z.; Ni, W.; Wang, J. *Langmuir* **2008**, *24*, 5233–5237.
49. Miller, M. M.; Lazarides, A. A. *J. Phys. Chem. B* **2005**, *109*, 21556–21565.
50. Lee, K.; El-Sayed, M. A. *J. Phys. Chem. B* **2006**, *110*, 19220–19225.

51. Chen, C.; Cheng, S.; Chau, L.; Wang, C. C. *Biosens. Bioelectron.* **2007**, *22*, 926–932.
52. Mock, J. J.; Smith, D. R.; Schultz, S. *Nano Lett.* **2003**, *3*, 485–491.
53. Sherry, L. J.; Jin, R.; Mirkin, C. A.; Schatz, G. C.; Duyne, R. P. V. *Nano Lett.* **2006**, *6*, 2060–2065.
54. Raschke, G.; Kowarik, S.; Franzl, T.; Sönnichsen, C.; Klar, T. A.; Feldmann, J.; Nichtl, A.; Kurzinger, K. *Nano Lett.* **2003**, *3*, 935–938.
55. Mayer, K. M.; Lee, S.; Liao, H.; Rostro, B. C.; Fuentes, A.; Scully, P. T.; Nehl, C. L.; Hafner, J. H. *ACS Nano* **2008**, *2*, 687–692.
56. Russier-Antoine, I.; Huang, J.; Benichou, E.; Bachelier, G.; Jonin, C.; Brevet, P. *Chem. Phys. Lett.* **2008**, *450*, 345–349.
57. Wang, L.; Li, J.; Song, S.; Li, D.; Fan, C. *J. Phys. D: Appl. Phys.* **2009**, *42*, 203001.
58. Sonnichsen, C.; Reinhard, B. M.; Liphardt, J.; Alivisatos, A. P. *Nat. Biotechnol.* **2005**, *23*, 741–745.
59. Elghanian, R.; Storhoff, J. J.; Mucic, R. C.; Letsinger, R. L.; Mirkin, C. A. *Science* **1997**, *277*, 1078–1081.
60. Dyadyusha, L.; Yin, H.; Jaiswal, S.; Brown, T.; Baumberg, J. J.; Booy, F. P.; Melvin, T. *Chem. Commun.* **2005**, 3201–3203.
61. Aslan, K.; Luhrs, C. C.; Perez-Luna, V. H. *J. Phys. Chem. B* **2004**, *108*, 15631–15639.
62. Millstone, J. E.; Park, S.; Shuford, K. L.; Qin, L.; Schatz, G. C.; Mirkin, C. A. *J. Am. Chem. Soc.* **2005**, *127*, 5312–5313.
63. Malicka, J.; Gryczynski, I.; Kusba, J.; Shen, Y.; Lakowicz, J. R. *Biochem. Bioph. Res. Co.* **2002**, *294*, 886–892.
64. Zhang, J.; Fu, Y.; Lakowicz, J. R. *J. Phys. Chem. C* **2007**, *111*, 50–56.
65. Roy, R.; Hohng, S.; Ha, T. *Nat. Methods* **2008**, *5*, 507–516.
66. Gersten, J.; Nitzan, A. *J. Chem. Phys.* **1981**, *75*, 1139–1152.
67. Sokolov, K.; Nida, D.; Descour, M.; Lacy, A.; Levy, M.; Hall, B.; Dharmawardhane, S.; Ellington, A.; Korgel, B.; Richards-Kortum, R. *Adv. Cancer Res.* **2006**, *96*, 299–344.
68. Uechi, I.; Yamada, S. *Anal. Bioanal. Chem.* **2008**, *391*, 2411–2421.
69. Wokaun, A.; Gordon, J. P.; Liao, P. F. *Phys. Rev. Lett.* **1982**, *48*, 957.
70. Gersten, J.; Nitzan, A. *J. Chem. Phys.* **1980**, *73*, 3023–3037.
71. Xu, H.; Aizpurua, J.; Käll, M.; Apell, P. *Phys. Rev. E* **2000**, *62*, 4318–4324.
72. Cline, M. P.; Barber, P. W.; Chang, R. K. *J. Opt. Soc. Am. B* **1986**, *3*, 15–21.
73. Sönnichsen, C.; Franzl, T.; Wilk, T.; Plessen, G. V.; Feldmann, J.; Wilson, O.; Mulvaney, P. *Phys. Rev. Lett.* **2002**, *88*, 077402.
74. Messinger, B. J.; Raben, K. U. V.; Chang, R. K.; Barber, P. W. *Phys. Rev. B* **1981**, *24*, 649–657.
75. Heilweil, E. J.; Hochstrasser, R. M. *J. Chem. Phys.* **1985**, *82*, 4762–4770.
76. Bek, A.; Jansen, R.; Ringler, M.; Mayilo, S.; Klar, T. A.; Feldmann, J. *Nano Lett.* **2008**, *8*, 485–490.
77. Dieringer, J. A.; Lettan, R. B.; Scheidt, K. A.; Duyne, R. P. V. *J. Amer. Chem. Soc.* **2007**, *129*, 16249–16256.
78. Wang, Y.; Guo, S.; Chen, H.; Wang, E. *J. Colloid Interface Sci.* **2008**, *318*, 82–87.
79. Doering, W. E.; Nie, S. *Anal. Chem.* **2003**, *75*, 6171–6176.
80. Hao, F.; Nehl, C. L.; Hafner, J. H.; Nordlander, P. *Nano Lett.* **2007**, *7*, 729–732.
81. Kumar, P. S.; Pastoriza-Santos, I.; Rodríguez-González, B.; García de Abajo, F. J.; Liz-Marzán, L. M. *Nanotechnology* **2008**, *19*, 015606.
82. Hrelescu, C.; Sau, T. K.; Rogach, A. L.; Jackel, F.; Feldmann, J. *Appl. Phys. Lett.* **2009**, *94*, 153113–3.
83. Rodríguez-Lorenzo, L.; Álvarez-Puebla, R. A.; Pastoriza-Santos, I.; Mazzucco, S.; Stéphan, O.; Kociak, M.; Liz-Marzán, L. M.; García de Abajo, F. J. *J. Am. Chem. Soc.* **2009**, *131*, 4616–4618.
84. Käll, M.; Xu, H.; Johansson, P. *J. Raman Spectrosc.* **2005**, *36*, 510–514.
85. Aslan, K.; Lakowicz, J. R.; Geddes, C. D. *Anal. Bioanal. Chem.* **2005**, *382*, 926–933.

86. Aslan, K.; Leonenko, Z.; Lakowicz, J. R.; Geddes, C. D. *J. Phys.Chem. B* **2005**, *109*, 3157–3162.
87. Li, C.; Male, K. B.; Hrapovic, S.; Luong, J. H. T. *Chem. Commun.* **2005**, 3924–3926.
88. Anger, P.; Bharadwaj, P.; Novotny, L. *Phys. Rev. Lett.* **2006**, *96*, 113002.
89. Kuhn, S.; Hakanson, U.; Rogobete, L.; Sandoghdar, V. *Phys. Rev. Lett.* **2006**, *97*, 017402.
90. Lakowicz, J. R.; Malicka, J.; Gryczynski, I.; Gryczynski, Z.; Geddes, C. D. *J. Phys. D: Appl. Phys.* **2003**, *36*, R240–R249.
91. Dulkeith, E.; Morteani, A. C.; Niedereichholz, T.; Klar, T. A.; Feldmann, J.; Levi, S. A.; Veggel, F. C. J. M. V.; Reinhoudt, D. N.; Möller, M.; Gittins, D. I. *Phys. Rev. Lett.* **2002**, *89*, 203002.
92. Dulkeith, E.; Ringler, M.; Klar, T. A.; Feldmann, J.; Javier, A. M.; Parak, W. J. *Nano Lett.* **2005**, *5*, 585–589.
93. Barnes, W. L. *J. Mod. Opt.* **1998**, *45*, 661.
94. Lakowicz, J. R. *Anal. Biochem.* **2005**, *337*, 171–194.
95. Lakowicz, J. R. *Anal. Biochem.* **2001**, *298*, 1–24.
96. Guzatov, D. V.; Klimov, V. V. *Chem. Phys. Lett.* **2005**, *412*, 341–346.
97. Mohamed, M. B.; Volkov, V.; Link, S.; El-Sayed, M. A. *Chem. Phys. Lett.* **2000**, *317*, 517–523.
98. Varnavski, O. P.; Mohamed, M. B.; El-Sayed, M. A.; Goodson, T. *J. Phys. Chem. B* **2003**, *107*, 3101–3104.
99. Eustis, S.; El-Sayed, M. *J. Phys. Chem. B* **2005**, *109*, 16350–16356.
100. Imura, K.; Nagahara, T.; Okamoto, H. *J. Phys. Chem. B* **2005**, *109*, 13214–13220.
101. Wang, H.; Huff, T. B.; Zweifel, D. A.; He, W.; Low, P. S.; Wei, A.; Cheng, J. *Proc. Natl. Acad. Sci. U. S. A.* **2005**, *102*, 15752–15756.
102. Bouhelier, A.; Bachelot, R.; Lerondel, G.; Kostcheev, S.; Royer, P.; Wiederrecht, G. P. *Phys. Rev. Lett.* **2005**, *95*, 267405.
103. Wilcoxon, J. P.; Martin, J. E.; Parsapour, F.; Wiedenman, B.; Kelley, D. F. *J. Chem. Phys.* **1998**, *108*, 9137–9143.
104. Wang, G.; Huang, T.; Murray, R. W.; Menard, L.; Nuzzo, R. G. *J. Am. Chem. Soc.* **2005**, *127*, 812–813.
105. You, C.; De, M.; Han, G.; Rotello, V. M. *J. Am. Chem. Soc.* **2005**, *127*, 12873–12881.
106. Zhang, C. X.; Zhang, Y.; Wang, X.; Tang, Z. M.; Lu, Z. H. *Anal. Biochem.* **2003**, *320*, 136–140.
107. Han, G.; You, C.; Kim, B.; Turingan, R. S.; Forbes, N. S.; Martin, C. T.; Rotello, V. M. *Angew. Chem. Int. Ed.* **2006**, *45*, 3165–3169.
108. Han, G.; Martin, C. T.; Rotello, V. M. *Chem. Biol. Drug Des.* **2006**, *67*, 78–82.
109. Han, G.; Ghosh, P.; Rotello, V. M. *Nanomedicine* **2007**, *2*, 113–123.
110. Zanchet, D.; Micheel, C. M.; Parak, W. J.; Gerion, D.; Alivisatos, A. P. *Nano Lett.* **2001**, *1*, 32–35.
111. Lévy, R.; Wang, Z.; Duchesne, L.; Doty, R. C.; Cooper, A. I.; Brust, M.; Fernig, D. G. *ChemBioChem* **2006**, *7*, 592–594.
112. Yu, C.; Irudayaraj, J. *Biophys. J.* **2007**, *93*, 3684–3692.
113. Love, J. C.; Estroff, L. A.; Kriebel, J. K.; Nuzzo, R. G.; Whitesides, G. M. *Chem. Rev.* **2005**, *105*, 1103–1170.
114. McNeil, S. E. *J. Leukocyte Biol.* **2005**, *78*, 585–594.
115. Li, X.; Huskens, J.; Reinhoudt, D. N. *J. Mater. Chem.* **2004**, *14*, 2954–2971.
116. Sperling, R. A.; Pellegrino, T.; Li, J.; Chang, W.; Parak, W. *Adv. Funct. Mater.* **2006**, *16*, 943–948.
117. Hermanson, G. T. *Bioconjugate Techniques*; Academic Press: San Diego, **1996**.
118. Angelatos, A. S.; Radt, B.; Caruso, F. *J. Phys. Chem.B* **2005**, *109*, 3071–3076.
119. Angelatos, A. S.; Katagiri, K.; Caruso, F. *Soft Matter* **2006**, *2*, 18–23.
120. Cho, J.; Caruso, F. *Chem. Mater.* **2005**, *17*, 4547–4553.
121. Templeton, A. C.; Wuelfing, W. P.; Murray, R. W. *Acc. Chem. Res.* **2000**, *33*, 27–36.
122. Oishi, M.; Nakaogami, J.; Ishii, T.; Nagasaki, Y. *Chem. Lett.* **2006**, *35*, 1046–1047.

123. Hong, R.; Han, G.; Fernandez, J. M.; Kim, B.; Forbes, N. S.; Rotello, V. M. *J. Am. Chem. Soc.* **2006**, *128*, 1078–1079.

124. Liz-Marzan, L. M.; Giersig, M.; Mulvaney, P. *Langmuir* **1996**, *12*, 4329–4335.

125. Katz, E.; Willner, I. *Angew. Chem. Int. Ed.* **2004**, *43*, 6042–6108.

126. De la Fuente, J. M.; Penadés, S. *BBA-Gen. Subjects* **2006**, *1760*, 636–651.

127. Tkachenko, A. G.; Xie, H.; Coleman, D.; Glomm, W.; Ryan, J.; Anderson, M. F.; Franzen, S.; Feldheim, D. L. *J. Am. Chem. Soc.* **2003**, *125*, 4700–4701.

128. Hainfeld, J. F.; Slatkin, D. N.; Smilowitz, H. M. *Phys. Med. Biol.* **2004**, *49*, N309–N315.

129. Connor, E.; Mwamuka, J.; Gole, A.; Murphy, C.; Wyatt, M. *Small* **2005**, *1*, 325–327.

130. Shukla, R.; Bansal, V.; Chaudhary, M.; Basu, A.; Bhonde, R. R.; Sastry, M. *Langmuir* **2005**, *21*, 10644–10654.

131. Pan, Y.; Neuss, S.; Leifert, A.; Fischler, M.; Wen, F.; Simon, U.; Schmid, G.; Brandau, W.; Jahnen-Dechent, W. *Small* **2007**, *3*, 1941–1949.

132. Oberdörster, G.; Stone, V.; Donaldson, K. *Nanotoxicology* **2007**, *1*, 2–25.

133. Liu, Y.; Shipton, M. K.; Ryan, J.; Kaufman, E. D.; Franzen, S.; Feldheim, D. L. *Anal. Chem.* **2007**, *79*, 2221–2229.

134. Kogan, M. J.; Olmedo, I.; Hosta, L.; Guerrero, A. R.; Cruz, L. J.; Albericio, F. *Nanomedicine.* **2007**, *2*, 287–306.

135. Vlerken, L. E.; Vyas, T. K.; Amiji, M. M. *Pharm. Res.* **2007**, *24*, 1405–1414.

136. Cognet, L.; Lounis, B. *Gold Bull.* **2008**, *41*, 139–146.

137. Govorov, A. O.; Richardson, H. H. *Nano Today* **2007**, *2*, 30–38.

138. Richardson, H. H.; Hickman, Z. N.; Govorov, A. O.; Thomas, A. C.; Zhang, W.; Kordesch, M. E. *Nano Lett.* **2006**, *6*, 783–788.

139. Pissuwan, D.; Valenzuela, S. M.; Cortie, M. B. *Trends Biotechnol.* **2006**, *24*, 62–67.

140. Hu, M.; Wang, X.; Hartland, G. V.; Salgueiriño-Maceira, V.; Liz-**Marzán**, L. M. *Chem. Phys. Lett.* **2003**, *372*, 767–772.

141. Govorov, A.; Zhang, W.; Skeini, T.; Richardson, H.; Lee, J.; Kotov, N. *Nanoscale Res. Lett.* **2006**, *1*, 84–90.

142. Stehr, J.; Hrelescu, C.; Sperling, R. A.; Raschke, G.; Wunderlich, M.; Nichtl, A.; Heindl, D.; Kurzinger, K.; Parak, W. J.; Klar, T. A.; Feldmann, J. *Nano Lett.* **2008**, *8*, 619–623.

143. Palpant, B.; Guillet, Y.; Rashidi-Huyeh, M.; Prot, D. *Gold Bull.* **2008**, *41*, 105–115.

144. Liu, G. L.; Kim, J.; Lu, Y.; Lee, L. P. *Nat. Mater.* **2006**, *5*, 27–32.

145. Boyer, D.; Tamarat, P.; Maali, A.; Lounis, B.; Orrit, M. *Science* **2002**, *297*, 1160–1163.

146. Berciaud, S.; Cognet, L.; Blab, G. A.; Lounis, B. *Phys. Rev. Lett.* **2004**, *93*, 257402.

147. Berciaud, S.; Lasne, D.; Blab, G. A.; Cognet, L.; Lounis, B. *Phys. Rev. B* **2006**, *73*, 045424.

148. Skirtach, A. G.; Dejugnat, C.; Braun, D.; Susha, A. S.; Rogach, A. L.; Parak, W. J.; Mohwald, H.; Sukhorukov, G. B. *Nano Lett.* **2005**, *5*, 1371–1377.

149. Skirtach, A. G.; Javier, A. M. O.; Kreft, O.; Köhler, K.; Alberola, A. P.; Möhwald, H.; Parak, W. J.; Sukhorukov, G. B. *Angew. Chem. Int. Ed.* **2006**, *45*, 4612–4617.

150. Skirtach, A.; Karageorgiev, P.; Geest, B. D.; Pazos-Perez, N.; Braun, D.; Sukhorukov, G. *Adv. Mater.* **2008**, *20*, 506–510.

151. Yao, C.; Rahmanzadeh, R.; Endl, E.; Zhang, Z.; Gerdes, J.; Huttmann, G. *J. Biomed. Opt.* **2005**, *10*, 064012.

152. Pitsillides, C. M.; Joe, E. K.; Wei, X.; Anderson, R. R.; Lin, C. P. *Biophys. J.* **2003**, *84*, 4023–4032.

153. Hirsch, L. R.; Stafford, R. J.; Bankson, J. A.; Sershen, S. R.; Rivera, B.; Price, R. E.; Hazle, J. D.; Halas, N. J.; West, J. L. *Proc. Natl. Acad. Sci. U. S. A.* **2003**, *100*, 13549–13554.

154. Letfullin, R. R.; Joenathan, C.; George, T. F.; Zharov, V. P. *Nanomedicine* **2006**, *1*, 473–480.

155. Zharov, V. P.; Galitovsky, V.; Viegas, M. *Appl. Phys. Lett.* **2003**, *83*, 4897–4899.

156. Khlebtsov, B.; Zharov, V.; Melnikov, A.; Tuchin, V.; Khlebtsov, N. *Nanotechnology* **2006**, *17*, 5167–5179.

157. Chen, J.; Wang, D.; Xi, J.; Au, L.; Siekkinen, A.; Warsen, A.; Li, Z.; Zhang, H.; Xia, Y.; Li, X. *Nano Lett.* **2007**, *7*, 1318–1322.

158. Hilger, I.; Andrä, W.; Bähring, R.; Daum, A.; Hergt, R.; Kaiser, W. A. *Invest. Radiol.* **1997**, *32*.
159. Hilger, A.; Cüppers, N.; Tenfelde, M.; Kreibig, U. *Eur. Phys. J. D* **2000**, *10*, 115–118.
160. Hilger, I.; Andrä, W.; Hergt, R.; Hiergeist, R.; Schubert, H.; Kaiser, W. A. *Radiology* **2001**, *218*, 570–575.
161. Andrä, W.; d'Ambly, C. G.; Hergt, R.; Hilger, I.; Kaiser, W. A. *J. Magn. Magn. Mater.* **1999**, *194*, 197–203.
162. Pan, D.; Lanza, G. M.; Wickline, S. A.; Caruthers, S. D. *Eur. J. Radiol.* **2009**, *70*, 274–285.
163. Hainfeld, J. F.; Powell, R. D. *J. Histochem. Cytochem.* **2000**, *48*, 471–480.
164. Cai, W.; Chen, X. *Small* **2007**, *3*, 1840–1854.
165. Sharma, P.; Brown, S.; Walter, G.; Santra, S.; Moudgil, B. *Adv. Colloid Interface Sci.* **2006**, *123–126*, 471–485.
166. Bruchez, M.; Moronne, M.; Gin, P.; Weiss, S.; Alivisatos, A. P. *Science* **1998**, *281*, 2013–2016.
167. Chan, W. C. W.; Maxwell, D. J.; Gao, X.; Bailey, R. E.; Han, M.; Nie, S. *Curr. Opin. Biotechnol.* **2002**, *13*, 40–46.
168. Åkerman, M. E.; Chan, W. C. W.; Laakkonen, P.; Bhatia, S. N.; Ruoslahti, E. *Proc. Nat. Acad. Sci. U. S. A.* **2002**, *99*, 12617–12621.
169. Gao, X.; Cui, Y.; Levenson, R. M.; Chung, L. W. K.; Nie, S. *Nat. Biotechnol.* **2004**, *22*, 969–976.
170. Kim, D. K.; Mikhaylova, M.; Wang, F. H.; Kehr, J.; Bjelke, B.; Zhang, Y.; Tsakalakos, T.; Muhammed, M. *Chem. Mater.* **2003**, *15*, 1343–4351.
171. Martina, M.; Fortin, J.; Menager, C.; Clement, O.; Barratt, G.; Grabielle-Madelmont, C.; Gazeau, F.; Cabuil, V.; Lesieur, S. *J. Am. Chem. Soc.* **2005**, *127*, 10676–10685.
172. Lee, J.; Huh, Y.; Jun, Y.; Seo, J.; Jang, J.; Song, H.; Kim, S.; Cho, E.; Yoon, H.; Suh, J.; Cheon, J. *Nat. Med.* **2007**, *13*, 95–99.
173. Cognet, L.; Tardin, C.; Boyer, D.; Choquet, D.; Tamarat, P.; Lounis, B. *Proc. Nat. Acad. Sci. U. S. A.* **2003**, *100*, 11350–11355.
174. Loo, C.; Lowery, A.; Halas, N.; West, J.; Drezek, R. *Nano Lett.* **2005**, *5*, 709–711.
175. Huang, X.; El-Sayed, I. H.; Qian, W.; El-Sayed, M. A. *J. Am. Chem. Soc.* **2006**, *128*, 2115–2120.
176. Ipe, B. I.; Yoosaf, K.; Thomas, K. G. *J. Am. Chem. Soc.* **2006**, *128*, 1907–1913.
177. Lewis, D. J.; Day, T. M.; MacPherson, J. V.; Pikramenou, Z. *Chem. Commun.* **2006**, 1433–1435.
178. Qian, X.; Peng, X.; Ansari, D. O.; Yin-Goen, Q.; Chen, G. Z.; Shin, D. M.; Yang, L.; Young, A. N.; Wang, M. D.; Nie, S. *Nat. Biotechnol.* **2008**, *26*, 83–90.
179. El-Sayed, I. H.; Huang, X.; El-Sayed, M. A. *Nano Lett.* **2005**, *5*, 829–834.
180. Thurn, K.; Brown, E.; Wu, A.; Vogt, S.; Lai, B.; Maser, J.; Paunesku, T.; Woloschak, G. *Nanoscale Res. Lett.* **2007**, *2*, 430–441.
181. Lewinski, N.; Colvin, V.; Drezek, R. *Small* **2008**, *4*, 26–49.
182. Debbage, P.; Jaschke, W. *Histochem. Cell Biol.* **2008**, *130*, 845–875.
183. Oyelere, A. K.; Chen, P. C.; Huang, X.; El-Sayed, I. H.; El-Sayed, M. A. *Bioconjugate Chem.* **2007**, *18*, 1490–1497.
184. Ding, H.; Yong, K.; Roy, I.; Pudavar, H. E.; Law, W. C.; Bergey, E. J.; Prasad, P. N. *J. Phys. Chem. C* **2007**, *111*, 12552–12557.
185. Yu, C.; Nakshatri, H.; Irudayaraj, J. *Nano Lett.* **2007**, *7*, 2300–2306.
186. Stone, J. W.; Sisco, P. N.; Goldsmith, E. C.; Baxter, S. C.; Murphy, C. J. *Nano Lett.* **2007**, *7*, 116–119.
187. Murphy, C. J.; Gole, A. M.; Stone, J. W.; Sisco, P. N.; Alkilany, A. M.; Goldsmith, E. C.; Baxter, S. C. *Acc. Chem. Res.* **2008**, *41*, 1721–1730.
188. Chen, J.; Saeki, F.; Wiley, B. J.; Cang, H.; Cobb, M. J.; Li, Z.; Au, L.; Zhang, H.; Kimmey, M. B.; Li; Xia, Y. *Nano Lett.* **2005**, *5*, 473–477.
189. Oldenburg, A. L.; Hansen, M. N.; Zweifel, D. A.; Wei, A.; Boppart, S. A. *Opt. Exp.* **2006**, *14*, 6724–6738.

190. Oldenburg, A. L.; Hansen, M. N.; Wei, A.; Boppart, S. A.; Achilefu, S.; Bornhop, D. J.; Raghavachari, R. In *Molecular Probes for Biomedical Applications II*; SPIE, **2008**; Vol. 6867, p. 68670E–10.
191. Troutman, T. S.; Barton, J. K.; Romanowski, M. *Opt. Lett.* **2007**, *32*, 1438–1440.
192. Fujimoto, J. G. *Nat. Biotechnol.* **2003**, *21*, 1361–1367.
193. Wang, Y.; Xie, X.; Wang, X.; Ku, G.; Gill, K. L.; O'Neal, D. P.; Stoica, G.; Wang, L. V. *Nano Lett.* **2004**, *4*, 1689–1692.
194. Wang, X.; Pang, Y.; Ku, G.; Xie, X.; Stoica, G.; Wang, L. V. *Nat. Biotechnol.* **2003**, *21*, 803–806.
195. Liao, C.; Huang, S.; Wei, C.; Li, P. *J. Biomed. Opt.* **2007**, *12*, 064006–9.
196. Eghtedari, M.; Oraevsky, A.; Copland, J. A.; Kotov, N. A.; Conjusteau, A.; Motamedi, M. *Nano Lett.* **2007**, *7*, 1914–1918.
197. Chamberland, D. L.; Agarwal, A.; Kotov, N.; Fowlkes, J. B.; Carson, P. L.; Wang, X. *Nanotechnology* **2008**, *19*, 095101–095101.
198. Copland, J. A.; Eghtedari, M.; Popov, V. L.; Kotov, N.; Mamedova, N.; Motamedi, M.; Oraevsky, A. A. *Mol. Imaging Biol.* **2004**, *6*, 341–349.
199. Agarwal, A.; Huang, S. W.; O'Donnell, M.; Day, K. C.; Day, M.; Kotov, N.; Ashkenazi, S. *J. Appl. Phys.* **2007**, *102*, 064701–4.
200. Park, J.; Estrada, A.; Sharp, K.; Sang, K.; Schwartz, J. A.; Smith, D. K.; Coleman, C.; Payne, J. D.; Korgel, B. A.; Dunn, A. K.; Tunnell, J. W. *Opt. Exp.* **2008**, *16*, 1590–1599.
201. Okamoto, H.; Imura, K. *J. Mater. Chem.* **2006**, *16*, 3920–3928.
202. Durr, N. J.; Larson, T.; Smith, D. K.; Korgel, B. A.; Sokolov, K.; Ben-Yakar, A. *Nano Lett.* **2007**, *7*, 941–945.
203. Tong, L.; Zhao, Y.; Huff, T. B.; Hansen, M. N.; Wei, A.; Cheng, J. X. *Adv. Mater.* **2007**, *19*, 3136–3141.
204. Huff, T. B.; Hansen, M. N.; Zhao, Y.; Cheng, J.; Wei, A. *Langmuir* **2007**, *23*, 1596–1599.
205. Debouttière, P.; Roux, S.; Vocanson, F.; Billotey, C.; Beuf, O.; Favre-Réguillon, A.; Lin, Y.; Pellet-Rostaing, S.; Lamartine, R.; Perriat, P.; Tillement, O. *Adv. Funct. Mater.* **2006**, *16*, 2330–2339.
206. Melancon, M. P.; Lu, W.; Li, C. *MRS Bull.* **2009**, *34*, 415–421.
207. Hainfeld, J. F.; Slatkin, D. N.; Focella, T. M.; Smilowitz, H. M. *Br. J. Radiol.* **2006**, *79*, 248–253.
208. Cai, Q.; Kim, S. H.; Choi, K. S.; Kim, S. Y.; Byun, S. J.; Kim, K. W.; Park, S. H.; Juhng, S. K.; Yoon, K. *Invest. Radiol.* **2007**, *42*, 797–806.
209. Kim, D.; Park, S.; Lee, J. H.; Jeong, Y. Y.; Jon, S. *Nanomed.-Nanotechnol.* **2007**, *3*, 352.
210. Chen, J.; Wiley, B.; Li, Z. Y.; Campbell, D.; Saeki, F.; Cang, H.; Au, L.; Lee, J.; Li, X.; Xia, Y. *Adv. Mater.* **2005**, *17*, 2255–2261.
211. Mallidi, S.; Larson, T.; Aaron, J.; Sokolov, K.; Emelianov, S. *Opt. Exp.* **2007**, *15*, 6583–6588.
212. Faulk, W. P.; Taylor, G. M. *Immunochemistry* **1971**, *8*, 1081–1083.
213. Nie, B.; Shortreed, M. R.; Smith, L. M. *Anal. Chem.* **2006**, *78*, 1528–1534.
214. Ryadnov, M. G.; Woolfson, D. N. *J. Am. Chem. Soc* **2004**, *126*, 7454–7455.
215. Levit-Binnun, N.; Lindner, A. B.; Zik, O.; Eshhar, Z.; Moses, E. *Anal. Chem.* **2003**, *75*, 1436–1441.
216. Lin, C. J.; Yang, T.; Lee, C.; Huang, S. H.; Sperling, R. A.; Zanella, M.; Li, J. K.; Shen, J.; Wang, H.; Yeh, H.; Parak, W. J.; Chang, W. H. *ACS Nano* **2009**, *3*, 395–401.
217. Hizoh, I.; Haller, C. *Invest. Radiol.* **2002**, *37*.
218. Haller, C.; Hizoh, I. *Invest. Radiol.* **2004**, *39*.
219. Jackson, P. A.; Rahman, W. N. W. A.; Wong, C. J.; Ackerly, T.; Geso, M. *Eur. J. Radiol.* **2010**, *75*, 104–109.
220. Lin, W.; Hyeon, T.; Lanza, G. M.; Zhang, M.; Meade, T. J. *MRS Bull.* **2009**, *34*, 441–448.
221. Caravan, P.; Ellison, J. J.; McMurry, T. J.; Lauffer, R. B. *Chem. Rev.* **1999**, *99*, 2293–2352.
222. Aime, S.; Fasano, M.; Terreno, E. *Chem. Soc. Rev.* **1998**, *27*, 19–29.
223. Albrecht-Buehler, G. *Cell* **1977**, *11*, 395–404.

224. Albrecht-Buehler, G. *Cell* **1977**, *12*, 333–339.
225. Albrecht-Buehler, G. *Sci. Am.* **1978**, *238*, 68–76.
226. Chen, J.; Helmold, M.; Kim, J.; Wynn, K.; Woodley, D. *Dermatology* **1994**, *188*, 6–12.
227. Liao, H.; Nehl, C. L.; Hafner, J. H. *Nanomedicine* **2006**, *1*, 201–208.
228. Zhao, J.; Zhang, X.; Yonzon, C. R.; Haes, A. J.; Van Duyne, R. P. *Nanomedicine* **2006**, *1*, 219–228.
229. Stuart, D. A.; Haes, A. J.; Yonzon, C. R.; Hicks, E. M.; Duyne, R. P. V. *IEE Proc. Nanobiotechnol.* **2005**, *152*, 13–32.
230. Willets, K. A.; Duyne, R. P. V. *Annu. Rev. Phys. Chem.* **2007**, *58*, 267–297.
231. Anker, J. N.; Hall, W. P.; Lyandres, O.; Shah, N. C.; Zhao, J.; Duyne, R. P. V. *Nat. Mater.* **2008**, *7*, 442–453.
232. McFadden, P. *Science* **2002**, *297*, 2075–2076.
233. Li, H.; Rothberg, L. *Proc. Nat. Acad. Sci. U. S. A.* **2004**, *101*, 14036–14039.
234. Yang, M.; Kostov, Y.; Bruck, H. A.; Rasooly, A. *Int. J. Food Microbiol.* **2009**, *133*, 265–271.
235. Yang, H.; Li, H.; Jiang, X. *Microfluid. Nanofluid.* **2008**, *5*, 571–583.
236. Simonian, A.; Good, T.; Wang, S.; Wild, J. *Anal. Chim. Acta* **2005**, *534*, 69–77.
237. Nicu, L.; Leichle, T. *J. Appl. Phys.* **2008**, *104*, 111101–16.
238. Otsuka, H.; Akiyama, Y.; Nagasaki, Y.; Kataoka, K. *J. Am. Chem. Soc.* **2001**, *123*, 8226–8230.
239. Haes, A.; Stuart, D. A.; Nie, S.; Van Duyne, R. *J. Fluoresc.* **2004**, *14*, 355–367.
240. Nam, J.; Stoeva, S. I.; Mirkin, C. A. *J. Am. Chem. Soc.* **2004**, *126*, 5932–5933.
241. Sato, K.; Hosokawa, K.; Maeda, M. *J. Am. Chem. Soc.* **2003**, *125*, 8102–8103.
242. Sato, K.; Hosokawa, K.; Maeda, M. *Nucleic Acids Res.* **2005**, *33*, e4.
243. Chah, S.; Hammond, M. R.; Zare, R. N. *Chem. Biol.* **2005**, *12*, 323–328.
244. De, M.; You, C.; Srivastava, S.; Rotello, V. M. *J. Am. Chem. Soc.* **2007**, *129*, 10747–10753.
245. GhoshMoulick, R.; Bhattacharya, J.; Mitra, C. K.; Basak, S.; Dasgupta, A. K. *Nanomed.: Nanotechnol.* **2007**, *3*, 208–214.
246. Aili, D.; Selegård, R.; Baltzer, L.; Enander, K.; Liedberg, B. *Small* **2009**, *5*, 2445–2452.
247. Bhattacharya, J.; Jasrapuria, S.; Sarkar, T.; GhoshMoulick, R.; Dasgupta, A. K. *Nanomed.: Nanotechnol.* **2007**, *3*, 14–19.
248. Tsai, C.; Yu, T.; Chen, C. *Chem. Commun.* **2005**, 4273–4275.
249. Guarise, C.; Pasquato, L.; De Filippis, V.; Scrimin, P. *Proc. Nat. Acad. Sci. U. S. A.* **2006**, *103*, 3978–3982.
250. Zhao, W.; Lam, J. C. F.; Chiuman, W.; Brook, M. A.; Li, Y. *Small* **2008**, *4*, 810–816.
251. Leuvering, J. H.; Thal, P. J.; Van Der Waart, M.; Schuurs, A. H. *J. Immunoassay* **1980**, *1*, 77–91.
252. Englebienne, P.; Hoonacker, A. V.; Valsamis, J. *Clin. Chem.* **2000**, *46*, 2000–2003.
253. Leuvering, J. H.; Goverde, B. C.; Thal, P. J.; Schuurs, A. H. *J. Immunol. Methods* **1983**, *60*, 9–23.
254. Leuvering, J. H.; Thal, P. J.; Schuurs, A. H. *J. Immunol. Methods* **1983**, *62*, 175–184.
255. Tanaka, M.; Matsuo, K.; Enomoto, M.; Mizuno, K. *Clin. Biochem.* **2004**, *37*, 27–35.
256. Thanh, N. T. K.; Rosenzweig, Z. *Anal. Chem.* **2002**, *74*, 1624–1628.
257. Dykman, L. A.; Bogatyrev, V. A.; Khlebtsov, B. N.; Khlebtsov, N. G. *Anal. Biochem.* **2005**, *341*, 16–21.
258. Takae, S.; Akiyama, Y.; Otsuka, H.; Nakamura, T.; Nagasaki, Y.; Kataoka, K. *Biomacromolecules* **2005**, *6*, 818–824.
259. Huang, C.; Chen, C.; Shiang, Y.; Lin, Z.; Chang, H. *Anal. Chem.* **2009**, *81*, 875–882.
260. Chen, Y.; Chen, S.; Chien, Y.; Chang, Y.; Liao, H.; Chang, C.; Jan, M.; Wang, K.; Lin, C. *ChemBioChem* **2005**, *6*, 1169–1173.
261. Aslan, K.; Zhang, J.; Lakowicz, J. R.; Geddes, C. D. *J. Fluoresc.* **2004**, *14*, 391–400.
262. Aslan, K.; Lakowicz, J. R.; Geddes, C. D. *Anal. Biochem.* **2004**, *330*, 145–155.
263. Aslan, K.; Lakowicz, J. R.; Geddes, C. D. *Anal. Chem.* **2005**, *77*, 2007–2014.

264. Pandey, P.; Singh, S. P.; Arya, S. K.; Gupta, V.; Datta, M.; Singh, S.; Malhotra, B. D. *Langmuir* **2007**, *23*, 3333–3337.
265. Liu, X.; Dai, Q.; Austin, L.; Coutts, J.; Knowles, G.; Zou, J.; Chen, H.; Huo, Q. *J. Am. Chem. Soc.* **2008**, *130*, 2780–2782.
266. Wilchek, M.; Bayer, E. A.; Livnah, O. *Immunol. Lett.* **2006**, *103*, 27–32.
267. Bayer, E. A.; Ben-Hur, H.; Wilchek, M. *Anal. Biochem.* **1986**, *154*, 367–370.
268. Lichstein, H. C.; Birnbaum, J. *Biochem.Bioph. Res. Co.* **1965**, *20*, 41–45.
269. Lazaridis, T.; Masunov, A.; Gandolfo, F. *Proteins* **2002**, *47*, 194–208.
270. Stayton, P. S.; Freitag, S.; Klumb, L. A.; Chilkoti, A.; Chu, V.; Penzotti, J. E.; To, R.; Hyre, D.; Le Trong, I.; Lybrand, T. P.; Stenkamp, R. E. *Biomol. Eng.* **1999**, *16*, 39–44.
271. González, M.; Bagatolli, L. A.; Echabe, I.; Arrondo, J. L. R.; Argaraña, C. E.; Cantor, C. R.; Fidelio, G. D. *J. Biol. Chem.* **1997**, *272*, 11288–11294.
272. Gould, E. A.; Buckley, A.; Cammack, N. *J. Virol. Methods* **1985**, *11*, 41–48.
273. Sastry, M.; Lala, N.; Patil, V.; Chavan, S. P.; Chittiboyina, A. G. *Langmuir* **1998**, *14*, 4138–4142.
274. White, K. A.; Rosi, N. L. *Nanomedicine* **2008**, *3*, 543–553.
275. Li, B.; Zhao, L.; Zhang, C.; Hei, X.; Li, F.; Li, X.; Shen, J.; Li, Y.; Huang, Q.; Xu, S. *Anal. Sci.* **2006**, *22*, 1367–1370.
276. Nath, N.; Chilkoti, A. *J. Fluoresc.* **2004**, *14*, 377–389.
277. Aslan, K.; Luhrs, C. C.; Perez-Luna, V. H. *J. Phys. Chem. B* **2004**, *108*, 15631–15639.
278. Kim, N. H.; Baek, T. J.; Park, H. G.; Seong, G. H. *Anal. Sci.* **2007**, *23*, 177–181.
279. Aslan, K.; Pérez-Luna, V. *Plasmonics* **2006**, *1*, 111–119.
280. Nath, N.; Chilkoti, A. *P. Soc. Photo-Opt. Ins.* **2002**, *4626*, 441–448.
281. Nath, N.; Chilkoti, A. *Anal. Chem.* **2004**, *76*, 5370–5378.
282. Nusz, G. J.; Marinakos, S. M.; Curry, A. C.; Dahlin, A.; Hook, F.; Wax, A.; Chilkoti, A. *Anal. Chem.* **2008**, *80*, 984–989.
283. Hernandez, F. J.; Dondapati, S. K.; Ozalp, V. C.; Pinto, A.; O'Sullivan, C. K.; Klar, T. A.; Katakis, I. *J. Biophotonics* **2009**, *2*, 227–231.
284. Sönnichsen, C.; Alivisatos, A. P. *Nano Lett.* **2005**, *5*, 301–304.
285. Orendorff, C. J.; Baxter, S. C.; Goldsmith, E. C.; Murphy, C. J. *Nanotechnology* **2005**, *16*, 2601–2605.
286. Selhuber-Unkel, C.; Zins, I.; Schubert, O.; Sönnichsen, C.; Oddershede, L. B. *Nano Lett.* **2008**, *8*, 2998–3003.
287. Wang, Z.; Lee, J.; Cossins, A. R.; Brust, M. *Anal. Chem.* **2005**, *77*, 5770–5774.
288. Wang, Z.; Levy, R.; Fernig, D. G.; Brust, M. *Bioconjugate Chem.* **2005**, *16*, 497–500.
289. Newman, G. R.; Jasani, B. *J. Pathol.* **1998**, *186*, 119–125.
290. Blab, G.; Cognet, L.; Berciaud, S.; Alexandre, I.; Husar, D.; Remacle, J.; Lounis, B. *Biophys. J.* **2006**, *90*, L13–L15.
291. Reichert, J.; Csaki, A.; Kohler, J. M.; Fritzsche, W. *Anal. Chem.* **2000**, *72*, 6025–6029.
292. Zhang, G.; Möller, R.; Kretschmer, R.; Csáki, A.; Fritzsche, W. *J. Fluoresc.* **2004**, *14*, 369–375.
293. Li, T.; Guo, L.; Wang, Z. *Biosens. Bioelectron.* **2008**, *23*, 1125–1130.
294. Moeremans, M.; Daneels, G.; Van Dijck, A.; Langanger, G.; De Mey, J. *J. Immunol. Methods* **1984**, *74*, 353–360.
295. Li, T.; Liu, D.; Wang, Z. *Biosens. Bioelectron.* **2009**, *24*, 3335–3339.
296. Sun, L.; Liu, D.; Wang, Z. *Anal. Chem.* **2007**, *79*, 773–777.
297. Qian, X.; Nie, S. M. *Chem. Soc. Rev.* **2008**, *37*, 912–920.
298. Vo-Dinh, T.; Yan, F.; Wabuyele, M. B. *J. Raman Spectrosc.* **2005**, *36*, 640–647.
299. Han, X.; Zhao, B.; Ozaki, Y. *Anal. Bioanal. Chem.* **2009**, *394*, 1719–1727.
300. Xu, S.; Ji, X.; Xu, W.; Zhao, B.; Dou, X.; Bai, Y.; Ozaki, Y. *J. Biomed. Opt.* **2005**, *10*, 031112–12.
301. Xu, S.; Ji, X.; Xu, W.; Li, X.; Wang, L.; Bai, Y.; Zhao, B.; Ozaki, Y. *Analyst* **2004**, *129*, 63–68.
302. Manimaran, M.; Jana, N. R. *J. Raman Spectrosc.* **2007**, *38*, 1326–1331.

303. Song, C.; Wang, Z.; Zhang, R.; Yang, J.; Tan, X.; Cui, Y. *Biosens. Bioelectron.* **2009**, *25*, 826–831.
304. Chen, J.; Lei, Y.; Liu, X.; Jiang, J.; Shen, G.; Yu, R. *Anal. Bioanal. Chem.* **2008**, *392*, 187–193.
305. Li, T.; Guo, L.; Wang, Z. *Anal. Sci.* **2008**, *24*, 907–910.
306. Zhong, W. *Anal. Bioanal. Chem.* **2009**, *394*, 47–59.
307. Matveeva, E. G.; Shtoyko, T.; Gryczynski, I.; Akopova, I.; Gryczynski, Z. *Chem. Phys. Lett.* **2008**, *454*, 85–90.
308. Demers, L. M.; Mirkin, C. A.; Mucic, R. C.; Reynolds, R. A.; Letsinger, R. L.; Elghanian, R.; Viswanadham, G. *Anal. Chem.* **2000**, *72*, 5535–5541.
309. Oh, E.; Lee, D.; Kim, Y.; Cha, S. Y.; Oh, D.; Kang, H. A.; Kim, J.; Kim, H. *Angew. Chem. Int. Ed.* **2006**, *45*, 7959–7963.
310. Lee, H.; Lee, K.; Kim, I. K.; Park, T. G. *Biomaterials* **2008**, *29*, 4709–4718.
311. Lee, H.; Lee, K.; Kim, I.; Park, T. G. *Adv. Funct. Mater.* **2009**, *19*, 1884–1890.
312. McCord, J. M. *Science* **1974**, *185*, 529–531.
313. Babior, B. M. *Blood* **1999**, *93*, 1464–1476.
314. Schumacker, P. T. *Cancer Cell* **2006**, *10*, 175–176.
315. Selvaraj, V.; Alagar, M. *Int. J. Pharm.* **2007**, *337*, 275–281.
316. Katz, E.; Willner, I.; Wang, J. *Electroanal.* **2004**, *16*, 19–44.
317. Pumera, M.; Sánchez, S.; Ichinose, I.; Tang, J. *Sensors Actuat. B-Chem.* **2007**, *123*, 1195–1205.
318. Yáñez-Sedeño, P.; Pingarrón, J. M, *Anal. Bioanal. Chem.* **2005**, *382*, 884–886.
319. Shlyahovsky, B.; Katz, E.; Xiao, Y.; Pavlov, V.; Willner, I. *Small* **2005**, *1*, 213–216.
320. Tang, L.; Zeng, G.; Shen, G.; Zhang, Y.; Li, Y.; Fan, C.; Liu, C.; Niu, C. *Anal. Bioanal. Chem.* **2009**, *393*, 1677–1684.
321. Du, D.; Chen, W.; Cai, J.; Zhang, J.; Tu, H.; Zhang, A. *J. Nanosci. Nanotechno.* **2009**, *9*, 2368–2373.
322. Jena, B. K.; Raj, C. R. *Anal. Chem.* **2006**, *78*, 6332–6339.
323. Xiao, Y.; Patolsky, F.; Katz, E.; Hainfeld, J. F.; Willner, I. *Science* **2003**, *299*, 1877–1881.
324. Zhao, J.; O'Daly, J. P.; Henkens, R. W.; Stonehuerner, J.; Crumbliss, A. L. *Biosens. Bioelectron.* **1996**, *11*, 493–502.
325. Wang, L.; Wang, E. *Electrochem. Commun.* **2004**, *6*, 225–229.
326. Miscoria, S. A.; Barrera, G.; Rivas, G. *Electroanal.* **2005**, *17*, 1578–1582.
327. Yabuki, S.; Mizutani, F. *Electroanal.* **1997**, *9*, 23–25.
328. Tang, M.; Chen, S.; Yuan, R.; Chai, Y.; Gao, F.; Xie, Y. *Anal. Sci.* **2008**, *24*, 487–491.
329. Willner, B.; Katz, E.; Willner, I. *Curr. Opin. Biotechnol.* **2006**, *17*, 589–596.
330. Pinijsuwan, S.; Rijiravanich, P.; Somasundrum, M.; Surareungchai, W. *Anal. Chem.* **2008**, *80*, 6779–6784.
331. Lin, J.; Qu, W.; Zhang, S. *Anal. Sci.* **2007**, *23*, 1059–1063.
332. Gu, H.; Sa, R.; Yuan, S.; Chen, H.; Yu, A. *Chem. Lett.* **2003**, *32*, 934–935.
333. Park, S.; Taton, T. A.; Mirkin, C. A. *Science* **2002**, *295*, 1503–1506.
334. Chen, P. C.; Mwakwari, S. C.; Oyelere, A. K. *Nanotechnol., Sci. Appl.* **2008**, *1*, 45–66.
335. Sahoo, S.; Parveen, S.; Panda, J. *Nanomed.-Nanotechnol.* **2007**, *3*, 20–31.
336. Prato, M.; Kostarelos, K.; Bianco, A. *Acc. Chem. Res.* **2008**, *41*, 60–68.
337. Champion, J. A.; Mitragotri, S. *Proc. Natl. Acad. Sci. U. S. A.* **2006**, *103*, 4930–4934.
338. Chithrani, B. D.; Ghazani, A. A.; Chan, W. C. W. *Nano Lett.* **2006**, *6*, 662–668.
339. Chithrani, B. D.; Ghazani, A. A.; Chan, W. C. W. *Nano Lett.* **2007**, *7*, 1542–1550.
340. Zhang, K.; Fang, H.; Chen, Z.; Taylor, J. A.; Wooley, K. L. *Bioconjugate Chem.* **2008**, *19*, 1880–1887.
341. Zhang, K.; Rossin, R.; Hagooly, A.; Chen, Z.; Welch, M. J.; Wooley, K. L. *J. Polym. Sci. Pol. Chem.* **2008**, *46*, 7578–7583.
342. Woodle, M. C.; Engbers, C. M.; Zalipsky, S. *Bioconjugate Chem.* **1994**, *5*, 493–496.
343. Rogers, W. J.; Basu, P. *Atherosclerosis* **2005**, *178*, 67–73.
344. Paciotti, G. F.; Kingston, D. G.; Tamarkin, L. *Drug Develop. Res.* **2006**, *67*, 47–54.

345. Niidome, T.; Yamagata, M.; Okamoto, Y.; Akiyama, Y.; Takahashi, H.; Kawano, T.; Kata-yama, Y.; Niidome, Y. *J. Controlled Release* **2006**, *114*, 343–347.

346. Gref, R.; Lück, M.; Quellec, P.; Marchand, M.; Dellacherie, E.; Harnisch, S.; Blunk, T.; Müller, R. H. *Colloid. Surface. B* **2000**, *18*, 301–313.

347. Gu, Y.; Cheng, J.; Lin, C.; Lam, Y. W.; Cheng, S. H.; Wong, W. *Toxicol. Appl. Pharm.* **2009**, *237*, 196–204.

348. Maysinger, D. *Org. Biomol. Chem.* **2007**, *5*, 2335–2342.

349. Ghosh, P.; Han, G.; De, M.; Kim, C. K.; Rotello, V. M. *Adv.Drug Deliver. Rev.* **2008**, *60*, 1307–1315.

350. Kim, C.; Ghosh, P.; Rotello, V. M. *Nanoscale* **2009**, *1*, 61–67.

351. De la Fuente, J. M.; Berry, C. C. *Bioconjugate Chem.* **2005**, *16*, 1176–1180.

352. Visaria, R. K.; Griffin, R. J.; Williams, B. W.; Ebbini, E. S.; Paciotti, G. F.; Song, C. W.; Bischof, J. C. *Mol. Cancer Ther.* **2006**, *5*, 1014–1020.

353. Oishi, M.; Nakaogami, J.; Ishii, T.; Nagasaki, Y. *Chem. Lett.* **2006**, *35*, 1046–1047.

354. Sahoo, S. K.; Labhasetwar, V. *Drug Discov. Today* **2003**, *8*, 1112–1120.

355. Maeda, H.; Wu, J.; Sawa, T.; Matsumura, Y.; Hori, K. *J. Controlled Release* **2000**, *65*, 271–284.

356. Maeda, H.; Fang, J.; Inutsuka, T.; Kitamoto, Y. *Int. Immunopharmacol.* **2003**, *3*, 319–328.

357. Greish, K. *J. Drug Target.* **2007**, *15*, 457–464.

358. Fang, J.; Sawa, T.; Maeda, H. *Adv. Exp. Med. Biol.* **2003**, *519*, 29–49.

359. Yang, P.; Sun, X.; Chiu, J.; Sun, H.; He, Q. *Bioconjugate Chem.* **2005**, *16*, 494–496.

360. North, R. J.; Havell, E. A. *J. Exp. Med.* **1988**, *167*, 1086–1099.

361. Brouckaert, P. G.; Leroux-Roels, G. G.; Guisez, Y.; Tavernier, J.; Fiers, W. *Int. J. Cancer* **1986**, *38*, 763–769.

362. Hieber, U.; Heim, M. E. *Oncology* **1994**, *51*, 142–153.

363. Cohenuram, M.; Saif, M. W. *JOP* **2007**, *8*, 4–15.

364. Rocha-Lima, C. M.; Soares, H. P.; Raez, L. E.; Singal, R. *Cancer Control* **2007**, *14*, 295–304.

365. Mendelsohn, J.; Baselga, J. *Semin. Oncol.* **2006**, *33*, 369–385.

366. Prudkin, L.; Wistuba, I. I. *Ann. Diagn. Pathol.* **2006**, *10*, 306–315.

367. Ahmed, S. M.; Salgia, R. *Respirology* **2006**, *11*, 687–692.

368. Paez, J. G.; Janne, P. A.; Lee, J. C.; Tracy, S.; Greulich, H.; Gabriel, S.; Herman, P.; Kaye, F. J.; Lindeman, N.; Boggon, T. J.; Naoki, K.; Sasaki, H.; Fujii, Y.; Eck, M. J.; Sellers, W. R.; Johnson, B. E.; Meyerson, M. *Science* **2004**, *304*, 1497–1500.

369. Arteaga, C. L. *J. Clin. Oncol.* **2001**, *19*, 32S–40S.

370. El-Sayed, I. H.; Huang, X.; El-Sayed, M. A. *Cancer Lett.* **2006**, *239*, 129–135.

371. Patra, C. R.; Bhattacharya, R.; Wang, E.; Katarya, A.; Lau, J. S.; Dutta, S.; Muders, M.; Wang, S.; Buhrow, S. A.; Safgren, S. L.; Yaszemski, M. J.; Reid, J. M.; Ames, M. M.; Mukherjee, P.; Mukhopadhyay, D. *Cancer Res.* **2008**, *68*, 1970–1978.

372. Kalderon, D.; Richardson, W. D.; Markham, A. F.; Smith, A. E. *Nature* **1984**, *311*, 33–38.

373. Liu, Y.; Franzen, S. *Bioconjugate Chem.* **2008**, *19*, 1009–1016.

374. Tkachenko, A. G.; Xie, H.; Liu, Y.; Coleman, D.; Ryan, J.; Glomm, W. R.; Shipton, M. K.; Franzen, S.; Feldheim, D. L. *Bioconjugate Chem.* **2004**, *15*, 482–490.

375. Joshi, H. M.; Bhumkar, D. R.; Joshi, K.; Pokharkar, V.; Sastry, M. *Langmuir* **2006**, *22*, 300–305.

376. Sudimack, J.; Lee, R. J. *Adv. Drug Deliver. Rev.* **2000**, *41*, 147–162.

377. Lu, Y.; Low, P. S. *Adv. Drug Deliver. Rev.* **2002**, *54*, 675–693.

378. Bhattacharya, R.; Patra, C. R.; Earl, A.; Wang, S.; Katarya, A.; Lu, L.; Kizhakkedathu, J. N.; Yaszemski, M. J.; Greipp, P. R.; Mukhopadhyay, D.; Mukherjee, P. *Nanomed.-Nano-technol.* **2007**, *3*, 224–238.

379. Roy, E. J.; Gawlick, U.; Orr, B. A.; Kranz, D. M. *Adv. Drug Deliver. Rev.* **2004**, *56*, 1219–1231.

380. Dixit, V.; Van Den Bossche, J.; Sherman, D. M.; Thompson, D. H.; Andres, R. P. *Bioconju-gate Chem.* **2006**, *17*, 603–609.

381. Shi, X.; Wang, S.; Meshinchi, S.; Antwerp, M. V.; Bi, X.; Lee, I.; Jr, J. B. *Small* **2007**, *3*, 1245–1252.
382. Vijayaraghavalu, S.; Raghavan, D.; Labhasetwar, V. *Curr. Opin. Investig. D.* **2007**, *8*, 477–484.
383. Ganesh, T. *Bioorgan. Med. Chem.* **2007**, *15*, 3597–3623.
384. Chen, Y.; Tsai, C.; Huang, P.; Chang, M.; Cheng, P.; Chou, C.; Chen, D.; Wang, C.; Shiau, A.; Wu, C. *Mol. Pharm.* **2007**, *4*, 713–722.
385. Podsiadlo, P.; Sinani, V. A.; Bahng, J. H.; Kam, N. W. S.; Lee, J.; Kotov, N. A. *Langmuir* **2008**, *24*, 568–574.
386. Li, J.; Wang, X.; Wang, C.; Chen, B.; Dai, Y.; Zhang, R.; Song, M.; Lv, G.; Fu, D. *ChemMedChem* **2007**, *2*, 374–378.
387. Shenoy, D.; Fu, W.; Li, J.; Crasto, C.; Jones, G.; DiMarzio, C.; Sridhar, S.; Amiji, M. *Int. J. Nanomedicine* **2006**, *1*, 51–57.
388. Jolesz, F. A.; Hynynen, K. *Cancer J.* **2002**, *8 Suppl 1*, S100–112.
389. Zharov, V. P.; Galitovskaya, E. N.; Johnson, C.; Kelly, T. *Laser. Surg. Med.* **2005**, *37*, 219–226.
390. Huang, W.-C.; Tsai, P.-J.; Chen, Y.-C. *Nanomedicine* **2007**, *2*, 777–787.
391. Zharov, V. P.; Mercer, K. E.; Galitovskaya, E. N.; Smeltzer, M. S. *Biophys. J.* **2006**, *90*, 619–627.
392. Norman, R. S.; Stone, J. W.; Gole, A.; Murphy, C. J.; Sabo-Attwood, T. L. *Nano Lett.* **2008**, *8*, 302–306.
393. Everts, M.; Saini, V.; Leddon, J. L.; Kok, R. J.; Stoff-Khalili, M.; Preuss, M. A.; Millican, C. L.; Perkins, G.; Brown, J. M.; Bagaria, H.; Nikles, D. E.; Johnson, D. T.; Zharov, V. P.; Curiel, D. T. *Nano Lett.* **2006**, *6*, 587–591.
394. Lowery, A. R.; Gobin, A. M.; Day, E. S.; Halas, N. J.; West, J. L. *Int. J. Nanomed.* **2006**, *1*, 149–154.
395. Bernardi, R.; Lowery, A.; Thompson, P.; Blaney, S.; West, J. *J. Neuro-Oncol.* **2008**, *86*, 165–172.
396. Gannon, C.; Patra, C.; Bhattacharya, R.; Mukherjee, P.; Curley, S. *J. Nanobiotechnol.* **2008**, *6*, 2.
397. Lapotko, D. O.; Lukianova, E.; Oraevsky, A. A. *Laser. Surg. Med.* **2006**, *38*, 631–642.
398. Lapotko, D. *Nanomedicine* **2009**, *4*, 253–256.
399. Chou, C.; Chen, C.; Wang, C. R. C. *J. Phys. Chem. B* **2005**, *109*, 11135–11138.
400. Huff, T. B.; Tong, L.; Zhao, Y.; Hansen, M. N.; Cheng, J.; Wei, A. *Nanomedicine* **2007**, *2*, 125–132.
401. Black, K. C.; Kirkpatrick, N. D.; Troutman, T. S.; Liping, X.; Vagner, J.; Gillies, R. J.; Barton, J. K.; Utzinger, U.; Romanowski, M. *Mol. Imaging* **2008**, *7*, 50–57.
402. Takahashi, H.; Niidome, T.; Nariai, A.; Niidome, Y.; Yamada, S. *Chem. Lett.* **2006**, *35*, 500–501.
403. Dickerson, E. B.; Dreaden, E. C.; Huang, X.; El-Sayed, I. H.; Chu, H.; Pushpanketh, S.; McDonald, J. F.; El-Sayed, M. A. *Cancer Lett.* **2008**, *269*, 57–66.
404. Skrabalak, S.; Chen, J.; Au, L.; Lu, X.; Li, X.; Xia, Y. *Adv. Mater.* **2007**, *19*, 3177–3184.
405. Skrabalak, S. E.; Au, L.; Lu, X.; Li, X.; Xia, Y. *Nanomedicine* **2007**, *2*, 657–668.
406. Pissuwan, D.; Valenzuela, S. M.; Miller, C. M.; Cortie, M. B. *Nano Lett.* **2007**, *7*, 3808–3812.
407. Roa, W.; Zhang, X.; Guo, L.; Shaw, A.; Hu, X.; Xiong, Y.; Gulavita, S.; Patel, S.; Sun, X.; Chen, J.; Moore, R.; Xing, J. Z. *Nanotechnology* **2009**, *20*, 375101.
408. Anal, A. K. *Recent Pat. Endocr., Metab. Immune Drug Discovery* **2007**, *1*, 83–90.
409. Sershen, S. R.; Westcott, S. L.; Halas, N. J.; West, J. L. *J. Biomed. Mater. Res.* **2000**, *51*, 293–298.
410. West, J. L.; Halas, N. J. *Curr. Opin. Biotechnol.* **2000**, *11*, 215–217.
411. Bikram, M.; Gobin, A. M.; Whitmire, R. E.; West, J. L. *J. Controlled Release* **2007**, *123*, 219–227.
412. Radt, B.; Smith, T.; Caruso, F. *Adv. Mater.* **2004**, *16*, 2184–2189.

413. Takahashi, H.; Niidome, Y.; Yamada, S. *Chem. Commun.* **2005**, 2247–2249.
414. Kitagawa, R.; Honda, K.; Kawazumi, H.; Niidome, Y.; Nakashima, N.; Yamada, S. *Jpn. J. Appl.Phys.* **2008**, *47*, 1374–1376.
415. Lin, F. Y. H.; Sabri, M.; Alirezaie, J.; Li, D.; Sherman, P. M. *Clin. Diagn. Lab. Immunol.* **2005**, *12*, 418–425.
416. Yakes, B. J.; Lipert, R. J.; Bannantine, J. P.; Porter, M. D. *Clin. Vaccine Immunol.* **2007**, *15*, 227–234.
417. Williams, D.; Ehrman, S.; Pulliam Holoman, T. *J. Nanobiotechnol.* **2006**, *4*, 3.
418. Huang, W.; Tsai, P.; Chen, Y. *Nanomedicine* **2007**, *2*, 777–787.
419. Gu, H.; Ho, P. L.; Tong, E.; Wang, L.; Xu, B. *Nano Lett.* **2003**, *3*, 1261–1263.
420. Dykman, L.; Sumaroka, M.; Staroverov, S.; Zaitseva, I.; Bogatyrev, V. *Biol. Bull.* **2004**, *31*, 75–79.
421. Nie, Z.; Liu, K. J.; Zhong, C.; Wang, L.; Yang, Y.; Tian, Q.; Liu, Y. *Free Radical Biol. Med.* **2007**, *43*, 1243–1254.
422. Faraday, M. *Phil. Trans. R. Soc. Lond.* **1857**, *147*, 145–181.
423. Murphy, C. J.; Sau, T. K.; Gole, A.; Orendorff, C. J. *MRS Bull.* **2005**, *30*, 349–355.
424. Murphy, C. J.; Sau, T. K.; Gole, A. M.; Orendorff, C. J.; Gao, J.; Gou, L.; Hunyadi, S. E.; Li, T. *J. Phys. Chem. B* **2005**, *109*, 13857–13870.
425. Ma, H.; Yin, B.; Wang, S.; Jiao, Y.; Pan, W.; Huang, S.; Chen, S.; Meng, F. *ChemPhysChem* **2004**, *5*, 68–75.
426. Chen, W.; Cai, W.; Zhang, L.; Wang, G.; Zhang, L. *J. Colloid Interface Sci.* **2001**, *238*, 291–295.
427. Mandal, M.; Ghosh, S. K.; Kundu, S.; Esumi, K.; Pal, T. *Langmuir* **2002**, *18*, 7792–7797.
428. Seino, S.; Kinoshita, T.; Nakagawa, T.; Kojima, T.; Taniguci, R.; Okuda, S.; Yamamoto, T. *J. Nanopart. Res.* **2008**, *10*, 1071–1076.
429. Gachard, E.; Remita, H.; Khatouri, J.; Keita, B.; Nadjo, L.; Belloni, J. *New J. Chem.* **1998**, *22*, 1257–1265.
430. Xia, Y.; Xiong, Y.; Lim, B.; Skrabalak, S. E. *Angew.Chem. Int. Ed.* **2008**, *48*, 60–103.
431. Gardea-Torresdey, J. L.; Parsons, J. G.; Gomez, E.; Peralta–Videa, J.; Troiani, H. E.; Santiago, P.; Yacaman, M. J. *Nano Lett.* **2002**, *2*, 397–401.
432. Sharma, N. C.; Sahi, S. V.; Nath, S.; Parsons, J. G.; Torresdey, J. L. G.; Pal, T. *Environ. Sci. Technol.* **2007**, *41*, 5137–5142.
433. Shankar, S. S.; Rai, A.; Ankamwar, B.; Singh, A.; Ahmad, A.; Sastry, M. *Nat. Mater.* **2004**, *3*, 482–488.
434. Chandran, S. P.; Chaudhary, M.; Pasricha, R.; Ahmad, A.; Sastry, M. *Biotechnol. Prog.* **2006**, *22*, 577–583.
435. Mukherjee, P.; Ahmad, A.; Mandal, D.; Senapati, S.; Sainkar, S. R.; Khan, M. I.; Ramani, R.; Parischa, R.; Ajayakumar, P. V.; Alam, M.; Sastry, M.; Kumar, R. *Angew. Chem. Int. Ed.* **2001**, *40*, 3585–3588.
436. Riddin, T. L.; Gericke, M.; Whiteley, C. G. *Nanotechnology* **2006**, *17*, 3482–3489.
437. Kumar, S. A.; Peter, Y.; Nadeau, J. L. *Nanotechnology* **2008**, *19*, 495101.
438. Klaus-Joerger, T.; Joerger, R.; Olsson, E.; Granqvist, C. *Trends Biotechnol.* **2001**, *19*, 15–20.
439. Govindaraju, K.; Basha, S.; Kumar, V.; Singaravelu, G. *J. Mater. Sci.* **2008**, *43*, 5115–5122.
440. Turkevich, J.; Stevenson, P.; Hillier, J. *Discuss. Faraday Soc.* **1951**, *11*, 55.
441. Frens, G. *Nature (London), Phys. Sci.* **1973**, *241*, 20–22.
442. Goia, D. V.; Matijevic, E. *Colloids Surf., A* **1999**, *146*, 139–152.
443. Sugunan, A.; Thanachayanont, C.; Dutta, J.; Hilborn, J. *Sci. Technol. Adv. Mater.* **2005**, *6*, 335–340.
444. Schellenberger, E. A.; Reynolds, F.; Weissleder, R.; Josephson, L. *ChemBioChem* **2004**, *5*, 275–279.
445. Morrow, B. J.; Matijevic, E.; Goia, D. V. *J. Colloid Interface Sci.* **2009**, *335*, 62–69.
446. Schmid, G.; West, H.; Malm, J.; Bovin, J.; Grenthe, C. *Chem. Eur. J.* **1996**, *2*, 1099–1103.
447. Wilcoxon, J. P.; Provencio, P. P. *J. Am. Chem. Soc.* **2004**, *126*, 6402–6408.
448. Brown, K. R.; Natan, M. J. *Langmuir* **1998**, *14*, 726–728.

449. Brown, K. R.; Walter, D. G.; Natan, M. J. *Chem. Mater.* **2000**, *12*, 306–313.
450. Sau, T. K.; Pal, A.; Jana, N.; Wang, Z.; Pal, T. *J. Nanopart. Res.* **2001**, *3*, 257–261.
451. Jana, N. R.; Gearheart, L.; Murphy, C. J. *Langmuir* **2001**, *17*, 6782–6786.
452. Schmid, G. *Chem. Rev.* **1992**, *92*, 1709–1727.
453. Niu, J.; Zhu, T.; Liu, Z. *Nanotechnology* **2007**, *18*, 325607.
454. Zhong, C.; Nioki, P. N.; Luo, J. *Controlled synthesis of highly monodispersed gold nanoparticles*, Patent Number: US 7524354, Issue Date: **2009**, 4, 28.
455. Hirai, H.; Aizawa, H. *J. Colloid Interface Sci.* **1993**, *161*, 471–474.
456. Stoeva, S.; Klabunde, K. J.; Sorensen, C. M.; Dragieva, I. *J. Am. Chem. Soc.* **2002**, *124*, 2305–2311.
457. Brust, M.; Walker, M.; Bethell, D.; Schiffrin, D. J.; Whyman, R. *J. Chem. Soc., Chem. Commun.* **1994**, 801–802.
458. Brust, M.; Fink, J.; Bethell, D.; Schiffrin, D. J.; Kiely, C. *J. Chem. Soc., Chem. Commun.* **1995**, 1655–1656.
459. Rothrock, A. R.; Donkers, R. L.; Schoenfisch, M. H. *J. Am. Chem. Soc.* **2005**, *127*, 9362–9363.
460. You, C.; Verma, A.; Rotello, V. M. *Soft Matter* **2006**, *2*, 190–204.
461. West, J. L.; Halas, N. J. *Annu. Rev. Biomed. Eng.* **2003**, *5*, 285–292.
462. Agrawal, A.; Huang, S.; Lin, A. W. H.; Lee, M.; Barton, J. K.; Drezek, R. A.; Pfefer, T. J. *J. Biomed. Opt.* **2006**, *11*, 041121.
463. Caruso, F. *Adv. Mater.* **2001**, *13*, 11–22.
464. Oldenburg, S. J.; Westcott, S. L.; Averitt, R. D.; Halas, N. J. *J. Chem.Phys.* **1999**, *111*, 4729–4735.
465. Graf, C.; van Blaaderen, A. *Langmuir* **2002**, *18*, 524–534.
466. Pham, T.; Jackson, J. B.; Halas, N. J.; Lee, T. R. *Langmuir* **2002**, *18*, 4915–4920.
467. Shi, W.; Sahoo, Y.; Swihart, M. T.; Prasad, P. N. *Langmuir* **2005**, *21*, 1610–1617.
468. Kah, J. C.; Phonthammachai, N.; Wan, R. C.; Song, J.; White, T.; Mhaisalkar, S.; Ahmad, I.; Sheppard, C.; Olivo, M. *Gold Bull.* **2008**, *41*, 23–36.
469. Laurent, S.; Forge, D.; Port, M.; Roch, A.; Robic, C.; Vander Elst, L.; Muller, R. N. *Chem. Rev.* **2008**, *108*, 2064–2110.
470. Lu, A.; Salabas, E.; Schüth, F. *Angew. Chem. Int. Ed.* **2007**, *46*, 1222–1244.
471. Mandal, M.; Kundu, S.; Ghosh, S. K.; Panigrahi, S.; Sau, T. K.; Yusuf, S.; Pal, T. *J. Colloid Interface Sci.* **2005**, *286*, 187–194.
472. Ji, X.; Shao, R.; Elliott, A. M.; Stafford, R. J.; Esparza-Coss, E.; Bankson, J. A.; Liang, G.; Luo, Z.; Park, K.; Markert, J. T.; Li, C. *J. Phys. Chem. C* **2007**, *111*, 6245–6251.
473. Nikoobakht, B.; El-Sayed, M. A. *Chem. Mater.* **2003**, *15*, 1957–1962.
474. Sau, T. K.; Murphy, C. J. *Langmuir* **2004**, *20*, 6414–6420.
475. Sau, T. K.; Murphy, C. J. *Philos. Mag.* **2007**, *87*, 2143–2158.
476. Lu, X.; Au, L.; McLellan, J.; Li, Z.; Marquez, M.; Xia, Y. *Nano Lett.* **2007**, *7*, 1764–1769.

Chapter 5
Biomedical Applications of Magnetic Particles

Evgeny Katz and Marcos Pita

Abstract Magnetic particles of nano- and micro-size functionalized with various biomolecules were synthesized and extensively used for many bioanalytical and biomedical applications. Different biosensors, including immunosensors and DNA sensors, were developed using functionalized magnetic particles for their operation *in vitro* and *in vivo*. Their use for magnetic targeting (drugs, genes, radiopharmaceuticals), magnetic resonance imaging, diagnostics, immunoassays, RNA and DNA purification, gene cloning, cell separation and purification has been developed. The present chapter summarizes the recent advances in the bioanalytical and biomedical applications of functional magnetic particles.

Keywords Magnetic particles • Nanoparticles • Magnetism • Biomedical applications • Nanotechnology • Nanobiotechnology

5.1 Introduction

Magnetic particles (microspheres, nanospheres and ferrofluids) are widely studied and applied in various fields of biology and medicine such as magnetic targeting (drugs, genes, radiopharmaceuticals), magnetic resonance imaging, diagnostics, immunoassays, RNA and DNA purification, gene cloning, cell separation and purification [1]. Also, magnetic nano-objects of complex topology, such as magnetic nanorods and nanotubes, were produced to serve as parts of various nano-devices, e.g. tunable fluidic channels for tiny magnetic particles, data storage devices in nanocircuits, and scanning tips for magnetic force microscopes [2].

Biomolecule-functionalized magnetic particles generally exist in a 'core-shell' configuration where biological species (cells, nucleic acids, proteins) are bound to the magnetic 'core' through organic linkers, often organized as a polymeric 'shell'

E. Katz (✉)
Department of Chemistry and Biomolecular Science, Clarkson University, Potsdam, NY, USA
e-mail: ekatz@clarkson.edu

M. Pita
Instituto de Catálisis y Petroleoquímica CSIC. C/Marie Curie 2, 28049 Madrid, Spain
e-mail: marcospita@icp.csic.es

E. Matijević (ed.), *Fine Particles in Medicine and Pharmacy,*
DOI 10.1007/978-1-4614-0379-1_5, © Springer Science+Business Media, LLC 2012

around the core [3, 4]. The efficacy of biomaterial binding to the primary organic shell surrounding the magnetic core was analyzed by various techniques (e.g. capillary electrophoresis with laser-induced fluorescence detection) [3]. Magnetic particles are extensively used as labeling units and immobilization platforms in various biosensing schemes [5], mainly for immunosensing and DNA analysis, as well as in environmental monitoring [6]. The results of these studies are outlined in the present chapter.

5.2 Synthesis of Biomolecule-Functionalized Magnetic Particles

The core of magnetic particles is usually composed of Fe_3O_4 or γ-Fe_2O_3 and its primary modification with an organic 'shell' can include the adsorption of an organic polymer [7] or the covalent attachment of a functionalized organosilane film [8]. Convenient syntheses of magnetic nanoparticles with controlled size, shape and magnetization were developed [9–14] and the synthesized magnetic particles were used for various biotechnological [12] and biomedical applications [15]. For example, size-controlled synthesis of magnetite (Fe_3O_4) nanoparticles was reported in organic solvents [16]. Atomic force microscopy and transmission electron microscopy were applied to characterize the size dispersion of biocompatible magnetic nanoparticles [17]. Particular attention was given to the synthesis of monodisperse and uniform particles [18]. Superparamagnetic iron oxide nanoparticles of controllable size (<20 nm) were prepared in the presence of reduced polysaccharides [19]. The polysaccharide shell increased the nanoparticles stability and provided functional groups for further functionalization by biomolecules. Structural and magnetic study of biocompatible superparamagnetic Fe_3O_4 nanoparticles was performed to optimize their use as labeling units in biomedical applications [20]. Highly crystalline monodisperse iron oxide (Fe_3O_4) nanoparticles with a continuous size spectrum of 6–13 nm were synthesized from the monodisperse Fe nanoparticles upon their controllable oxidation [21]. Iron oxide/polystyrene (core/shell) magnetic particles with the cross-linked shell were synthesized [22]. Cross-linking of polymeric chains in the organic shell could additionally stabilize the shell structure protecting the magnetic core from physical and chemical decomposition. Bimagnetic FePt-MFe$_2$O$_4$ (M = Fe, Co) core-shell nanoparticles were synthesized with the magnetic properties tunable by varying the chemical composition and thickness of the coating materials [23]. Also, various magnetic materials were examined as an alternative to iron oxide particles to be used for different biomedical and bioanalytical applications [24]. For example, ferromagnetic FeCo nanoparticles demonstrated superior properties that make them promising candidates for magnetically assisted bioseparation and analysis.

Reversible photo-switching of the magnetization of γ-Fe_2O_3 nanoparticles was achieved by the functionalization of the magnetic core with photoisomerizable units included in the organic shell [25]. The organic shell was generated by the assembly of the photoisomerizable amphiphilic 8-[4-{4-butoxy-phenyl(azo)-phenoxy}octan-1-ol] (**1**) and *n*-octylamine (**2**), which dilutes the photoisomerizable units and provides free space in the shell for their photo-isomerization. The azo-component of the shell upon ultraviolet (UV) irradiation ($\lambda = 360$ nm) was isomerized from the *trans*-state (1a) to the *cis*-state (1b) and returned back when illuminated with visible light ($\lambda = 400$–700 nm) (Fig. 5.1a). The reversible photoisomerization of the azo-component of the shell was followed by the absorbance at $\lambda = 360$ nm (Fig. 5.1b). Upon UV illumination, the intensity of the absorbance band at 360 nm in the spectrum of the functionalized magnetic particles decreased and the intensity of the band at 480 nm increased, indicating the *trans*-to-*cis* photoisomerization. The illumination with visible light resulted in the opposite change of the spectrum indicating the *cis*-to-*trans* photoisomerization. The change of the dipole moment of the shell upon the reversible photoisomerization between the *trans*-state (1a) and the *cis*-state (1b) resulted in the reversible change of the magnetization of the γ-Fe$_2$O$_3$ core (Fig. 5.1c). The relationship between the electronic polarization (including the charge or the dipole moments) and the magnetization is well established, however the exact mechanism of the phenomenon is still not fully described [26–28]. The magnetic nanoparticles with photochemically controlled magnetization represent novel potential labeling tools for biomaterials.

Coating of magnetic nanoparticles (e.g. Fe$_3$O$_4$) by thin films of noble metals (such as Au or Ag) results in the enhanced chemical stability of the magnetic core [29]. Gold-coated iron nanoparticles with a specific magnetic moment of 145 emu g^{-1} and a coercivity of 1664 Oe were prepared and tested for biomedical applications [30]. Also, Au-coated nanoparticles with magnetic Co cores were prepared for biomedical applications with the controlled size (5–25 nm; ± 1 nm) and tailored morphologies (spheres, discs with specific aspect ratio of 5×20 nm) [31]. The gold-shell allows further modification of the magnetic nanoparticles with biomolecules using the self-assembly method. Gold-coated composite nanoparticles of different diameters (50, 70 and 100 nm) were prepared by the reduction of AuCl$_4^-$ ions with hydroxylamine in the presence of Fe$_3$O$_4$ nanoparticles as seeds [32]. The Au shell was used to modify the magnetic nanoparticles by antibodies (rabbit anti-HIVp24 IgG or goat anti-human IgG) through a simple self-assembling procedure. The generated antibody-functionalized Au-coated magnetic nanoparticles were applied in immunoassay, providing easy separation/purification steps in a standard direct sandwich ELISA method. Also, Au-coated magnetic nanoparticles were used to construct thin films with the controlled structure and properties upon application of dithiols as molecular linkers between the Au shells [33].

Silica particles carrying encapsulated magnetic nanoparticles were used for proteins binding to the outer-sphere of the inorganic beads and for biocatalysis [34, 35]. Particles with a magnetic core and mesoporous silica shell were used as magnetically transported matrices with the pores filled by biomolecules (e.g. drugs) [36]. Also, iron oxide magnetic nanoparticles (γ-Fe$_2$O$_3$ 20 nm or Fe$_3$O$_4$ 6–7 nm) were coated by a SiO$_2$ shell (thickness of 2–5 nm) using wet chemical synthesis and used for immobilization of biomolecules [37, 38]. Silica nanotubes were synthesized in alumina template and the inner surfaces of the nanotubes were modified with Fe$_3$O$_4$ magnetic nanoparticles [39]. The resulting magnetic nanotubes were applied for magnetic-field-assisted bioseparation, biointeraction, and drug delivery. Iron oxide magnetic

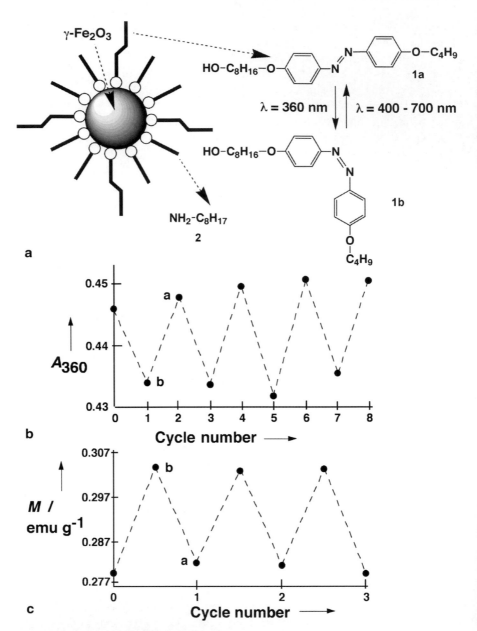

Fig. 5.1 a Magnetic nanoparticles functionalized with a mixed organic shell composed of a photoisomerizable azo-component (**1a/1b**) and a diluter *n*-octylamine (**2**). **b** The reversible absorbance changes (at λ = 360 nm) in the spectrum of the functionalized magnetic particles upon reversible photoisomerization of the azo-component: (*a*) the *trans*-states (**1a**) generated by visible illumination (λ = 400–700 nm); (*b*) the *cis*-states (**1b**) generated by UV illumination (λ = 360 nm). **c** The reversible change of the magnetization of the γ-Fe$_2$O$_3$ core upon illumination of the functionalized magnetic nanoparticles with visible light (*a*) and UV light (*b*). (Adapted from [25], Figs. 2b and 4b, with permission.)

nanoparticles (5–15 nm) were coated with inorganic fluorescent shell composed of ytterbium and erbium co-doped sodium yttrium fluoride ($NaYF_4/Y/Er$) that provides an efficient infrared-to-visible up-conversion [40]. The magnetic/fluorescent hybrid nanoparticles were coated with SiO_2 to immobilize biomolecules (e.g. streptavidin). The generated multi-functional particles provided magnetic separation, fluorescence detection and bioaffinity association with complementary biomolecules. In another approach, the magnetic Fe_3O_4 nanoparticles (8.5 nm) were coated with a controlled number of polyelectrolyte layers using layer-by-layer deposition of positively charged polyallylamine and negatively charged polystyrene sulfonate [41]. Negatively charged thioglycolic acid-capped CdTe nanoparticles were electrostatically bound to the positively charged polyallylamine exterior layer in the polyelectrolyte shell of the magnetic nanoparticles. The hybrid assemblies were prepared with the different numbers of the polyelectrolyte layers separating the magnetic core nanoparticle and the satellite fluorescent CdTe nanoparticles. Also, the method allowed the generation of several layers of the CdTe nanoparticles separated from the magnetic core and from each other by the polyelectrolyte layers. Also, Co-CdSe core-shell magnetic-fluorescent assemblies were generated by controlled deposition of the CdSe semiconductor layer on the pre-formed magnetic Co core in a non-aqueous solution using dimethylcadmium as an organic precursor [42]. The generated fluorescent-active/magnetic nanocomposite particles were suggested as versatile labels for biomolecules, which demonstrate advantages of the both fluorescent reporting part and magnetic separating/transporting part of the assembly. Further developments to improve the nano-hybrid assembly properties will be directed to the formation of a spacer layer between the magnetic core and the fluorescent shell to inhibit the quenching of the photoexcited semiconductor by the metal core.

Another approach to generate polyfunctional nano-assemblies includes formation of hetero-dimeric structures composed of two different nanoparticles (e.g. magnetic and metal or semiconductor) linked one to another as siamese twins (dumbbell-like bifunctional particles) [43–45]. Magnetic nanoparticles, Fe_3O_4 or FePt, (8 nm) coated with surfactant were dispersed in an organic solvent (e.g. dichlorobenzene) and added to an aqueous solution of Ag^+ salt [43]. Ultrasonication of the aqueous/organic system resulted in the formation of micelles with the magnetic nanoparticles self-assembled on the liquid/liquid interface (Fig. 5.2a). Defects in the surfactant shell allowed catalytic reduction of Ag^+ ions at Fe^{2+} sites on the surface of the magnetic nanoparticles faced to the aqueous phase to yield seeding for a Ag nanoparticle. Further reduction of Ag^+ ions on the Ag seed resulted in the growth of the Ag nanoparticle on a side of the magnetic nanoparticle as depicted in a TEM image (Fig. 5.2c). Similarly a Ag nanoparticle was grown on a side of a FePt magnetic nanoparticle (Fig. 5.2d). The size of the generated Ag nanoparticle was dependent on the allowed time-interval of the catalytic growing process. The generated hetero-dimers provided a platform for the directed functionalization of the magnetic and metallic parts of the assembly with different organic or bioorganic molecules using the difference of the surface properties of the two parts of the dimer. For example, the Ag nanoparticle in the dimeric hybrid was functionalized with thiolated porphyrin (3), while the Fe_3O_4 magnetic nanoparticle was modified by a biotin derivative (4) using dopamine units

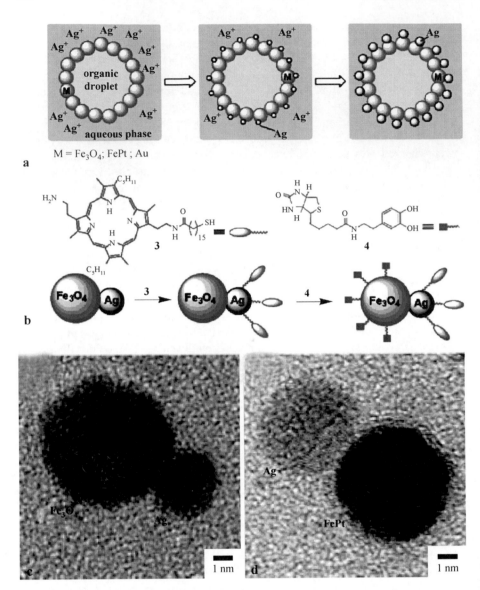

Fig. 5.2 a Assembly of Fe₃O₄-Ag hetero-dimeric nanoparticles in an aqueous/organic micell-system. **b** Directed functionalization of the Fe₃O₄ nanoparticle and Ag nanoparticle with different functional units, such as dopamine-derivatized biotin and thiol-derivatized porphyrin, respectively. **c, d** TEM images of Fe₃O₄-Ag and FePt-Ag hetero-dimers. (Adapted from [43], Scheme 1 and Figs. 5.1d, 5.1e, with permission. Copyright (2005) American Chemical Society)

as anchor groups (Fig. 5.2b). In another approach, FePt nanoparticles were coated with amorphous CdS layer to form a metastable core-shell structure, in which the CdS shell was transformed into a crystalline state upon heating [44]. Because of the incompatibility of the lattices of FePt and CdS the core-shell nanoparticles were transformed into heterodimers composed of interconnected CdS and FePt nanoparticles, which have sizes less than 10 nm and exhibit both superparamagnetism and fluorescence, providing excellent means for labeling of biomaterials. In a different approach, separately synthesized superparamagnetic γ-Fe_2O_3 nanoparticles (ca. 11.8 nm) and semiconductive CdSe nanoparticles (ca. 3.5 nm) were mixed and encapsulated together in a silica shell yielding a complex multifunctional assembly that preserved the unique magnetic property of γ-Fe_2O_3 and fluorescent property of CdSe quantum dots [46]. The silica shell served as panzer keeping together the functional nano-components and also provided chemically accessible sites for further functionalization of the hybrid system with organic molecules.

Various organic shells (e.g. aminosiloxane, dextran or dimercaptosuccinic acid) were generated around magnetic particles [47–49]. Organic functional groups introduced in the outer-layer of magnetic particles allowed coupling of biomolecules to the organic 'shell' [50, 51]. Several synthetic approaches have been applied to couple biomolecules to magnetic particles and to use the hybrid assemblies for various biochemical, bioanalytical and biomedical applications [52]. For example, dextran-coated magnetic particles were further covalently modified with polyclonal IgG anti-horseradish peroxidase antibody and used for the capturing of horseradish peroxidase from a crude protein extract from *Escherichia coli* [52]. Magnetite nanoparticles were silanized and further covalently modified with polyamidoamine dendrimer (PAMAM) [53]. The amino groups of PAMAM were used for the covalent binding of streptavidin. The streptavidin loading was found to be up to 3.4-fold higher comparing to the direct binding of streptavidin to the polysiloxane-modified magnetite nanoparticles. This was attributed to the increase of the organic shell diameter and the increase of the number of the amino groups, which are responsible for the covalent binding of streptavidin. Dopamine was suggested as a robust anchor group to link biomaterials to magnetic particles [54]. Dopamine ligands bind to iron oxide magnetic nanoparticles through coordination of the dihydroxylphenyl units with Fe^{+2} surface sites providing amino groups for further covalent attachment of various biomolecules. Preparation and characterization of magnetic nanoparticles coated by organic shells for covalent immobilization of proteins (e.g. bovine serum albumin) [8, 55, 56] or enzymes (e.g. HRP or lipase) [57–60] was described recently. The immobilization of enzymes on magnetic particles yields biocatalytically active particles. In one example, alcohol dehydrogenase was covalently bound to Fe_3O_4 magnetic particles, and the immobilized enzyme demonstrated high biocatalytic activity [61, 62]. Enzyme-modified magnetic nanoparticles could also demonstrate bioelectrocatalytic activities, while being in contact with an electrode surface [60]. Reversible association of proteins was achieved with negatively charged polyacrylic-shell/Fe_3O_4-core magnetic nanoparticles [63]. The binding of proteins was controlled by electrostatic interactions between the proteins and the negatively charged shells. The protein molecules, positively charged at low pH values (pH < pI, isoelectric point), were electrostatically attracted to the negatively charged shells, while at higher pH values (pH > pI) the negatively charged protein molecules were repelled from the functionalized magnetic particles. The reversible association/dissociation of the proteins to and from the magnetic nanoparticles, respectively, was used to collect, purify and transport proteins. Magnetic particles were functionalized with carbohydrate oligomers to yield multivalent binding of the magnetic labels to proteins or cells *via* specific carbohydrate-protein interactions, thus allowing their imaging [64]. DNA molecules were reacted with a mixture of Fe^{2+}/Fe^{3+} ions that were electrostatically associated with DNA chains [65]. These iron ions were then chemically reacted to give Fe_3O_4 magnetic particles associated with the DNA molecules. The labeled DNA could hybridize with complementary oligonucleotides, and the magnetic particles linked to the DNA molecules allowed the separation of the labeled DNA from non-labeled strands.

The cooperative assembly of magnetic nanoparticles in the presence of amino acid-based polymers was studied, and it was demonstrated that electrostatic interactions between block co-polypeptides and nanoparticles could control the organization of these components [66]. The addition of polyaspartic acid initiated the aggregation of maghemite nanoparticles into clusters, without the formation of a precipitate. The addition of the block co-polypeptide poly(EG_2-Lys)$_{100}$-*b*-poly(Asp)$_{30}$ induced a controlled organization of magnetic nanoparticles, possibly through the formation of micelles with cores consisting of the nanoparticles electrostatically bound to the polyaspartic acid end of the block co-polypeptide. In this case, the poly(EG_2-Lys) ends of the copolymers would form the micelle shell and, as such, both stabilize the clusters and control their size. By altering the composition of the block co-polypeptide, it should be possible to control both the size and stability of the resulting dispersed nanoparticle clusters, thereby greatly expanding the potential applications and usefulness of these cooperatively assembled nanocomposites.

5.3   Application of Magnetic Particles for Separation and Purification of Biomolecules and Cells

Most of the applications of biomaterial-magnetic particle hybrid conjugates involve the concentration, separation, regeneration, mechanical translocation and targeting of biomolecules, such as proteins, DNA/RNA, and of cells.

The following systems exemplify applications of the functionalized magnetic particles for separation, purification and transportation of bioanalytes to simplify analytical procedures and enhance the sensitivity and specificity of the analysis. Magnetic microparticles were coated with polymeric organic shells composed of poly(2-hydroxyethyl methacrylate) or poly(glycidyl methacrylate), and then different proteins, such as RNAse A, DNAse I, anti-*Salmonella* and proteinase K, were immobilized on the polymer-coated magnetic microparticles [67]. The protein-functionalized magnetic particles were applied as biocatalytic carries in the degradation of bacterial RNA, chromosomal and plasmid DNA, magnetic separation of *Salmonella* cells or degradation of their intracellular inhibitors. Phospholipid-coated magnetic nanoparticles with a mean magnetite core size of 8 nm were used for the recovery and separation of proteins from protein mixtures [4]. Fe_3O_4 particles coated with co-polymerized methacrylate-divinylbenzene were further derivatized with ethylenediamine to yield amino-functionalized magnetic micro-particles (1–5 µm) [68]. Protein A was covalently bound to the amino groups in the organic shells of the magnetic particles and then used for bioaffinity coupling with antibodies. The protein A-functionalized magnetic particles demonstrated the efficient immunoaffinity purification of monoclonal antibodies IgG2a (22 mg per gram of magnetic particles) from mouse ascites. In another approach, magnetic nanoparticles were functionalized with aptamer units specific to lysozyme allowing its separation, purification and further electrochemical analysis [69]. Streptavidin-functionalized magnetic beads were reacted with biotinylated anti-lysozyme aptamer, 5'-ATC TAC GAA TTC ATC AGG GCT AAA GAG TGC AGA GTT ACT TAG-3', and then they were reacted with the target lysozyme protein to yield an affinity complex. The complex of lysozyme protein with the aptamer-functionalized magnetic beads was magnetically separated from a mixture with other proteins and amino acids. Then purified lysozyme was released from the complex by alkaline treatment and analyzed electrochemically with a detection limit of 350 fmol (7 nM). Alternatively, streptavidin-functionalized magnetic beads were modified with a DNA primer (**5**), then reacted with a DNA analyte (**6**) and with a DNA-signaling probe carrying a dye (**7**) to yield a "sandwich" system (Fig. 5.3) [70]. The resulting hybrid system was magnetically separated and purified, and then the DNA assembly was dissociated to release the purified DNA analyte and DNA-signaling probe. These components were re-hybridized and reacted with a cationic conjugated polymer (**8**) to yield the electrostatically assembled DNA/polymer system. The generated system revealed the fluorescence resonance energy transfer (FRET) between the conjugated polymer and the dye molecules associated with the DNA-signaling probe. Thus, lighting-up the dye molecules in the DNA system was the read-out signal for the DNA analyte. Obviously, the use of the magnetic particles in this system is limited by the isolation and purification steps of the analytical procedure.

Multi-component magnetic nanorods were synthesized for magnetically controlled separation of His-tagged biomolecules [71]. Nanorods (ca. 330 nm diameter and ca. 10 μm length) of controlled composition with a central domain made of Ni and short end domains made of Au were prepared by electrochemical synthesis in micro-porous alumina template [72, 73]. Ni domain providing ferromagnetic properties of nanorods was used for specific binding of His-tagged biomolecules, whereas Au end domains prevented etching of the nanorods upon chemical dissolution of the template membrane. The Au parts of the multi-component nanorods were passivated with 11-mercaptoundecyl-tri(ethylene glycol) to prevent non-specific adsorption of proteins. A suspension of the nanorods was challenged with a mixture of His-tagged and untagged proteins labeled with different fluorescent dyes (His-tagged ubiquitin labeled with Alexa 568 red dye and anti-rabbit IgG without a His-tag labeled with Alexa 488 green dye) (Fig. 5.4a). The His-tagged ubiquitin was associated with the Ni domains of the nanorods and magnetically separated from the mixture, leaving the untagged anti-rabbit IgG in the solution. Then the His-tagged ubiquitin was released from the Ni surface by washing with eluent buffer pH 2.8, regenerating the bare surface of the Ni domains. The method was extended to the separation of His-tag-specific antibodies (Fig. 5.4b). The Ni domains of the nanorods were coated with polyhistidine (His$_6$) and then reacted with a mixture of antibodies (Alexa 568-labeled anti-poly-His IgG and Alexa 488-labeled anti-human IgG). The His-specific antibody was associated with the His$_6$-coated Ni domains, magnetically separated from the mixture of antibodies, and then it was washed off

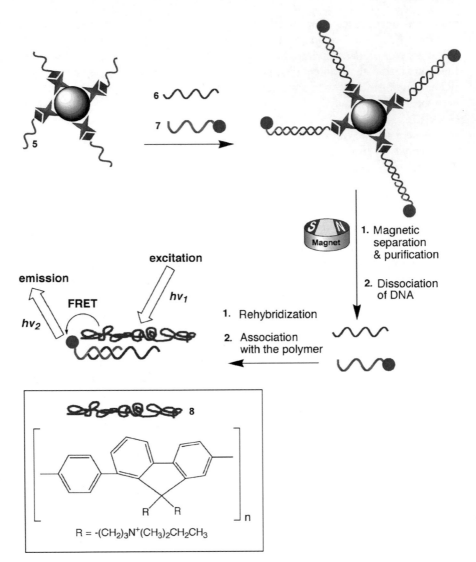

Fig. 5.3 DNA analysis based on the FRET in the polymer/DNA assembly with the isolation and purification steps assisted by magnetic particles

from the Ni surface with the eluent buffer pH 2.8 (this buffer is known to release His$_6$ from the Ni surface and to decrease the affinity interaction between the His$_6$ and the antibody). The method represents a general route for the magnetic separation of His-tagged or His-specific proteins from their mixtures with other biomolecules providing an alternative to chromatographic columns. Au parts of the nanorods, which are presently used as protecting ends of the nanorods, may be further used as supports for various functional components. Multifunctional magnetic

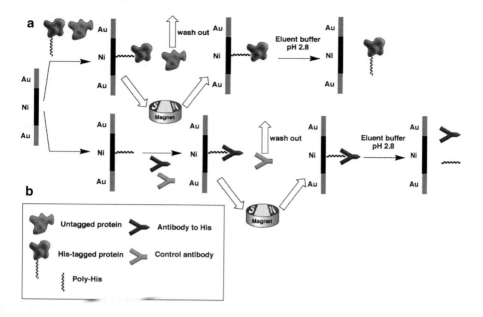

Fig. 5.4 a Separation of His-tagged proteins from untagged proteins using magnetic Ni-nanorods. **b** Separation of antibodies specific to polyhistidine (anti-poly-His) from other antibodies using magnetic Ni-nanorods. (Adapted from [71], Scheme 1, with permission).

nanorods composed of Au and Ni domains and carrying various biomolecules on the Au and Ni parts could also be used for the trapping and precise positioning in the external magnetic field [74], thus opening a new way to the design of nanocircuitry and construction of nanobiodevices.

A magnetic nanocapturer has been developed for the separation and collection of trace amounts of DNA/mRNA from mixtures and cancer cells. The nanocapturer was fabricated using a magnetic nanoparticle as a magnetic carrier, which was functionalized with a molecular beacon as a DNA probe for gene recognition and collection. Isolation, purification and further PCR amplified analysis of DNA from pathogenic bacteria *Listeria monocytogenes* were performed with the use of magnetic nanoparticles as transporting units [75].

Antibodies were adsorbed on synthetic Fe_3O_4 magnetic particles and were then used for specific binding to cells and further separation of the cells by an external magnetic field [76, 77]. Superparamagnetic polystyrene microscopic beads impregnated with magnetic particles were covalently modified with polyclonal anti-*E. coli* O157 antibodies and used for affinity capturing of *E. coli* O157:H7 bacteria [78]. The bacterial cells captured by the magnetic beads were analyzed using immunosensing with fluorescent quantum dots. The application of the magnetic beads as a platform for the immunosensing of bacteria provided means for their separation, purification and further analysis. For example, IgG-functionalized magnetic nanoparticles were employed as effective affinity probes for selective concentration of target bacteria from sample solutions followed by their analysis [79]. Magnetic nanoparticles (FePt, ca. 4 nm diameter) functionalized with vancomycin (Van) were used to selectively capture Gram-positive bacteria cells through molecular recognition between Van and the terminal peptide, D-Ala-D-Ala, on the surface of the cells. Unexpectedly, the Van-modified magnetic particles were also found to associate with Gram-negative bacteria such as *E. coli* [80, 81]. The captured *E. coli* cells were then purified and transported with the use of an external magnet. Bacterial magnetic particles containing protein A in the biological shell were genetically engineered and anti-mouse IgG antibodies were bound through their F fragments to the protein A units on the surface of the magnetic particles [82]. The antibody-functionalized magnetic particles were used for specific binding to mononuclear cells from peripheral blood and their separation.

5.4   Application of Magnetic Nanoparticles as Magnetic Labels for Biomolecules and Cells

Intrinsic magnetic properties of nanoparticles allowed their applications as labels for biomolecules. Several approaches were developed in biosensing with the use of magnetic nanoparticles based on their magnetic properties: (i) Application of magnetic nanoparticles as labeling units and their detection with the use of highly sensitive magnetometers. (ii) Analysis of the relaxation properties of magnetic nanoparticles in alternative magnetic fields upon binding of biomolecules. (ii) The nuclear magnetic resonance (NMR) analysis of an aqueous environment affected by magnetic nanoparticles associated or dissociated in course of biorecognition or biocatalytic processes. The later approach became a background for the NMR imaging that found extensive use in medicine and *in vivo* bioanalysis.

Rapid developments in magnetoelectronics [83] brought into reality highly sensitive magnetic field sensors. These sensors, primarily developed for electronics (e.g. as information storage devices in computer technology), recently became used for detection of magnetic nanoparticles linked as labeling units to biomolecules, offering various immunosensors and DNA sensors [84–87]. Superconducting quantum interference devices (SQUIDs), the world's most sensitive magnetic flux detectors, were applied as analytical tools in the sensing of biomolecules labeled with magnetic nanoparticles [88], e.g. for DNA sensing [89] and immunosensing [90]. However, as it requires expensive cryogenic devices, practical application of biosensors based on SQUIDs is very limited. Relative magnetic permeability measurements of biomolecule-assemblies labeled with magnetic nanoparticles performed with the use of relatively simple induction coil-based magnetometers were also applied for biosensing. For example, superparamagnetic nanoparticles functionalized with histone H1 were specifically attached to a DNA analyte (calf thymus DNA or plasmid DNA) and used as magnetic tracers for quantitative DNA analysis based on measurements of relative magnetic permeability [91]. However, the simplicity of the used instrumentation does not provide sufficiently high sensitivity of the analysis.

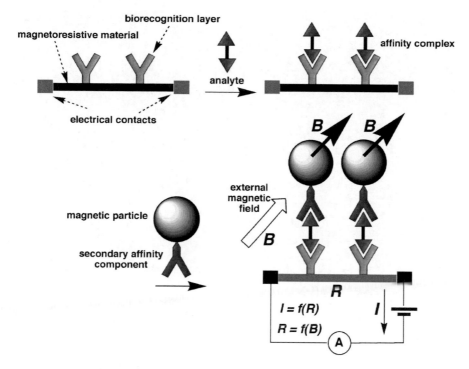

Fig. 5.5 Biosensors based on magnetoresitive materials that provide magneto-electronic transduction of biorecognition events of biomolecules labeled by magnetic nanoparticles

Practically useful, highly sensitive magnetometers based on magnetoresistive materials were recently developed and used in biosensing devices. The effect of the resistance change in the presence of a magnetic field (magnetoresistive effect) discovered by William Thomson [92] in 1856 became practically useful only at the end of twentieth century with the advances in solid state technology allowing fabrication of thin films of ferromagnetic materials (e.g. $Ni_{80}Fe_{20}$) showing strong magnetoresistive effect. Recently discovered effect of the antiferromagnetic interlayer exchange coupling, the giant magnetoresistive (GMR) effect and the anisotropic magnetoresistive effect (AMR) are used in various schemes of the magnetometers based on magnetoresistive materials (e.g. planar Hall sensors, AMR ring sensors [93], GMR sensors [94–96], spin valves [97–100]).

DNA sensors and immunosensors based on these magnetometers consist of a magnetoresistive thin film support functionalized with target biomolecules (oligonucleotides, antigens or antibodies) (Fig. 5.5) [84]. Coupling of analyte complementary biomolecules (complementary DNA, antibodies or antigens), followed by further binding of secondary complementary biomolecules (oligonucleotide, anti-antibody) labeled with magnetic nanoparticles results in the change of the electrical resistance of the magnetoresistive support. Current passing through the magnetoresistive support is alternated in the presence of the magnetically labeled biomole-

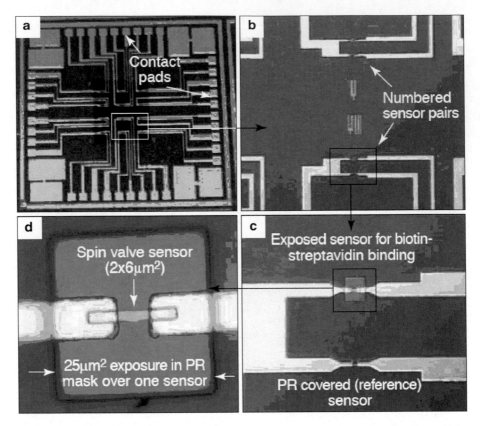

Fig. 5.6 A magnetoresistive-based biochip designed with 12 pairs of spin valve sensors, used for single or differential signal measurements. **a** The 8×8 mm^2 chip has the sensor pairs fabricated in the central area, covered with photoresist mask, the sensor connections running to contact pads arranged around the outer edges of the chip. **b** Each sensor pair has one input line and two separate output lines. **c** One sensor of each pair is exposed to on-chip biochemistry (active sensor) via an exposure in the photoresist. **d** A single 2×6 mm^2 spin valve with contacts within a 25×25 mm^2 exposure in the photoresist mask. (Adapted from [84], Fig. 5.5, with permission.)

cules, thus allowing the electronic transduction of the biorecognition event. Usually, the biomolecules are labeled with non-remanent paramagnetic nanoparticles (iron oxide) having small random magnetic moments. Thus, an external magnetic field is applied to align the magnetic moments of the nanoparticles, yielding the change of the resistance in the magnetoresistive support. The biosensors were integrated into complex magnetosensitive microchips for the multi-target high throughput analysis (Fig. 5.6) [95, 96, 101–103]. The sensitivity provided by the magnetometers (e.g. using a miniaturized silicon Hall sensor [104]) allows detection of a single magnetic nanoparticle, thus allowing, in principle, the detection of a single biomolecular recognition event. It should be noted that the magnetically assisted transport of DNA-functionalized magnetic particles to the sensing interface also results in the acceleration of the DNA analysis [105].

Binding of biomolecules to magnetic nanoparticles results in the increase of their hydrodynamic radius, thus resulting in the change of their Brownian relaxation time upon excitation in an alternative magnetic field. The impedance analysis of the complex magnetic susceptibility of a fluid containing magnetic nanoparticles provides the means to follow the binding of biomolecules to the magnetic nanoparticles. Magnetic nanoparticles (mean diameter 130 nm) were functionalized with a monoclonal antibody to prostate specific antigen (PSA) [106]. Reacting the antibody-modified magnetic nanoparticles with PSA resulted in the formation of the affinity complex consisting of the magnetic nanoparticles, anti-PSA antibody and PSA. The analysis of the complex magnetic susceptibility showed the change of the Brownian relaxation time after the modification of the nanoparticles with the anti-PSA antibody and further change after the affinity binding of PSA reflecting the respective increase of the hydrodynamic radius of the species. The method proved its applicability to the design of various sensors for biorecognition events (e.g. immunosensors) [107, 108]. Similarly, superparamagnetic γ-Fe_2O_3 nanoparticles (50 nm) functionalized with antibodies specific to *Listeria monocytogenes* bacteria were used for the analysis of the bacterial cells [109]. Upon application of an external magnetic field the magnetic dipoles of the nanoparticles were aligned with the magnetic field direction. Afterwards the external magnetic field was switched off and the relaxation time of the magnetic dipoles of the nanoparticles was measured. While the unbound magnetic nanoparticles showed fast Brownian relaxation, the magnetic nanoparticles bound to the bacterial cells were unable to move and the relaxation time was much longer, providing means for the detection of the magnetically labeled bacteria. Further developments of the method were directed to the optical analysis of the rotational mobility of magnetic nanoparticles functionalized with biomolecules and fluorescent dyes [110]. Optical analysis of the rotational relaxation of magnetic particles was successfully applied to develop immunosensors [111]. The binding of polyclonal antibodies to the antigen-functionalized magnetic particles resulted in their cross-linking, thus decreasing the rotation mobility of the particles. This process was followed by optical means allowing immunosensing based on the analysis of magneto-optical relaxation curves.

Magnetic nanoparticles were extensively used to follow biorecognition processes and enzymatic biocatalytic reactions using magnetic resonance techniques, such as nuclear magnetic resonance (NMR) and magnetic resonance imaging (MRI) [112]. The cores of iron oxide nanoparticles $(Fe_2O_3)_n/(Fe_3O_4)_m$ are superparamagnetic and they are magnetized in an external magnetic field generating large magnetic dipoles. The formation of these dipoles results in local magnetic field gradients and yields non-homogeneity in the external magnetic field. Precession of water protons in the environment with the non-homogeneous magnetic field proceeds at an off-resonance frequency, dephasing their spins and increasing the relaxation time (1/T2). The structure and physical properties of the monodisperse magnetic nanoparticles causing this effect were well characterized [113]. The NMR-effect generated by the magnetic nanoparticles was extensively experimentally studied and theoretically modeled [114, 115]. The effect of the magnetized nanoparticles on the relaxation time of water protons is enhanced when the magnetic nanoparticles are aggregated and diminished upon their dissociation [112]. The association and dissociation of biocompatible magnetic nanoparticles was read out by magnetic resonance techniques using various concentrations of the nanoparticles at different temperatures [116]. The magnetic nanoparticles were attached to various biomaterials, such as oligonucleotides, proteins (antigens/antibodies) and viruses, and the association of the magnetic nanoparticles, resulted from the biorecognition events, was followed by magnetic resonance techniques. Enzymatic reactions resulting in the biocatalyzed disaggregation of the biomolecular assemblies labeled with the magnetic nanoparticles were also followed by magnetic resonance techniques [112].

DNA analysis was performed using NMR transduction of the hybridization process using oligonucleotides labeled with magnetic nanoparticles [112, 117, 118]. Superparamagnetic $(Fe_2O_3)_n/(Fe_3O_4)_m$ nanoparticles (3 nm) coated with a cross-linked dextran shell functionalized with amino groups were used as labels [119]. The magnetic nanoparticles were activated by N-succinimidyl 3-(2-pyridyldithio) propionate (**9**) as a linker providing further binding of thiolated oligonucleotides with the formation of disulfide bonds (Fig. 5.7). Two kinds of oligonucleotide-functionalized magnetic nanoparticles were generated by covalent binding of two different thiolated oligonucleotides HS-$(CH_2)_3$-CGC-ATT-CAG-GAT (**11**) and TCT-CAA-CTC-GTA-$(CH_2)_3$-SH (**10**), which are complementary to different domains of the analyte oligonucleotide 3'-GCG-TAA-GTC-CTA-AGA-GTT-GAG-CAT-5' (**12**). Mixing the **10**-functionalized and **11**-functionalized magnetic nanoparticles, carrying in average three oligonucleotide chains per nanoparticle, with the analyte oligonucleotide (**12**) resulted in the hybridization of the oligonucleotides linked to the magnetic nanoparticles with the complementary domains of the analyte oligonucleotide, yielding clusters of 140 ± 16 nm each containing four to five magnetic nanoparticles. The aggregation of the magnetic nanoparticles bridged by the hybridized oligonucleotides resulted in significant changes of the spin-spin relaxation time (T2) of protons in the neighboring water molecules (Fig. 5.8a, curve *a*). The observed changes of T2 were dependent on the concentration of the analyte oligonucleotide, thus allowing its quantitative analysis (Fig. 5.8a, inset). No changes of T2 were observed when a non-complementary oligonucleotide, 5'-ATG-CTA-AAT-GAC-GAC-TGC-CCA-CAT-3', was added in a control experiment (Fig. 5.8a, curve *b*). To prove that the observed changes of T2 originate from the hybridization process, the reversible melting of the double-stranded oligonucleotide upon elevation of temperature and re-hybridization upon the temperature decrease were performed, and the respective reversible changes of T2 were followed (Fig. 5.8b). In an additional control experiment, the addition of dithiothreitol (DTT, **13**) to the system resulted in the splitting of the disulfide bonds between the oligonucleotides and the magnetic nanoparticles, resulting in the irreversible disaggregation of the magnetic nanoparticles that cancelled the T2 changes during the temperature cycling (Fig. 5.8b). The selectivity of the DNA analysis was studied using the DNA analyte targeting a GFP gene sequence and its single-base mutants containing T, C or G bases instead of A base [118]. The magnetic nanoparticles functionalized with oligonucleotides complementary to the half-domains of the DNA analyte were added to the system. The observed changes of T2 in the presence of the target

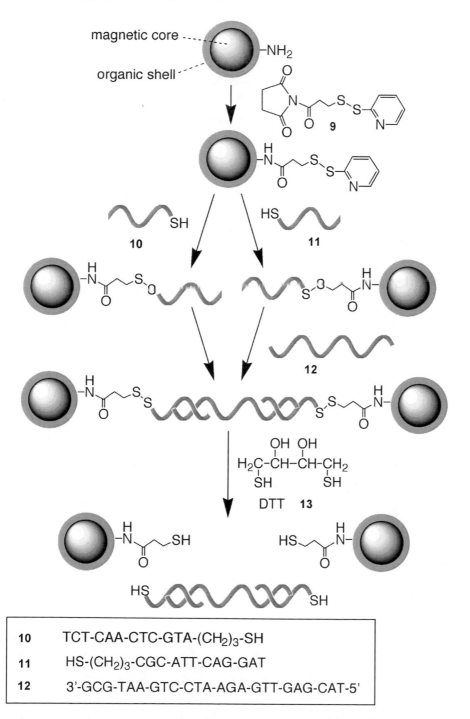

Fig. 5.7 DNA analysis using magnetic nanoparticles as labels allowing the NMR-transduction of the hybridization process

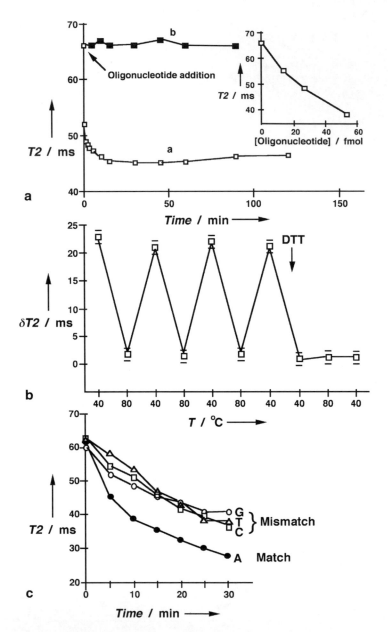

Fig. 5.8 a Changes of T2 upon addition the (**10**)- and (**11**)-oligonucleotide-functionalized magnetic nanoparticles to: (*a*) Complementary analyte oligonucleotide (**12**). (*b*) Non-complementary
oligonucleotide. Inset: The calibration plot of the T2 changes upon addition of the complementary
analyte oligonucleotide (**12**) at various concentrations. **b** Reversible changes of T2 upon dehybrid-
ization and hybridization of the oligonucleotides labeled with the magnetic nanoparticles caused
by the increase and decrease of temperature, respectively. The addition of dithiothreitol (DTT)
to the system results in the irreversible disaggregation of the magnetic clusters linked through
the olgonucleotides. **c** Changes of T2 in the presence of the fully complementary oligonucleotide analyte and single-base mismatched target oligonucleotides. (Adapted from [116], Scheme 1,
Figs. 5.3 and 5.4; [117], Fig. 5.3a, with permission.)

oligonucleotides with a single-base mismatch were smaller than that for the fully complementary DNA analyte (Fig. 5.8c), thus allowing the single-base mismatches DNA analysis.

The NMR-based method allows the DNA and RNA analysis in turbid media and in whole-cell lysates without purification of the samples. The selective high-throughput DNA analysis was performed for lysates of cell lines for GFP mRNA expression using the NMR analysis as the detection tool. The sensitivity level and the mismatch selectivity of the currently non-optimized NMR analysis of the non-purified cell lysates were comparable with the traditional fluorescence-based methods for the purified RNA [118]. The NMR-based method enables the imaging of the gene expression *in vivo* [120]. The existing technology and instrumentation enable the high-throughput analysis of thousands of samples simultaneously and rapidly [121].

Similarly to the DNA analysis, the immunosensing based on the NMR analysis of the aggregation of the magnetic nano-labels was developed [112, 118]. Super-paramagnetic $(Fe_2O_3)_n/(Fe_3O_4)_m$ nanoparticles were covalently modified with avidin, yielding conjugates that carry an average of two avidin units per nanoparticle, and used as generic reagent for labeling of any biotinylated antibodies. Biotinylated anti-green fluorescent protein (GFP) polyclonal antibody was then attached to yield a GFP-sensitive immunosensor. The addition of the GFP analyte to the anti-GFP antibody labeled with magnetic nanoparticles resulted in the cross-linking and aggregation of the magnetic nanoparticles, yielding the change of T2 detected by the NMR analysis. A similar approach was used to design an enantioselective immunosensor [122]. The magnetic nanoparticles were labeled with a derivative of D-phenylalanine. The addition of antibodies specific to D-amino acids resulted in the cross-linking and aggregation of the magnetic nanoparticles, which led to a decrease of more than 100 ms in the T2 relaxation time. The analysis of D-phenyl-alanine impurities in samples of L-phenylalanine was performed by the competitive immunoassay. The addition of a mixture of the enatiomers to the assembly of magnetic nanoparticles cross-linked with the D-amino acids-specific antibodies led to the disaggregation of the magnetic nanoparticles due to the competitive binding of the antibodies to the D-phenylalanine. This process resulted in the increase of T2 relaxation time, thus providing the detection of D-phenylalanine. The rate and magnitude of the change in the T2 relaxation time was dependent on the concentration of D-phenylalanine in the sample, whereas the pure L-phenylalanine did not result in any change of T2.

Modification of the magnetic nanoparticles with virus-surface-specific antibodies allowed immunosensing of viruses with the NMR analysis as a transduction tool [112, 123]. The superparamagnetic $(Fe_2O_3)_n/(Fe_3O_4)_m$ nanoparticles were functionalized with anti-adenovirus 5 (ADV-5) or anti-herpes simplex virus 1 (HSV-1) antibodies. The magnetic labels were cross-linked upon addition to the system the respective viral species, adenovirus (ADV-5) or herpes simplex virus (HSV-1), yielding an aggregated system detected by measuring the changes of the water protons T2 relaxation times. The method allowed the detection of five viral species as lower sensitivity limit. The analyzed model viruses represent a model for other more pathogenic viruses, and are currently used as viral vectors in gene-therapy studies. The viral-specific magnetic nanoparticles can be further developed into viral-specific imaging agents for the magnetic resonance imaging of distribution of viruses and other pathogens *in vivo*. For example, the magnetic nanoparticles modified by anti-human E-selectin (CD62E) F(ab')$_2$ fragments were used for the magnetic resonance imaging of inducible E-selectin expression in human endothelial cell culture [124], that is important for early diagnostics of tumor angiogenesis, arthritis, inflammations and atherosclerosis.

Since enzyme-catalyzed reactions can lead to the formation of various assemblies or alternatively can result in the disassembling of organic materials, these processes could be followed by the NMR analysis with the use of the magnetic nano-labels, providing the means to analyze the respective enzyme activity [112]. Serotonin-functionalized magnetic nanoparticles were reacted with hydrogen peroxide in the presence of myeloperoxidase, an enzyme implicated with inflammation and coronary diseases, to yield the cross-linked magnetic matrix due to the enzyme catalyzed oxidation of serotonin leading to its dimerization (Fig. 5.9a) [125]. The formation of the biocatalytically generated magnetic matrix was followed by the change of T2 relaxation time, thus allowing the analysis of the enzyme activity in the sample (Fig. 5.9b). Protease activity was assayed using avidin-functionalized magnetic nanoparticles and di-biotin linkers with a protease-cleavable peptide sequence in the spacer [126]. The addition of a di-biotin derivative, biotin-GPAR-LAIK-biotin, which includes a specific peptide sequence cleavable by trypsin, to the avidin-functionalized magnetic nanoparticles results in their cross-linking and yields the respective change of the T2 relaxation time. In the presence of tyrosin, the peptide sequence is cut and the produced mono-biotin derivative is not capable to cross-link the avidin-functionalized magnetic nanoparticles, thus inhibiting the T2 change (Fig. 5.10a). Similarly, magnetic nano-assemblies cleavable by renin and matrix metalloprotease-2 were developed and used for the assay of these proteases [126]. The analysis of the protease activity with the use of the magnetic nanoparticles opens the way for the magnetic resonance imaging of various proteases, playing key roles in diverse diseases, *in vivo*.

Similarly, to the assay of the protease activity, the analysis of DNA-cleaving enzymes was reported [127]. Two kinds of magnetic nanoparticles modified with complementary single-stranded DNA, 3'-AATGCGGGATCCTACGAG-5' and 5'-TTACGCCCTAGGATGCTC-3', were prepared to yield upon hybridization the double-stranded DNA sequence recognized by *BamH*I restriction endonuclease (Fig. 5.10b). The hybridization process produced the magnetic nanoparticles assembly and resulted in the change of the T2 relaxation time. The ds-DNA linker was cut upon the addition of *BamH*I restriction endonuclease, thus generating the monomeric magnetic nanoparticles and returning the T2 relaxation time to its initial value with the rate proportional to the enzyme activity. The addition of other endonucleases, such as *EcoR*I, *Hind*III, and *Dpn*I, to the *BamH*I-sensitive nanoassembly did not cause the change of T2, thus resulting in the specificity of the assay.

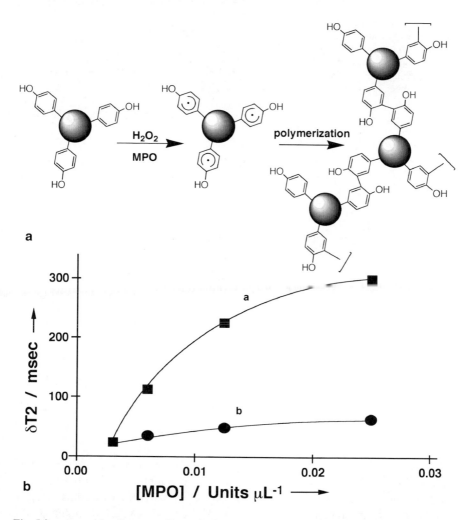

Fig. 5.9 a Assembly of a magnetic matrix from serotonin-functionalized magnetic nanoparticles biocatalyzed by myeloperoxidase (MPO). **b** Calibration plot showing the changes of T2 relaxation time as a function of the myeloperoxidase activity: (a) in the presence of H_2O_2, (b) in the absence of H_2O_2 in the control samples. (Adapted from [124], Scheme 1 and Fig. 5.1a, with permission. Copyright (2004) American Chemical Society)

Telomerase activity was assayed with the use of the magnetic nanoparticles assembled on the telomeric DNA sequence [128]. The magnetic nanoparticles were functionalized with the oligonucleotide, 5'-CCC-TAA-CCC-TAA-3', that is complementary to the telomeric repeat sequence generated by telomerase. The telomerization process was performed in the system consisting of a telomer-primer, 5'-AAT-CCG-TCG-AGC-AGA-GTT-3', the mixture of nucleotides (dNTP-mix)

Fig. 5.10 The enzyme activity assay with the use of the magnetic nanoparticles crosslinked by the enzyme-cleavable spacer: **a** Protease assay. **b** Restriction endonuclease assay

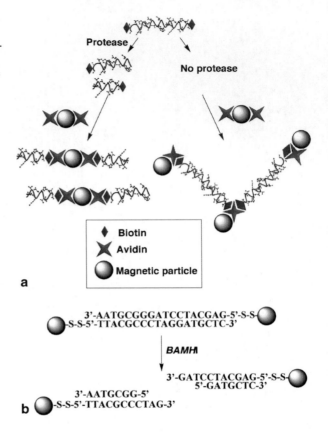

dATP, dGTP, dUTP, and a cell extract containing telomerase prepared from various cell lines (B16 melanoma, CaPan-2 pancreas carcinoma, MCF-7 breast cancer, PC3 prostate cancer, ovcar-5 ovarian carcinoma, H-4-II-E hepatocellular carcinoma, Lewis lung carcinoma, 9L glioblastoma, HeLa, E6-1 Jurkat lymphoma, and rin 5F rat insulinoma). Telomerase catalyzed elongation of the telomer primer by the telomeric repeat units TTAGGG (Fig. 5.11a). The oligonucleotide-functionalized magnetic nanoparticles were bound to the elongated oligonucleotide yielding the change of the T2 relaxation time (Fig. 5.11b, curve *a*). It should be noted that the oligonucleotide bound to the magnetic nanoparticles was not complementary to the telomer-primer and the magnetic nanoparticles were not associated with the telomer-primer in the absence of the telomeric repeats generated in the presence of telomerase (Fig. 5.11b, curve *b*). The method allowed the telomerase assay in various cancer cells, as well as the analysis of the telomerase inhibition and activation processes.

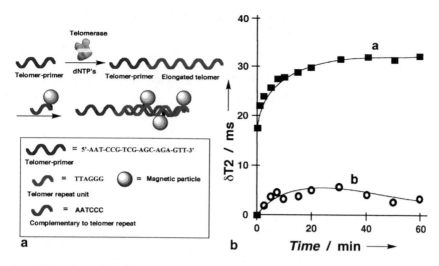

Fig. 5.11 a Assembly of the magnetic nanoparticles on the telomeric repeat units generated by telomerase. **b** The change of T2 relaxation time after addition of the oligonucleotide-functionalized magnetic nanoparticles to: (a) the oligonucleotide consisting of 54 telomeric repeat units generated by telomerase; (b) the telomeric-primer without the telomeric repeat units. (Adapted from [127], Fig. 5.1b, with permission.)

5.5   Conclusions and Perspectives

The state-of-the-art in the preparation, characterization and application of bio-molecule-functionalized magnetic particles and other related micro-/nano-objects allows efficient performance of various *in vitro* and *in vivo* biosensors. Many of these biosensors are directed to biomedical applications. For example, $CoFe_2O_4$-core/Au-shell nanoparticles have been successfully used to design a biosensor for foot-and-mouth viral disease [129]. In this system a biomimetic oligo peptide-nucleic acid (PNA) was assembled on the gold surface of the nanoparticles and then hybridized with a complementary DNA sequence. Detection was performed by incubation with an intercalating fluorescence probe, Rhodamine 6G. The present example demonstrates powerful applicability of the biomolecular-functionalized magnetic nanoparticles in biomedical biosensors. Many other examples outlined in the review demonstrated various aspects of magnetic particles applications in medicine and related areas.

Future developments in this area will be directed to further complexity increase of the systems scaling up their structural and functional properties. Two examples might be mentioned to give some ideas about the possible research directions: (i) Self-assembling of magnetic nanoparticles in the form of magnetic nanowires was achieved recently and the system was considered as a platform for future biosensor/bioelectronic devices [130]. In this system the particles will operate as an aggregated community controlled by an external magnetic field rather than a dispersion of individual species. (ii) Magnetic nanoparticles can be associated with functional switchable interfaces providing a platform for multi-enzyme signal processing system [131]. The present approach will allow functional integration of nano-structured species, functional interfaces and signal-processing biomolecular systems resulting in the novel bioelectronic devices for biosensing and bioactuation. Tremendous changes in the performance of biomedical systems could be expected soon based on the rapid developments in this exciting research area.

References

1. *Scientific and Clinical Applications of Magnetic Carriers*, U. Häfeli, W. Schütt, J. Teller, M. Zborowski (Eds.), Plenum Press, New York, **1997**.
2. Z. Liu, D. Zhang, S. Han, C. Li, B. Lei, W. Lu, J. Fang, C. Zhou, J. Am. Chem. Soc. **2005**, *127*, 6-7.
3. F.-H. Wang, T. Yoshitake, D.-K. Kim, M. Muhammed, B. Bjelke, Jan Kehr, *J. Nanoparticle Res.* **2003**, *5*, 137-146.
4. S. Bucak, D.A. Jones, P.E. Laibinis, T.A. Hatton, *Biotechnol. Prog.* **2003**, *19*, 477-484.
5. L. Stanciu, Y.H. Won, M. Ganesana, S. Andreescu, *Sensors* **2009**, 9, 2976-2999.
6. S. Andreescu, J. Njagi, C. Ispas, M.T. Ravalli, *J. Environmental Monitoring* **2009**, *11*, 27-40.
7. S. Yu, G.M. Chow, *J. Mater. Chem.* **2004**, *14*, 2781-2786.
8. X. Liu, J. Xing, Y. Guan, G. Shan, H. Liu, *Colloids Surfaces A* **2004**, *238*, 127-131.
9. N.R. Jana, Y. Chen, X. Peng, *Chem. Mater.* **2004**, *16*, 3931-3935.
10. J. Lai, K.V.P.M. Shafi, A. Ulman, K. Loos, Y. Lee, T. Vogt, W.-L. Lee, N.P. Ong, *J. Phys. Chem. B* **2005**, *109*, 15-18.
11. D.L. Huber, *Small* **2005**, *1*, 482-501.
12. P. Tartaj, M.P. Morales, T. González-Carreño, S. Veintemillas-Verdaguer, C.J. Serna, *J. Magnet. Magnet. Mater.* **2005**, 290-291, 28-34.
13. L. Shen, P. E. Laibinis, T.A. Hatton, *Langmuir* **1999**, *15*, 447-453.
14. T. Hyeon, *Chem. Commun.* **2003**, 927-934.
15. A.K. Gupta, M. Gupta, *Biomaterials* **2005**, *26*, 3995-4021.
16. S. Sun, H. Zeng, *J. Am. Chem. Soc.* **2002**, *124*, 8204-8205.
17. L.P. Silva, Z.G.M. Lacava, N. Buske, P.C. Morais R.B. Azevedo, *J. Nanoparticle Res.* **2004**, *6*, 209-213.
18. E. Matijević, *Langmuir* **1994**, *10*, 8-16.
19. K.G. Paul, T.B. Frigo, J.Y. Groman, E.V. Groman, *Bioconjugate Chem.* **2004**, *15*, 394-401.
20. L.F. Gamarra, G.E.S. Brito, W.M. Pontuschka, E. Amaro, A.H.C. Parma, G.F. Goya, *J. Magnet.Magnet. Mater.* **2005**, *289*, 439-441.
21. J. Park, E. Lee, N.-M. Hwang, M. Kang, S.C. Kim, Y. Hwang, J.-G. Park, H.-J. Noh, J.-Y. Kim, J.-H. Park, T. Hyeon, *Angew. Chem. Int. Ed.* **2005**, *44*, 2872-2877.
22. G. Li, J. Fan, R. Jiang, Y. Gao, *Chem. Mater.* **2004**, *16*, 1835-1837.
23. H. Zeng, S. Sun, J. Li, Z.L. Wang, J.P. Liu, *Appl. Phys. Lett.* **2004**, *85*, 792-794.
24. A. Hütten, D. Sudfeld, I. Ennen, G. Reiss, K. Wojczykowski, P. Jutzi, *J. Magnet. Magnet. Mater.* **2005**, *293*, 93-101.
25. R. Mikami, M. Taguchi, K. Yamada, K. Suzuki, O. Sato, Y. Einaga, *Angew. Chem. Int. Ed.* **2004**, *43*, 6135-6139.
26. V.K. Sharma, F. Waldner, *J. Appl. Phys.* **1977**, *48*, 4298-4302.
27. J. Baker-Jarvis, P. Kabos, *Phys. Rev. E* **2001**, *64*, 056127-1-056127-14.
28. S. Mørup, E. Tronc, *Phys. Rev. Lett.* **1994**, *72*, 3278-3281.
29. M. Mandal, S. Kundu, S.K. Ghosh, S. Panigrahi, T.K. Sau, S.M. Yusuf, T. Pal, *J. Colloid Interface Sci.* **2005**, *286*, 187-194.

30. M. Chen, S. Yamamuro, D. Farrell, S.A. Majetich, *J. Appl. Phys.* **2003**, *93*, 7551-7553.
31. Y. Bao, K.M. Krishnan, *J. Magnet. Magnet. Mater.* **2005**, *293*, 15-19.
32. Y. Cui, Y. Wang, W. Hui, Z. Zhang, X. Xin, C. Chen, *Biomed. Microdevices* **2005**, *7*, 153-156.
33. L. Wang, J. Luo, M.M. Maye, Q. Fan, Q. Rendeng, M.H. Engelhard, C. Wang, Y. Lin, C.-J. Zhong, *J. Mater. Chem.* **2005**, *15*, 1821-1832.
34. X. Gao, K.M.K. Yu, K.Y. Tam, S.C. Tsang, *Chem. Commun.* **2003**, 2998-2999.
35. H.-H. Yang, S.-Q. Zhang, X.-L. Chen, Z.-X. Zhuang, J.-G. Xu, X.-R. Wang, *Anal. Chem.* **2004**, *76*, 1316-1321.
36. W. Zhao, J. Gu, L. Zhang, H. Chen, J. Shi, *J. Am. Chem. Soc.* **2005**, *127*, 8916-8917.
37. S.H. Yun, C.W. Lee, J.S. Lee, C.W. Seo, E.K. Lee, *Mater. Sci. Forum* **2004**, *449*, 1033-1036.
38. Y.P. He, S.Q. Wang, C.R. Li,Y.M. Miao, Z.Y. Wu, B.S. Zou, *J. Phys. D* **2005**, *38*, 1342-1350.
39. S.J. Son, J. Reichel, B. He, M. Schuchman, S.B. Lee, *J. Am. Chem. Soc.* **2005**, *127*, 7316-7317.
40. H. Lu, G. Yi, S. Zhao, D. Chen, L.-H. Guo, J. Cheng, *J. Mater. Chem.* **2004**, *14*, 1336-1341.
41. X. Hong, J. Li, M. Wang, J. Xu, W. Guo, J. Li, Y. Bai, T. Li, *Chem.Mater.* **2004**, *16*, 4022-4027.
42. H. Kim, M. Achermann, L.P. Balet, J.A. Hollingsworth, V.I. Klimov, *J. Am. Chem. Soc.* **2005**, *127*, 544-546.
43. H. Gu, Z. Yang, J. Gao, C.K. Chang, B. Xu, *J. Am. Chem. Soc.* **2005**, *127*, 34-35.
44. H. Gu, R. Zheng, X.-X. Zhang, B. Xu, *J. Am. Chem. Soc.* **2004**, *126*, 5664-5665.
45. H. Yu, M. Chen, P.M. Rice, S.X, Wang, R.L. White, S. Sun, *Nano Lett.* **2005**, *5*, 379-382.
46. D.K. Yi, S.T. Selvan, S.S. Lee, G.C. Papaefthymiou, D. Kundaliya, J.Y. Ying, *J. Am. Chem. Soc.* **2005**, *127*, 4990-4991.
47. P.C. Morais, J.G. Santos, L.B. Silveira, C. Gansau, N. Buske, W.C. Nunes, J.P. Sinnecker, *J. Magnet.Magnet. Materials* **2004**, *272-276*, 2328-2329.
48. A. del Campo, T. Sen, J.-P. Lellouche, I.J. Bruce, *J. Magnet.Magnet. Mater.* **2005**, *293*, 33-40.
49. I.J. Bruce, T. Sen, *Langmuir* **2005**, *21*, 7029-7035.
50. T. Tanaka, T. Matsunaga, *Anal. Chem.* **2000**, *72*, 3518-3522.
51. C.R. Martin, D.T. Mitchell, *Anal.Chem.* **1998**, *70*, 322A-327A.
52. M. Fuentes, C. Mateo, J.M. Guisán, R. Fernández-Lafuente, *Biosens. Bioelectron.* **2005**, *20*, 1380-1387.
53. F. Gao, B.-F. Pan, W.-M. Zheng, L.-M. Ao, H.-C. Gu, *J. Magnet. Magnet. Mater.* **2005**, *293*, 48-54.
54. C. Xu, K. Xu, H. Gu, R. Zheng, H. Liu, X. Zhang, Z. Guo, B. Xu, *J. Am. Chem. Soc.* **2004**, *126*, 9938-9939.
55. Z.G. Peng, K. Hidajat, M.S. Uddin, *J. Colloid Interface Sci.* **2004**, *271*, 277-283.
56. M. Mikhaylova, D.K. Kim, C.C. Berry, A. Zagorodni, M. Toprak, A.S.G. Curtis, M. Muhammed, *Chem. Mater.* **2004**, *16*, 2344-2354.
57. A. Dyal, K. Loos, M. Noto, S.W. Chang, C. Spagnoli, K.V.P.M. Shafi, A. Ulman, M. Cowman, R.A. Gross, *J. Am. Chem. Soc.* **2003**, *125*, 1684-1685.
58. M. Ma, Y. Zhang, W. Yu, H. Shen, H. Zhang, N. Gu, *Colloids Surfaces A* **2003**, *212*, 219-226.
59. S.-H. Huang, M.-H. Liao, D.-H. Chen, *Biotechnol. Prog.* **2003**, *19*, 1095-1100.
60. D. Cao, P. He, N. Hu, *Analyst* **2003**, *128*, 1268-1274.
61. M. Shinkai, H. Honda, T. Kobayashi, *Biocatalysis* **1991**, *5*, 61-69.
62. M.-H. Liao, D.-H. Chen, *Biotechnol.Lett.* **2001**, *23*, 1723-1727.
63. M.-H. Liao, D.-H. Chen, *Biotechnol. Lett.* **2002**, *24*, 1913-1917.
64. X.-L. Sun, W. Cui, C. Haller, E.L. Chaikof, *ChemBioChem* **2004**, *5*, 1593-1596.
65. S. Mornet, A. Vekris, J. Bonnet, E. Duguet, F. Grasset, J.-H. Choy, J. Portier, *Mater. Lett.* **2000**, *42*, 183-188.
66. L.E. Euliss, S.G. Grancharov, S. O'Brien, T.J. Deming, G.D. Stucky, C.B. Murray, G.A. Held, *Nano Lett.* **2003**, *3*, 1489-1493.
67. D. Horák, B. Rittich, A. Španová, M.J. Benes, *Polymer* **2005**, *46*, 1245-1255.
68. X. Liu, Y. Guan, Y. Yang, Z. Ma, X. Wu, H. Liu, *J. Appl. Polymer Sci.* **2004**, *94*, 2205-2211.

69. A.-N. Kawde, M.C. Rodriguez, T.M.H. Lee, J. Wang, *Electrochem.Commun.* **2005**, *7*, 537-540.

70. H. Xu, H. Wu, F. Huang, S. Song, W. Li, Y. Cao, C. Fan, *Nucleic Acids Res.* **2005**, *33*, paper e83.

71. K.-B. Lee, S. Park, C.A. Mirkin, *Angew. Chem. Int. Ed.* **2004**, *43*, 3048-3050.

72. C.R. Martin, *Science* **1994**, *266*, 1961-1966.

73. D. Routkevitch, T. Bigioni, M. Moskovits, J.M. Xu, *J. Phys. Chem.* **1996**, *100*, 14037-14047.

74. D.H. Reich, M. Tanase, A. Hultgren, L.A. Bauer, C.S. Chen, G.J. Meyer, *J. Appl.Phys.* **2003**, *93*, 7275-7280.

75. G. Amagliani, G. Brandi, E. Omiccioli, A. Casiere, I.J. Bruce, M. Magnani, *Food Microbiol.* **2004**, *21*, 597-603.

76. S.V. Sonti, A. Bose, *J. Colloid Inter. Sci.* **1995**, *170*, 575-585.

77. J. Roger, J.N. Pons, R. Massart, A. Halbreich, J.C. Bacri, *Eur. Phys. J. Appl.* **1999**, *5*, 321-325.

78. X.-L. Su, Y. Li, *Anal. Chem.* **2004**, *76*, 4806-4810.

79. K.-C. Ho, P.-J. Tsai, Y.-S. Lin, Y.-C. Chen, *Anal. Chem.* **2004**, *76*, 7162-7168.

80. H. Gu, P.-L. Ho, K.W.T. Tsang, C.-W. Yu. B. Xu, *Chem. Commun.* **2003**, 1966-1967.

81. H. Gu, P.-L. Ho, K.W.T. Tsang, L. Wang, B. Xu, *J. Am. Chem. Soc.* **2003**, *125*, 15702-15703.

82. M. Kuhara, H. Takeyama, T. Tanaka, T. Matsunaga, *Anal.Chem.* **2004**, *76*, 6207-6213.

83. G.A. Prinz, *Science* **1998**, *282*, 1660-1663.

84. D.L. Graham1, H.A. Ferreira, P.P. Freitas, *Trends Biotechnol.* **2004**, *22*, 455-462.

85. J. Richardson, A. Hill, R. Luxton, P. Hawkins, *Biosens. Bioelectron.* **2001**, *16*, 1127-1132.

86. J. Richardson, P. Hawkins, R. Luxton, *Biosens. Bioelectron.* **2001**, *16*, 989-993.

87. S.G. Grancharov, H. Zeng, S. Sun, S.X. Wang, S.O'Brien, C.B. Murray, J.R. Kirtley, G.A. Held, *J. Phys. Chem. B* **2005**, *109*, 13030-13035.

88. S.-K. Lee, W.R. Myers, H.L. Grossman, H.-M. Cho, Y.R. Chemla, J. Clarke, *Appl. Phys. Lett.* **2002**, *81*, 3094-3096.

89. S. Katsura, T. Yasuda, K. Hirano, A. Mizuno, S. Tanaka, *Supercond. Sci. Technol.* **2001**, *14*, 1131-1134.

90. K. Enpuku, A. Ohba, K. Inoue, T.Q. Yang, *Physica C* **2004**, *412-414*, 1473-1479.

91. D. Abrahamsson, K. Kriz, M. Lub, D. Kriz, *Biosens. Bioelectron.* **2004**, *19*, 1549-1557.

92. W. Thomson, *Proc. R. Soc. London A* **1857**, *8*, 546-550.

93. M.M. Miller, G.A. Prinz, S.-F. Cheng, S. Bounnak, *Appl. Phys. Lett.* **2002**, *81*, 2211-2213.

94. R.L. Edelstein, C.R. Tamanaha, P.E. Sheehan, M.M. Miller, D.R. Baselt, L.J. Whitman, R.J. Colton, *Biosens. Bioelectron.* **2000**, *14*, 805-813.

95. M.M. Miller, P.E. Sheehan, R.L. Edelstein, C.R. Tamanaha, L. Zhong, S. Bounnak, L.J. Whitman, R.J. Colton, *J. Magnet. Magnet. Mater.* **2001**, *225*, 138-144.

96. J.C. Rife, M.M. Miller, P.E. Sheehan, C.R. Tamanaha, M. Tondra, L.J. Whitman, *Sens. Actuat. A* **2003**, *107*, 209-218.

97. L. Lagae, R. Wirix-Speetjens, J. Das, D. Graham, H. Ferreira, P.P.F. Freitas, G. Borghs, J. De Boeck, *J. Appl. Phys.* **2002**, *91*, 7445-7447.

98. D.L. Graham, H. Ferreira, J. Bernardo, P.P. Freitas, J.M.S. Cabral, *J. Appl. Phys.* **2002**, *91*, 7786-7788.

99. H.A. Ferreira, D.L. Graham, P.P. Freitas, J.M.S. Cabral, *J. Appl. Phys.* **2003**, *93*, 7281-7286.

100. G. Li, V. Joshi, R.L. White, S.X. Wang, J.T. Kemp, C. Webb, R.W. Davis, S. Sun, *J. Appl. Phys.* **2003**, *93*, 7557-7559.

101. D.R. Baselt, G.U. Lee, M. Natesan, S.W. Metzger, P.E. Sheehan, R.J. Colton, *Biosens. Bioelectron.* **1998**, *13*, 731-739.

102. D.L. Graham, H.A. Ferreira, P.P. Freitas, J.M.S. Cabral, *Biosens. Bioelectron.* **2003**, *18*, 483-488.

103. M. Megens, M. Prins, *J. Magnet. Magnet. Mater.* **2005**, *293*, 702-708.

104. P.-A. Besse, G. Boero, M. Demierre, V. Pott, R. Popovic, *Appl. Phys. Lett.* **2002**, *80*, 4199-4201.
105. D.L. Graham, H.A. Ferreira, N. Feliciano, P.P. Freitas, L.A. Clarke, M.D. Amaral, *Sens. Actuat. B* **2005**, *107*, 936-944.
106. A.P. Astalan, F. Ahrentorp, C. Johansson, K. Larsson, A. Krozer, *Biosens. Bioelectron.* **2004**, *19*, 945-951.
107. C. Wilhelm, F. Gazeau, J. Roger, J.N. Pons, M.F. Salis, R. Perzynski, J.C. Bacri, *Phys. Rev. E* **2002**, *65*, 031404-1-031404-9.
108. S.H. Chung, A. Hoffmann, S.D. Bader, C. Liu, B. Kay, L. Makowski, L. Chen, *Appl. Phys. Lett.* **2004**, *85*, 2971-2973.
109. H.L. Grossman, W.R. Myers, V.J. Vreeland, R. Bruehl, M.D. Alper, C.R. Bertozzi, J. Clarke, *Proc. Nat. Acad. Sci. US* **2004**, *101*, 129-134.
110. C.J. Behrend, J.N. Anker, B.H. McNaughton, M. Brasuel, M.A. Philbert, R. Kopelman, *J. Phys. Chem. B* **2004**, *108*, 10408-10414.
111. G. Glöckl, V. Brinkmeier, K. Aurich, E. Romanus, P. Weber, W. Weitschies, *J. Magnet. Magnet. Mater.* **2005**, *289*, 480-483.
112. J.M. Perez, L. Josephson, R. Weissleder, *ChemBioChem* **2004**, *5*, 261-264.
113. T. Shen, R. Weissleder, M. Papisov, A. Bogdanov, T.J. Brady, *Magnet. Resonance Med.* **1993**, *29*, 599-604.
114. P. Gillis, F. Moiny, R.A. Brooks, *Magnet. Resonance Med.* **2002**, *47*, 257-263.
115. R.A. Brooks, *Magnet. Resonance Med.* **2002**, *47*, 388-391.
116. J.G. Santos, L.B. Silveira, C. Gansau, N. Buske, P.C. Morais, *J. Magn. Magn. Mater.* **2004**, *272-276*, 2330-2331.
117. L. Josephson, J.M. Perez, R. Weissleder, *Angew. Chem. Int. Ed.* **2001**, *40*, 3204-3206.
118. J.M. Perez, L. Josephson, T. O'Loughlin, D. Högelmann, R. Weissleder, *Nature Biotechnol.* **2002**, *20*, 816-820.
119. P. Wunderbaldinger, L. Josephson, R. Weissleder, *Acad. Radiol.* **2002**, *9*, S304-S306.
120. D. Högemann, L. Josephson, R. Weissleder, J.P. Basilion, *Bioconjugate Chem.* **2000**, *11*, 941-946.
121. D. Högemann, V. Ntziachristos, L. Josephson, R. Weissleder, *Bioconjugate Chem.* **2002**, *13*, 116-121.
122. A. Tsourkas, O. Hofstetter, H. Hofstetter, R. Weissleder, L. Josephson, *Angew. Chem. Int. Ed.* **2004**, *43*, 2395-2399.
123. J.M. Perez, F.J. Simeone, Y. Saeki, L. Josephson, R. Weissleder, *J. Am. Chem. Soc.* **2003**, *125*, 10192-10193.
124. H.W. Kang, L. Josephson, A. Petrovsky, R. Weissleder, A. Bogdanov, Jr. *Bioconjugate Chem.* **2002**, *13*, 122-127.
125. J.M. Perez, F.J. Simeone, A. Tsourkas, L. Josephson, R. Weissleder, *Nano Lett.* **2004**, *4*, 119-122.
126. M. Zhao, L. Josephson, Y. Tang, R. Weissleder, *Angew. Chem. Int. Ed.* **2003**, *42*, 1375-1378.
127. J.M. Perez, T. O'Loughin, F.J. Simeone, R. Weissleder, L. Josephson, *J. Am. Chem. Soc.* **2002**, *124*, 2856-2857.
128. J. Grimm, J.M. Perez, L. Josephson, R. Weissleder, *Cancer Res.* **2004**, *64*, 639-643.
129. M. Pita, J. M. Abad, C. Vaz-Dominguez, C. Briones, E. Mateo-Martí, J. A. Martín-Gago, M. P. Morales, V. M. Fernández. *J. Coll. Interf. Sci.* **2008**, *321*, 484-492.
130. J. Jimenez, R. Sheparovych, M. Pita, A. Narvaez Garcia, E. Dominguez, S. Minko, E. Katz, J. Phys. Chem. C **2008**, 112, 7337-7344.
131. M. Pita, T.K. Tam, S. Minko, E. Katz, *ACS Appl. Mater. Interfaces* **2009**, 1, 1166-1168.

Chapter 6
Degradable Polymer Particles in Biomedical Applications

Broden G. Rutherglen and Devon A. Shipp

Abstract Polymers are widely used in therapeutic and drug delivery applications, with applications including wound closure, tissue repair, bone augmentation and drug delivery. Of such polymers, those that are degradable (or resorbable) are of great interest. Furthermore, colloidal polymers are often used in biomedical and pharmaceutics, particularly in encapsulation, targeting, and controlled release systems. Regardless of the application, the molecular and architectural design of the colloidal system is of great importance in the performance of the material. This article reviews degradable polymers and particles made from them, with particular attention paid to particles made from polyanhydrides and their potential use in the therapeutic agent delivery applications. We initially review various biodegradable polymers, molecular architectures and outline their fundamental chemistry. Finally, we examine modes of polyanhydride particle synthesis, characterization, and use in therapeutic delivery applications.

Keywords Biodegradable • Erosion • Degradation • Drug delivery • Particle • Polyanhydrides • Polymer colloid

6.1 Introduction and Scope of Review

Polymers are widely used in therapeutic and drug delivery applications (see, for example, Duncan et al. [1] and references therein), with applications including wound closure, tissue repair, bone augmentation and drug delivery [2]. Within these applications, degradable (or resorbable) polymers are of great interest. Furthermore, colloidal polymers are often used in biomedical and pharmaceutics, particularly in encapsulation, targeting, and controlled release systems. Regardless of the application, the molecular and architectural design of the colloidal system is of great importance in the performance of the material. For example, chain length, crystallinity, branching, charge, composition, and hydrophilicity/hydrophobicity all have significant effects on the material properties of the polymer colloid, such as deg-

D. A. Shipp (✉)
Department of Chemistry & Biomolecular Science, Clarkson University,
8 Clarkson Avenue, Box 5810, Potsdam, NY 13699, USA
e-mail: dshipp@clarkson.edu

E. Matijević (ed.), *Fine Particles in Medicine and Pharmacy,*
DOI 10.1007/978-1-4614-0379-1_6, © Springer Science+Business Media, LLC 2012

Fig. 6.1 Schematic of various adjustable parameters of polymer colloid and bio-properties and bioapplications they affect

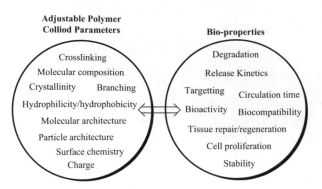

radation and release kinetics. The connection between the many polymer colloid parameters and bioapplication properties is outlined in Fig. 6.1.

This article reviews degradable polymers and particles made from them, with particular attention paid to particles made from polyanhydrides and their potential use in the therapeutic agent delivery applications. We initially review various biodegradable polymers, molecular architectures and outline their fundamental chemistry. Finally, we examine modes of polyanhydride particle synthesis, characterization, and use in therapeutic delivery applications.

6.1.1   Therapeutic Polymers: Particles and Assemblies

Finne-Wistrand and Albertsson [3] have recently reviewed the area of biodegradable polymer colloids, which includes spheres and capsules, and covers a wide range of polymer-based particulates of 10 nm and larger in size. The area has been fruitful for nearly 40 years, with the original work in producing drug release from polymers dating back to the 1960s. Polymer particles can be made from emulsion polymerization (or a related method such as dispersion polymerization) or linear polymers that are made in a first step, then followed by a particle-forming process (such as solvent evaporation or phase inversion). There are many factors that affect drug release from polymer particles. These include composition, architecture, molecular weight, crystallinity, porosity, and the size distribution of particles. Furthermore, other parameters such as pH, light or heat can be used to induce changes in the polymer materials (i.e. they are stimuli responsive) that result in enhanced release rates.

Particles can come in various architectures, with simple homogeneous particles, coreshell morphologies, and brush-like surfaces. In many cases, and particularly for core–shell structures, the use of surfactants to stabilize the particles can lead to interference in the drug delivery process by altering degradation or the size distribution, for example. Surfactants are also often difficult to remove. Particular domains of the particles can be cross-linked; for example, degradable shell cross-linked nanoparticles have been made by Wooley et al. [4, 5] via the self-assembly of block copolymers with subsequent cross-linking of the shell. In such cases, it is impor-

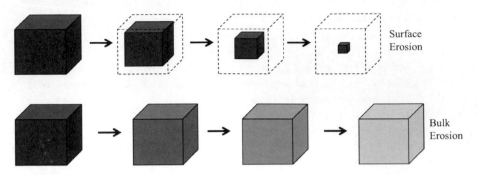

Scheme 6.1 A schematic diagram of surface and bulk erosion. Degree of shading represents material properties

Before discussing the various types of degradable polymers (based on molecular structure), definitions of erosion and degradation are needed. Degradation is defined as the breaking of covalent or non-covalent bonds within the polymer chains. On the other hand, erosion characterizes the macroscopic scale or physical breakdown, which is classified as the sum of all processes leading to the loss of mass from the polymer matrix [9]. An example of non-degradative erosion would be solvation.

Furthermore, there are two types of erosion: surface and bulk. Surface erosion is also referred to as heterogeneous erosion, where erosion is limited to the surface of the polymer only. This is due to the faster rate of degradation on the surface of the polymer relative to rate of bond cleavage of the polymer chains buried within the sample. The difference in rate is primarily due to (a) the presence in the polymer of highly reactive functionality toward water, and (b) the difference in local water concentration—if the material is hydrophobic enough to restrict water uptake into the bulk of the material then the local concentration of water is very high at the surface and very low (or zero) within the material. By inhibiting water from penetrating into the bulk of the material and having highly reactive hydrolysable groups at the surface, the material will be depleted only at the surface. In contrast, bulk erosion, which is also termed "homogeneous erosion" in some publications, the typical rate at which the bond cleavage occurs is much slower in comparison to surface eroding polymers, thus allowing penetration of water throughout the bulk of the material which leads to the uniform erosion of the polymer from within [9, 10]. Scheme 6.1 displays a schematic description of surface and bulk erosion that were described above. Degradable poly(α-hydroxy acids), such as PLA, PGA, and PLGA, all undergo bulk erosion.

6.2.1 Polyesters

While there have been many materials examined in the field of degradable polymers, to date the most widely investigated and used degradable polymers are poly-

Scheme 6.2 Ring opening
polymerization of *PGA* and
PLA respectively [11]

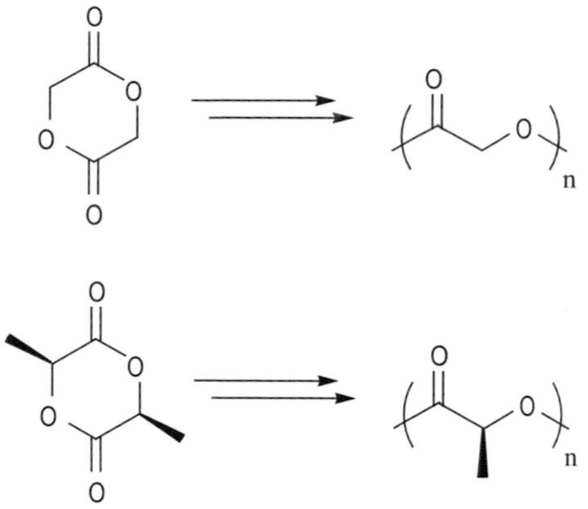

esters, particularly those belonging to the poly(α-hydroxy acid) family. These include poly(lactic acid) (PLA), poly(glycolic acid) (PGA), copolymers of these two
(producing PLGA), poly(orthoesters) and poly(β-amino esters) [11]. The notable
degrading functionality of this class of polymer is hydrolytically unstable ester moieties present in the backbone.

6.2.1.1   Poly(Lactic Acid), (PLA), Poly(Glycolic Acid) (PGA) and Their Copolymers

Scheme 6.2 shows how PLA, PGA and copolymers are created by ring opening
polymerizations of lactones, which is the most common method of synthesis [11].

These types of degradable polymers have been widely used due to their high
biocompatibility and variable degradation rates. Products already available from
these polymers include sutures, staples, screws, plates, and drug delivery devices [11]. The first patent for resorbable sutures was made as far back as 1967 by
Schmitt and Polistina [12]. Even though these polymers are highly biocompatible they have some drawbacks in terms of their use in the biomaterials field,
thus there are ongoing efforts to create better and more suitable materials for
various applications. Some of these disadvantages are that they are crystalline,
hydrophobic, and the mechanical properties are difficult to tune. The crystallinity causes several problems, including inconsistent swelling, increased hard-
ness with low resilience, and low diffusion coefficients which hinders the water
penetration to the ester bonds thus resulting in a material which could degrade
anywhere from 6 months to many years [13–16]. This is often at odds with the
requirements in the areas of soft tissue regeneration and drug delivery systems,
where it is often necessary for the material to be soft yet capable of withstanding
dynamic environments, and able to undergo surface erosion [17, 18].

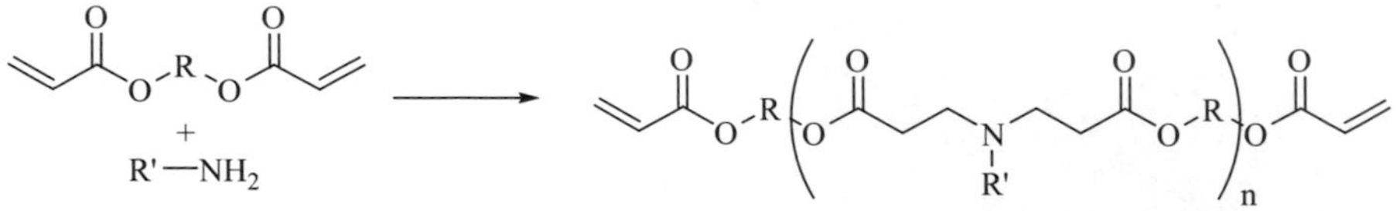

Scheme 6.3 An example of the synthesis of acrylate-terminated poly(β-amino esters)

6.2.1.2 Poly(Ortho Esters)

Early work on poly(ortho esters) by Heller et al. [19] showed that these polymers undergo surface erosion and potentially can provide zero-order drug release kinetics. Such polymers are primarily made via the addition of polyols to diketene acetals. However, degradation of poly(ortho esters) occurs only under acidic conditions; this can be both an advantage and disadvantage. The benefit is that acidic or basic active ingredients into the polymer matrix can be used to tune degradation rates. On the other hand, this makes the system more complex and may lead to the operation of various degradation/erosion mechanisms.

6.2.1.3 Poly(β-Amino Esters)

Recent work by Langer et al. [20–23] has shown that poly(β-amino esters) have great promise as drug delivery vehicles. Cross-linked acrylated poly(β-amino esters) degrade via hydrolysis of their ester backbones into simple amino acids, diols, and poly(acrylic acid), compounds that are easily removed by the body [23].

Acrylate-terminated poly(β-amino esters) have been developed more recently, and a large variety of cross-linking agents developed by combinatorial approaches [20, 21]. These polymers are made by the β-addition of diamines to diacrylates in a step-growth polymerization (Scheme 6.3). Since there is a wide range of such monomers available, this chemical diversity leads to a significant range of both degradation profiles and mechanical properties. Furthermore, their synthesis is simple and quite tolerant of other functionalities, such as ethers, alcohols and tertiary amines. As is typical for step-growth polymers, the molecular weight of such polymers is controlled by the stoichiometry of the starting monomers, allowing oligomers to be produced easily. Oligomers with acrylate end-groups have been used in the synthesis of biodegradable hydrogels as crosslinkers [24].

6.2.2 Polyanhydrides

This class of polymer is among the most reactive and hydrolytically unstable that have been developed for biomedical applications [25]. Polyanhydrides were first reported to be synthesized in the early twentieth century by the melt condensation

Scheme 6.4 General structure of polyanhydrides

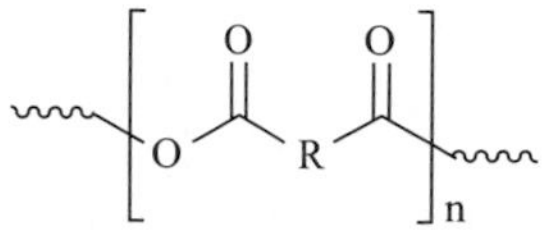

method between diacids, but they did not become widely studied until about three decades ago. The main reason why this class of polymer entered into the spotlight was the uniqueness of their degradation/erosion mechanism—they undergo surface erosion. Due to this characteristic they are the only surface eroding polymer that has been approved for applications by the Food and Drug Administration [25]. Scheme 6.4 shows the general polyanhydride structure.

The earliest publication concerning this class of polymer was conducted by Bucher and Slade in 1909, where they synthesized anhydrides from isophthalic and terephthalic acids [26]. Two decades later, Carothers and Hill prepared aliphatic polyanhydrides; polymers containing adipic and sebacic anhydride were produced by heating the individual acids into excess acetyl anhydride [27, 28]. The next to look at these polymers was Conix [29] in the late 1950s. He studied and reported polyanhydrides resulting from aromatic acids, mainly derived from poly[bis(p-carboxyphenoxy)alkane anhydrides]. Scheme 6.5 shows common monomers used by Conix, Hill and Carothers, and others. Another pioneer in the polyanhydride field was Yoda [30] in the early 1960s. He produced a new family of polyanhydrides that were heterocyclic crystalline compounds that were synthesized from various five member heterocyclic dibasic acids, in which he polymerized with acetic anhydride at 200–300°C in nitrogen and under vacuum.

The first to take real advantage of the polyanhydrides hydrolytically unstable nature was R. Langer, a pioneer in the sustained drug release community. During the mid 1980s Langer and coworkers began studying methods of polyanhydride synthesis, ultimately working towards increasing the molecular weights in an efficient manner. Particularly, they worked with poly(bis(p-carboxyphenoxy)propane) (poly(CPP)), poly(sebacic acid) (poly(SA)) anhydrides and copolymers of the two. One significant outcome of Langer's work is the Gliadel® wafer implant [31], which is made from the poly(sebacic acid-co-1,3-bis(p-carboxyphenoxy)propane) and administers an anti-tumor agent as it erodes to the brain cancer patient. Many other works concerning polyanhydrides have been reported, for example the production of poly(anhydrides-co-imides) which increases mechanical properties [32].

In the late 1990s Langer et al. developed a method of photopolymerization and cross-linking of polyanhydrides [33]. The desire for higher strength polymers with moldability, surface erosion, and controlled rates of degradation characteristics for orthopedic applications such as bone regeneration and augmentation were becoming evident. The study by Anseth et al. [33] reported the development of photopolymerizable methacrylated anhydride monomers and oligomers in which unite high strength, controlled degradation, and photoprocessibility into one system. They reported cross-linked networks with degradation times from 1 week to almost a year that retained 90% of the tensile modulus at 40% mass loss. These dimethacry-

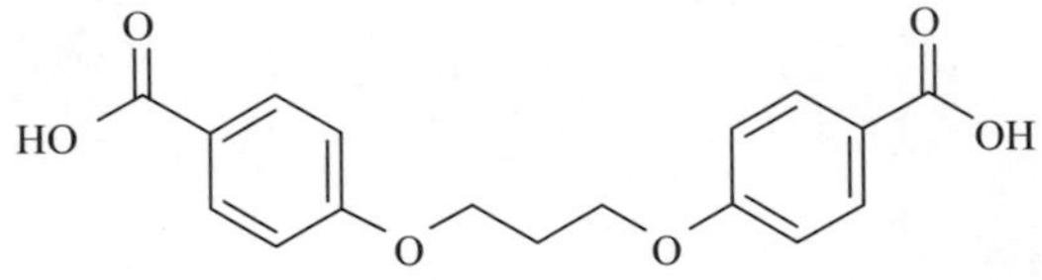

Bis(*p*-carboxyphenoxy)propane (CPP)

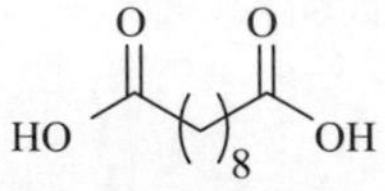

Sebacic acid (SA)

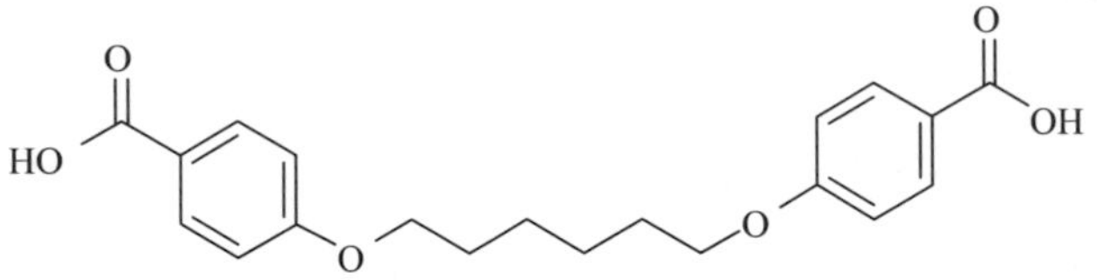

Bis(*p*-carboxyphenoxy)hexane (CPH)

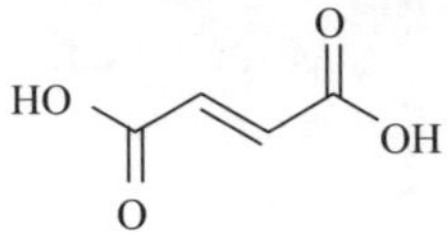

Fumaric acid (FA)

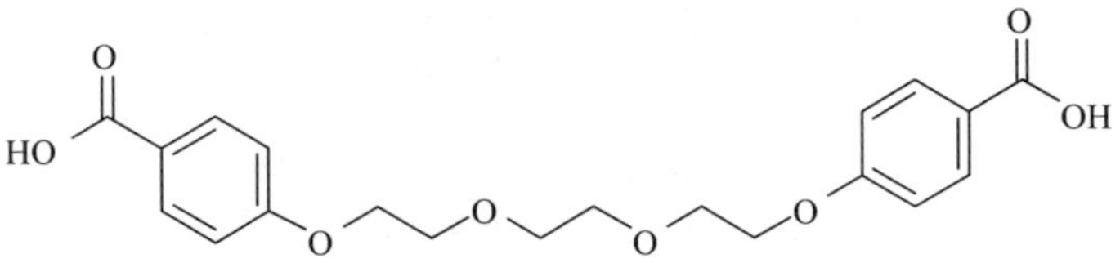

bis(*p*-carboxyphenoxy)triethylene glycol (CPTEG)

Scheme 6.5 Structures of common monomers used to make polyanhydrides

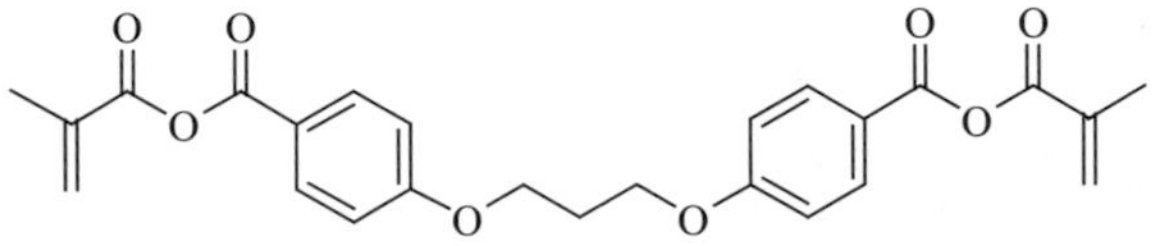

MCPP

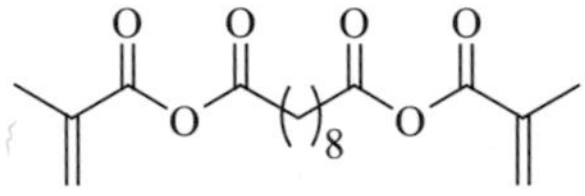

MSA

Scheme 6.6 Structures of *MCPP* and *MSA*

lated anhydride monomers were based on diacid molecules of SA, CPP and CPH, (forming MSA, MCPP and MCPH, respectively—see Scheme 6.6). Due to the high concentration of double bonds and the multifunctional nature of the monomer the formation of highly cross-linked, degradable networks were achieved upon minutes of photopolymerization. The tensile and compressive strengths of poly(MSA) and poly(MCPH) were reported and quite comparable to trabecular bone (5–10 MPa for each), but significantly lower than cortical bone (80–150 MPa and 130–220 MPa), respectively. They also conducted *in vivo* studies with the photocurable polymers in rats in which a 2 mm defect in the tibia was made and then filled with the photoreactive monomers and cured using 1 s pulses of UV light at 10 W/cm^2. Minimal damage to the surrounding tissue occurred, the polymer network adhered well and

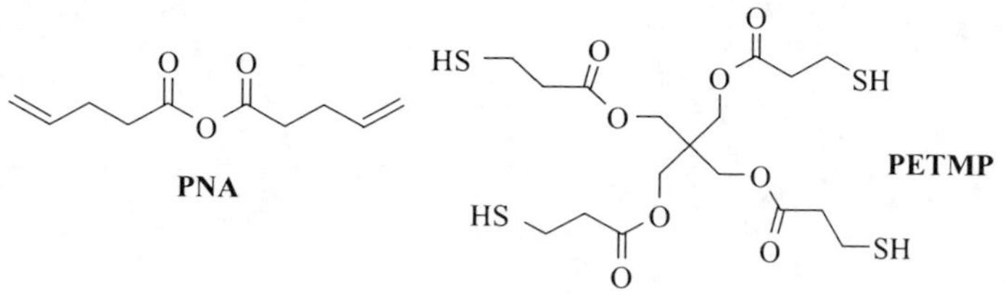

Scheme 6.7 The structures of 4-pentenoic anhydride (*PNA*) and pentaerythritol tetrakis(3-mercaptopropionate) (*PETMP*)

the absence of inflammatory cells suggests that the rat's system tolerated the network [33]. Other studies were conducted on similar systems, and from these it was concluded that these methacrylated anhydrides have characteristics suitable for use in orthopedic applications [34, 35].

We recently have shown that using commercially available monomers, thiol-ene polymerization yields degradable polyanhydrides that, when crosslinked, appear to undergo surface erosion [36]. Scheme 6.7 shows the key monomers used. The main components are an anhydride containing a multi-ene compound (PNA) and a multi-thiol compound (PETMP). It should be noted that the anhydride functionality could, in principle, be contained within the thiol entity, although some of our initial examinations of this avenue indicates the slow formation of a thioester if the two functional groups are present together [37]. Furthermore, a linear non-crosslinked polymer will be formed if both monomers are di-functional, with end-groups being determined by the relative quantities of the monomers.

Time-lapse photographs of a crosslinked polyanhydride cube sample after being immersed in deionized water for various amounts of time at room temperature are shown in Fig. 6.2 [36]. These series of photographs clearly show that the cube retains the basic shape during the erosion process. The sample also remains firm throughout, even though the cube decreases in size. Thus, the development of thiol-ene chemistry in the preparation of polyanhydrides has realized new materials that are photocurable and have controllable erosion rates. Furthermore, it is expected that using thiol-ene chemistry to make polyanhydride network polymers will provide significant flexibility in tailoring characteristics such as crosslink density, functionality and hydrophilicity/hydrophobicity. Such materials are likely to have significant potential in drug delivery and tissue engineering applications.

6.3 Preparation of Polyanhydride Micro- and Nano-particles

The following discussion focuses on the primary methods that have been developed to make polyanhydride micro- and nano-particles. These are hot melt encapsulation, phase inversion/precipitation, emulsions, and spray drying/cryogenic atomization.

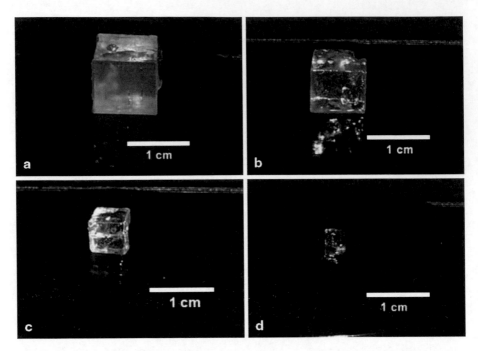

Fig. 6.2 Crosslinked thiol-ene polyanhydride cube (made with 1:1 functional group stoichiometry of PETMP and PNA, with 0.1 wt.% 1-hydroxycyclohexyl phenyl ketone as the photoinitiator) after **a** 0 h, **b** 18 h, **c** 48 h, **d** 72 h in deionized water. (Reproduced from Shipp et al. [36] by permission of The Royal Society of Chemistry (copyright 2009))

A discussion about synthesis, characterization and biomedical use of the particles is given within each section.

6.3.1   Hot Melt Encapsulation

The hot melt encapsulation method requires that the polymer (e.g. in this case, polyanhydrides) be heated above the melt temperature. The powdered active ingredient (i.e. the payload), having been sieved to a size of 50 μm or less, is added to the melt. With continued heating, the mixture is added to a non-miscible solvent (silicon oil is often used) and stirred until a stable emulsion is achieved. Upon cooling, the polymer solidifies, forming the particles with the active drug ingredient within. The particles can then be washed with suitable non-solvents and dried, typically giving free-flowing powders. While this approach may give a wide range of sizes and microspheres with dense and smooth surfaces, the high temperatures may lead to degradation of the drug or denaturation of proteins, depending on the active ingredient used. Furthermore, in order to make a melt with appropriate viscosity, molecular weights of the polymer should be in the range of 1,000–50,000.

Mathiowitz and Langer [38] were the first to develop hot-melt encapsulation with polyanhydrides, and indeed the first polyanhydride particles. Made from poly(CPP-SA), the microspheres with diameters of tens of microns up to 2,000 μm could be made, the size was controlled by the stirring rate during encapsulation and being independent of additional model payloads (dye and insulin). Using SEM, the mode of erosion appeared to be via surface erosion. When loaded with a payload, it was found that for the smaller microspheres (less than 50 μm) the rate of payload release was dependent on the size of the microspheres—the smaller the microsphere the faster the release rate. This proved that the microspheres were undergoing surface erosion, which in turn controlled the release of the payload. If erosion occurred via bulk erosion then the rate of release would be independent of microsphere size. The insulin-containing microspheres were injected into diabetic rates—this resulted in normaglycemia for 3–4 days.

6.3.2   Phase Inversion/Precipitation

Phase inversion preparation of polyanhydride particles simply involves the dissolution of the polymer into a good solvent and then the dropwise addition of the solution to a large excess of a non-solvent. The solvent that dissolves the polymer should be miscible with the non-solvent such that upon addition of the polymer solution to the non-solvent, the solvent quickly disperses into the non-solvent leaving the polymer to collapse, aggregate and so precipitate. Under appropriate conditions, particles of sub-micron to micron dimensions can be formed. Surfactants may be added to ensure particle stability.

Mathiowitz et al. [39] examined the use of poly(fumaric acid-co-sebacic acid) (poly(FA-SA)) as a oral drug delivery system. The poly(FA-SA) microspheres, with a 20:80 ratio of FA:SA and molecular weights of 500–2,000, were produced by dissolving the polymer into methylene chloride (a good solvent) which was then added to petroleum ether (a non-solvent). Particle sizes ranged from 0.1–5.0 μm. Examples are shown in Fig. 6.3. The poly(FA-SA) microspheres were loaded with insulin, plasmid DNA and dicumarol (an anticoagulant) and injected into rat models. The microspheres were found to be bioadhesive, and could continue contact with intestinal mucus and cell linings. These properties improved the absorption of all three model payloads.

Fu and Wu [40] performed a light scattering study on poly(SA) particles made by the phase inversion method. In their process, the poly(SA) was dissolved in tetrahydrofuran (THF) and then added to water containing surfactant. The surfactants used were sodium laurylsulfate (SDS), polyoxyethylene20 sorbitan monostearate (Tween 60), sorbitan monolaurate (Span 20), and polyoxyethylene20 sorbitan monolaurate (Tween 20). It was found that particles with hydrodynamic radii around 200 nm were made, and the degradation of the polymer was close to zero order (up to 75% weight loss) under a variety of conditions, including different temperatures, surfactants, surfactant concentration and pH.

Fig. 6.3 SEM image of
poly(FA-SA) particles
prepared by dispersing 4%
(w/v) polymer solution in
methylene chloride into a
100-fold excess of petroleum
ether. Scale bar = 5 μm.
(Reproduced from Mathiow-
itz et al. [39] by permission
of Macmillan Publishers Ltd
(copyright 1997))

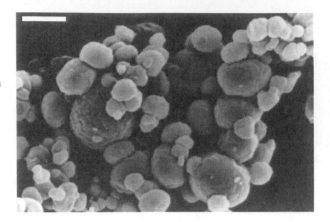

6.3.3   Emulsion Methods

Emulsions are useful in preparing polymer particles, particularly those containing
drugs, in part because the procedure can be done at room temperature. This removes
the negative effects of high temperatures on degradation of the polymer carrier or
biologicals, such as proteins. There is also a wide range of surfactants and information regarding their use already available. Oil-in-water (O/W) single emulsions
are used for hydrophobic polymers, whereas oil-in-water-in-oil (O/W/O) double
emulsions are normally used for hydrophilic polymers. There are also reports of
solid-in-oil-in-oil (S/O/O) double emulsions where the aim was to reduce premature
hydrolysis of the degradable linkage in the polymer chain by avoiding the use of
water in the emulsification [41].

Procedurally, O/W emulsions of the polymer can be achieved by dissolution of
the polymer in an organic, water-immiscible solvent (e.g. chloroform), which is
then added to an aqueous system containing a stabilizer (e.g. surfactant), followed
by high energy emulsification (e.g. through the use of sonication or homogenizers).
The polymer particles can be finally formed by chain crosslinking and/or solvent
removal.

In any of these approaches, there are several critical aspects of particle formation
that need to be evaluated and optimized. Firstly, stabilization of the emulsions is of
obvious importance, particularly when the final product is being loaded with drug
molecules. Without appropriate stabilization several unwanted outcomes may occur, including aggregation, agglomeration and loss of the active ingredient, which
may happen at any time in the particle formation process, but most often occurs
during emulsification. If appropriately stabilized, the particles may be crosslinked
in order to preserve the individual particles and allow the solvents to be removed.
The crosslinking chemistry must be compatible with both the polymer carrier and
the drug. Other factors that affect particle formation and stability include solvent
types and ratios, emulsification methodology and energy.

Fig. 6.4 SEM image of poly(SA) nanoparticles made with the O/W emulsion technique. (Reproduced from Lee and Chu [42] by permission of John Wiley & Sons (copyright 2008))

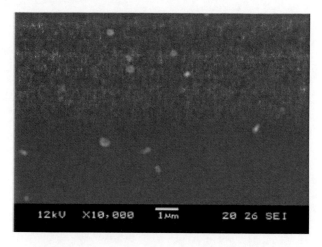

Lee and Chu [42] examined the degradation behavior of CPP/SA copolymers in nanoparticle and disc forms. The nanoparticles were formed using the O/W emulsion technique, followed by removal of the dichloromethane solvent. Poly(vinyl alcohol) (PVA) was used as a stabilizer. Figure 6.4 shows a SEM image of nanoparticles made from poly(SA) (50 mg/ml), using 2.5 wt/vol.% of PVA. The average particle sizes depended on the amount of PVA added; 1.0 w/v% PVA gave 540 nm, 2.5 w/v% gave 436 nm, and 5.0 w/v% gave 423 nm. When only 0.5 w/v% PVA was added the poly(SA) precipitated. It was noted that the nanoparticles could be freeze-dried and resuspended in phosphate buffered saline (PBS) solution (pH = 7.4, 37°C) without aggregation over a 10 h period, after which the average particle size increased significantly, up to several microns in diameter. However, this was not necessarily a major problem since the particles typically degraded in a similar time-frame, depending on the composition. For example, copolymers of SA:CPP (80:20) were shown to have eroded by approximately 80% (based on % monomer release; PBS pH = 7.4, 37°C) after 24 h. The degradation/erosion rates of the nanoparticles were compared to discs (8 mm diameter, 1 mm thickness) of the same composition. It was found that the degradation/erosion rates were much faster for the nanoparticles; this was attributed to the much larger surface area available for hydrdolysis of the anhydride moiety.

Pfeifer et al. [43] synthesized micro- and nanoparticles of copolymers based on SA and poly(L-lactic acid). The resulting polymers therefore contained both ester and anhydride moieties, and were expected to display a mixture of erosion mechanisms: bulk erosion due to the esters and surface erosion due to the anhydrides. The copolymers were synthesized according to Scheme 6.8. The two pathways used to make the copolymers provided control over the molecular weights of the final polymer, which could also influence the degradation of the particles in addition to the composition (e.g. ester vs. anhydride content). The particles were generated using a W/O/W double emulsion technique, using PVA as a stabilizer. The particles were characterized for size (using SEM, Coulter microparticle analyzer and light

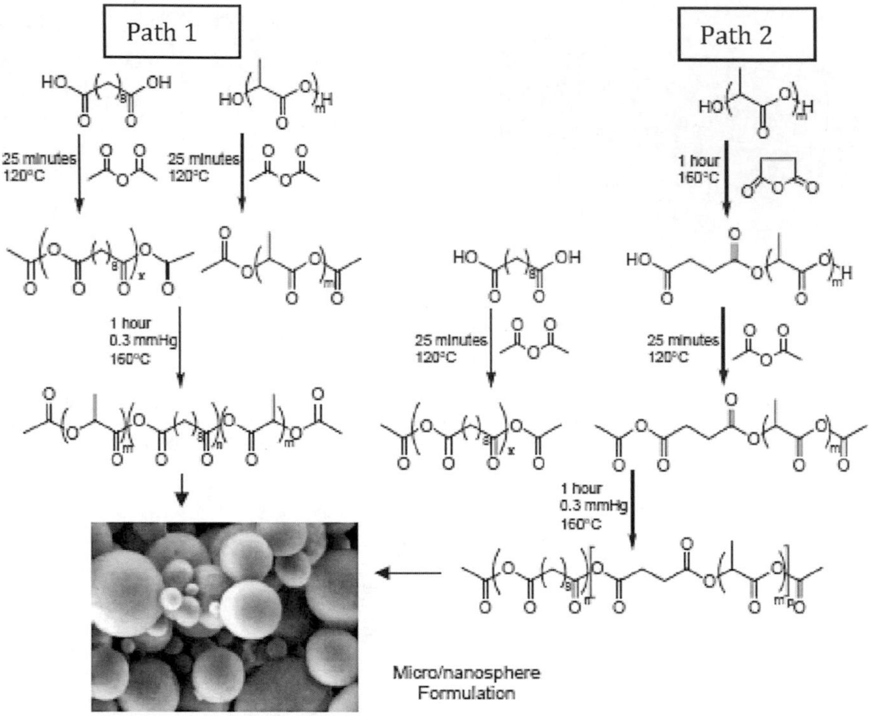

Scheme 6.8 Synthetic schemes for poly(ester-anhydride) copolymers leading to microsphere formation. (Reproduced from Pfeifer et al. [43] by permission of Elsevier Ltd (copyright 2005))

scattering) and zeta potential. The diameters for the microparticles were between 3–10 μm, while the nanoparticle diameters ranged from 300–500 nm, and all had negative zeta potentials under the conditions measured (ranging from −20 mV to approximately −35 mV in water). It was found, however, that upon freeze-drying the particles tended to aggregate, particularly those with larger polyanhydrides content. This was overcome through the use of a cryoprotectant, D(+)-trehalose.

Figure 6.5 shows SEM images of microparticles with various compositions (all ester, 50:50 ester:anhydride and all anhydride) before and after exposure to PBS solution. The spherical shape of the particles is clearly evident before dispersion in the PBS solution. However, only the particles containing polyanhydrides appear to become smaller after degradation, with the 100% polyanhydrides microspheres eroding to a much greater extent (17% over 3 days) than those made with 0% or 50% polyanhydrides (no noticeable erosion in 3 days). The degraded particles also appear to become more oblong in shape

Pfeifer et al. [43] also examined the functionalization of the polyanhydrides particles with cystamine groups, which were then subsequently reduced to yield surface thiol entities. This is outlined in Scheme 6.9. The degree of functionalization was

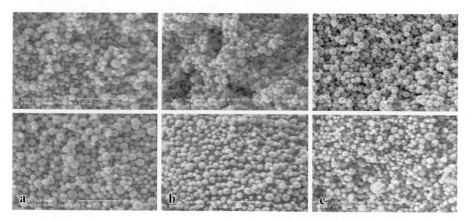

Fig. 6.5 SEM images of poly(ester-anhydride) microparticles made with the W/O/W double emulsion technique. *Top row*: initial microspheres. *Bottom row*: PBS degraded microspheres (in PBS solution for 3 days). **a** 0% polyanhydrides, **b** 50% polyanhydrides, **c** 100% polyanhydrides. (Reproduced from Pfeifer et al. [43] by permission of Elsevier Ltd (copyright 2005))

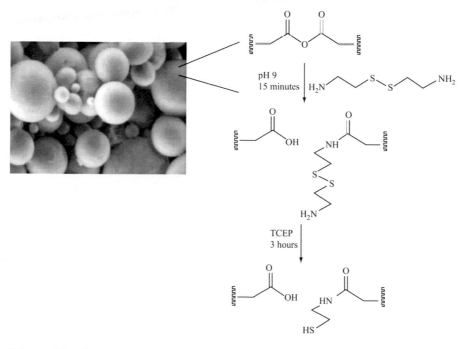

Scheme 6.9 Polyanhydride surface modification scheme with cystamine. (Reproduced from Pfeifer et al. [43] by permission of Elsevier Ltd (copyright 2005))

followed quantitatively by Ellman's reagent (5,5'-dithiobis(2-nitrobenzoic acid)), and it was found that the microparticles were functionalized at 0.20–35 µmol/g and the nanoparticles at 0.6–100 µmol/g.

Kipper et al. [44] studied the synthesis of poly(CPH) and poly(SA) microparticles as the carrier of tetanus toxoid (TT) model antigen. The primary aim was to load the polyanhydride spheres with TT, examine their release profiles and study their ability to induce an *in vivo* antigen-specific immune response. *In vivo* tests were performed on mice models. Figure 6.6 shows the SEM images of poly(CPH-SA) copolymer particles loaded with TT. Spheres in the range of 10–60 μm diameter can be seen, along with some small amount of loose surface material. Some small divots on the surface (as shown in the Fig. 6.6 by the arrows) indicate the fragility of the spheres. Figure 6.7 shows the release profile of TT from microspheres made up of 20:80 and 50:50 poly(CPH:SA). These indicate nearly zero-order release kinetics, with the more hydrophobic 50:50 polymer showing slower release data. They concluded that, in addition to prolonging the release of the TT, the polyanhydride/TT microsphere single dose delivery system stimulated a mature immune response that could be selectively tuned without supplementary adjuvants.

6.3.4 Spray Drying and Cryogenic Atomization

These methods are simple and widely applied in many fields, spray drying in particular [45]. In simplistic terms, these techniques utilize phase separation and solvent evaporation by passing a solution containing the active ingredient and excipient, along with other entities such as stabilizing agents, binders, flavors, etc., though a spray nozzle. The cooling and evaporation processes that occur once the solution expands after it passes through the nozzle can result in dried micron-sized particles. The spray may be directed into compressed or cryogenic liquids, such as liquid CO_2, liquid nitrogen or argon. The cryogenic atomization approach can yield higher loadings of the active ingredient and avoids high temperatures, which may denature proteins or degrade drugs [45].

For the production of polyanhydride particles, both spray drying and cryogenic atomization have been used to create micron-sized particles that encapsulate drugs or proteins. Spray drying to make polyanhydride microspheres was first reported by Mathiowitz et al. [46] The microspheres were made from homopolymers and copolymers of SA, CPP and fumaric acid, and were loaded (5–7%) with dyes or a protein (bovine somatotropin, STH) for release studies. Scanning electron microscopy (SEM) was used to evaluate the microsphere morphology, while x-ray diffraction and differential scanning calorimetry (DSC) were used to examine the crystallinity. It was found that less crystalline polymers gave smoother microspheres, while the spray drying process decreased the crystallinity of all polymers. An SEM image of poly(CPP-SA) (20:80) containing the protein, release kinetics and degradation data are shown in Fig. 6.8. The rough surface morphology is clearly evident, and the release kinetics indicates that for these microspheres the payload release appears independent of the microsphere erosion. In most cases studied, only 15% of the polymer microspheres were eroded within the first 4 h, while 60–70% of the

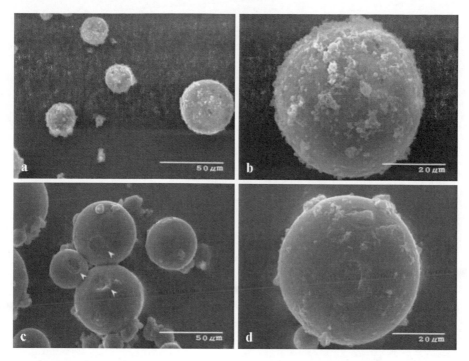

Fig. 6.6 TT-loaded poly(CPH-SA) microspheres. 20:80 copolymer microspheres showing small polymer particles flocculated on their surfaces [**a** and **b**] and 50:50 microspheres with small circular divots [indicated by the *arrowheads* in **c**] formed when the loosely adhered clumps of microspheres break apart [**c** and **d**]. (Reproduced from Kipper et al. [44] by permission of John Wiley & Sons (copyright 2006))

Fig. 6.7 *In vitro* TT release profiles for TT-loaded poly-(CPH-SA) 20:80 (●) and 50:50 (▲) microspheres. (Reproduced from Kipper et al. [44] by permission of John Wiley & Sons (copyright 2006))

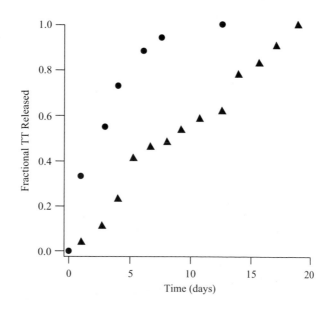

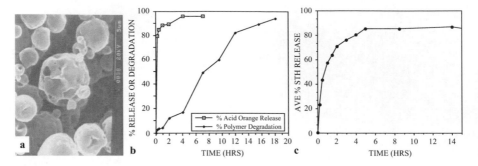

Fig. 6.8 a SEM image of poly(CPP-SA) (20:80) microspheres loaded with bovine somatotropin (STH), **b** acid orange release from poly(CPP-SA) (20:80) microspheres and polymer degradation, and **c** STH release from poly(CPP-SA) (20:80) microspheres. (Adapted from Mathiowitz et al. [46] by permission of John Wiley & Sons (copyright 1992))

protein/dye was released in this same time. It was concluded that payload release was dominated by diffusional processes rather than mass loss via polymer erosion.

Using cryogenic atomization, Narasimhan et al. [41] made microparticles of copolymers synthesized from CPH, SA and CPTEG monomers (see Scheme 6.5 for structures). In this work they systematically investigated the effect on polymer structure and particle fabrication method on the release kinetics of ovalbumin protein from polyanhydride microparticles. The size of the particles for both the CPH:SA and CPTEG:CPH copolymers ranged from 5–25 μm (Fig. 6.9), although they have obviously broad particle size distributions. Protein release kinetics from these particles showed that after an initial burst of protein, there was a sustained release profile for each sample. For the amphiphilic copolymers, those containing CPTEG, the initial burst of protein release was less than for the CPH:SA copolymers. This was attributed to the larger differences in hydrophobicities between the protein and the CPH:SA copolymers; this incompatibility leads to non-uniform distribution of the protein in the microsphere, with higher concentrations at the surface thereby leading to greater protein release at early stages of degradation. The stability of the protein is also greater with the CPTEG:CPH copolymer.

6.4   Conclusions and Outlook

Among the multitude of biomaterials, polymer colloids are unique in the breadth of potential applications. The ability to control size, architecture, surface functionality, and degradation profiles, along with other properties discussed above, allows polymer colloids to be used in many biomedicine and pharmaceutical applications. Furthermore, with the advent of instrumentation that allows for analysis of materials as low as the sub-nanometer range, it is likely that the use of nanoscale colloidal materials will continue to gain momentum. Another fertile area will be to examine

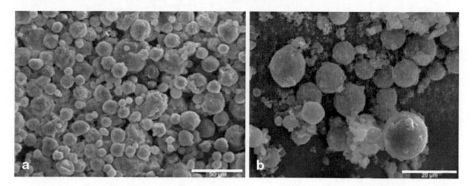

Fig. 6.9 Example polyanhydrides microspheres made by cryogenic atomization. **a** 50:50 CPH:SA, scale bar represent 50 μm. **b** 80:20 CPTEG:CPH, scale bar represent 20 μm. (Reproduced from Lopac et al. [41] by permission of John Wiley & Sons (copyright 2009))

the interactions between synthetic colloids and biological entities, such as cells and their substructures (e.g. proteins). Research in this area, while having progressed significantly in the past decade or so (see, for example, a recent *MRS Bulletin* devoted to materials for regenerative medicine [47]), is still in its infancy and with many unanswered questions remaining it is anticipated that it will be an active field for many years to come.

Acknowledgements   We thank the Center for Advanced Materials Processing, CAMP (a New York State Center for Advanced Technology, funded through NYSTAR) at Clarkson University, and the Department of Chemistry and Biomolecular Science at Clarkson University for financial support.

References

1. Duncan, R.; Ringsdorf, H.; Satchi-Fainaro, R. *Adv. Poly. Sci.* **2006**, *192*, 1-8.
2. Li, S. *J. Biomed. Mater. Res.* **1998**, *48*, 342-353.
3. Finne-Wistrand, A.; Albertson, A.-C. *Annu. Rev. Mater. Res.* **2006**, *36*, 369-395.
4. Zhang, Q.; Remsen, E. E.; Wooley, K. L. *J. Am. Chem. Soc.* **2000**, *122*, 3642-3651.
5. Wooley, K. L. *J. Polym. Sci. Part A: Polym. Chem.* **2000**, *38*, 1397-1407.
6. Nishiyama, N.; Kataoka, K. *Adv. Poly. Sci.* **2006**, *193*, 67-101.
7. Malmsten, M. *Soft Matter* **2006**, *2*, 760-769.
8. Nasongkla, N.; Bey, E.; Ren, J.; Ai, H.; Khemtong, C.; Guthi, J. S.; Chin, S.-F.; Sherry, A. D.; Boothman, D. A.; Gao, J. *Nano Lett.* **2006**, *6*, 2427-2430.
9. Gopferich, A.; Tessmar, J. *Advanced Drug Delivery* **2002**, *54*, 911-931.
10. Goepferich, A. *Macromolecules* **1997**, *30*, 2598-2604.
11. Uhrich, K. E.; Cannizzaro, S.M.; Langer, R.S.; Shakesheff, K.M. *Chem. Rev.* **1999**, *99*, 3181-3198.
12. Schmitt, E. E.; Polistina, R. A. *U.S. Patent 3297033* **1967**.
13. Vert, M.; Li, S.; Spenlehauer, G.; Guerin, P. *Biomaterials* **1994**, *15*, 1209-1213.
14. Slager J.; Domb, A. *J. Adv. Drug Delivery Rev.* **2003**, *55*, 549-583.
15. Alexis, F. *Polymer Int.* **2005**, *54*, 36-46.

16. Bigg, D. M. *Adv. Polymer Technol.* **2005**, *24*, 69-82.
17. Amsben, B. G.; Misra, G.; Gu, F.; Younes, M. H. *Biomacromolecules* **2004**, *5*, 2479-2486.
18. Peppas, N.; Langer, R. *Science* **1994**, 1715-1720.
19. Heller, J. *Biomaterials* **1990**, *11*, 659-665.
20. Anderson, D. G.; Lynn, D. M.; Langer, R. *Angew. Chem. Int. Ed.* **2003**, *42*, 3153-3158.
21. Anderson, D. G.; Tweedie, C. A.; Hossain, N.; Navarro, S. M.; Brey, D. M.; Van Vliet, K. J.; Langer, R.; Burdick, J. A. *Adv. Mater.* **2006**, *18*, 2614-2618.
22. Lynn, D. M.; Anderson, D. G.; Putnam, D.; Langer, R. *J. Am. Chem. Soc.* **2001**, *123*, 8155-8156.
23. Lynn, D. M.; Langer, R. *J. Am. Chem. Soc.* **2000**, *122*, 10761-10768.
24. McBath, R. A.; Shipp, D. A. *Polym. Chem.* **2010**, *1*, 860-865.
25. Tamada, J.; Langer, R. *J. Biomater. Sci. Polym. Ed.* **1992**, *3*, 315-353.
26. Bucher, J. E.; Slade, W. C. *J. Am. Chem. Soc.* **1909**, *31*, 1319-1321.
27. Hill, J.; Carothers, W. *J. Am. Chem. Soc.* **1932**, *54*, 1569-1578.
28. Hill, J. *J. Am. Chem. Soc.* **1930**, *52*, 4110-4114.
29. Conix, A. *J. Polymer Sci. Part A* **1958**, *29*, 343-353.
30. Yoda, N. *Makromol. Chem.* **1962**, *55*, 174-190.
31. Kattia, D. S.; Lakshmi, S.; Langer, R.; Laurencin C.T. *Adv. Drug Deliv. Rev.* **2002**, *54*, 933-961.
32. Uhrich, K. E.; Gupta, A.; Thomas, T.; Laurencin, C.T.; Langer, R. *Macromolecules* **1995**, *28*, 2184-2193.
33. Anseth, K. S.; Shastri, V. R.; Langer, R. *Nat. Biotechnol.* **1999**, *17*, 156-159.
34. Young, J. S.; Gonzales, K. D.; Anseth, K. S. *Biomaterials* **2000**, *21*, 1181-1188.
35. Weiner, A. A.; Shuck, D. M.; Bush, J. R.; Shastri, V. P. *Biomaterials* **2007**, *28*, 5259-5270.
36. Shipp, D. A.; McQuinn, C. W.; Rutherglen, B. G.; McBath, R. A. *Chem. Commun.* **2009**, 6415-6417.
37. Rutherglen, B. G.; McBath, R. A.; Huang, Y. L.; Shipp, D. A. *Macromolecules* **2010**, 10297-10303.
38. Mathiowitz, E.; Langer, R. *J. Control. Release* **1987**, *5*, 13-22.
39. Mathiowitz, E.; Jacob, J. S.; Jong, Y. S.; Carino, G. P.; Chickering, D. E.; Chaturved, P.; Santos, C. A.; Vijayaraghavan, K.; Montgomery, S.; Bassett, M.; Morrell, C. *Nature* **1997**, *386*, 410-414.
40. Fu, J.; Wu, C. *J. Polym. Sci. Part B: Polym. Phys.* **2001**, *39*, 703-708.
41. Lopac, S. K.; Torres, M. P.; Wilson-Welder, J. H.; Wannemuehler, M. J.; Narasimhan, B. *J. Biomed. Mater. Res. B* **2009**, *91B*, 938-947.
42. Lee, W. C.; Chu, I. M. *J. Biomed. Mater. Res. B* **2008**, *84B*, 138-146.
43. Pfeifer, B. A.; Burdick, J. A.; Langer, R. *Biomaterials* **2005**, *26*, 117-124.
44. Kipper, M. J.; Wilson, J. H.; Wannemuehler, M. J.; Narasimhan, B. *J. Biomed. Mater. Res. A* **2006**, *76A*, 798-810.
45. Barot, M.; Modi, D. M.; Parikh, J. R. *Drug Deliv. Tech.* **2006**, *6*, 54-60.
46. Mathiowitz, E.; Bernstein, H.; Giannos, S.; Dor, P.; Turek, T.; Langer, R. *J. Appl. Polym. Sci.* **1992**, *45*, 125-134.
47. Shastri, V. P.; Lendlein, A. *MRS Bulletin* **2010**, *35*, 571-575.

Chapter 7
Synthesis of Inorganic Nanoparticles Using Protein Templates

Artem Melman

Abstract The synthesis of nanoparticles with controlled composition, size and shape has long been of scientific and technological interest. Despite efforts invested to this problem, selective preparation of tailor made particles by conventional methods constitutes a considerable challenge. In contrast, highly selective fabrication of monodisperse functional nanosized particles occurs continuously in every living cell. The process of building of inorganic and hybrid organic–inorganic architectures on templates of biopolymers through biomineralization is also very common in biological systems and provides the level of control that has not been closely achieved in conventional technology. Any biomineralization is achieved through controlled nucleation of metal cations by functional groups in amino acid constituting proteins. Study of the biomineralization in the last two decades and attempts to mimic the process have been highly successful although details are far from being completely understood

Keywords Protein cages • Templated synthesis • Protein templates • Ferritin • Dps proteins • HSP proteins • Viruses • Hollow nanoparticles • Bimetallic nanoparticles

7.1 Introduction

The synthesis of nanoparticles with controlled composition, size and shape has long been of scientific and technological interest. Despite efforts invested to this problem, selective preparation of tailor made particles by conventional methods constitutes a considerable challenge. In contrast, highly selective fabrication of monodisperse functional nanosized particles occurs continuously in every living cell. The process of building of inorganic and hybrid organic–inorganic architectures on templates of biopolymers through biomineralization is also very common in biological systems and provides the level of and control that has not been closely achieved in conventional technology. Any biomineralization is achieved through controlled nucleation of metal cations by functional groups in amino acid constitut-

A. Melman (✉)
Department of Chemistry & Biomolecular Science, Clarkson University,
8 Clarkson Avenue, SC, Box 5810, Potsdam, NY 13699, USA
e-mail: amelman@clarkson.edu

E. Matijević (ed.), *Fine Particles in Medicine and Pharmacy,*
DOI 10.1007/978-1-4614-0379-1_7, © Springer Science+Business Media, LLC 2012

Fig. 7.1 Schematic representation of protein nanocage;
templating of inorganic
nanoparticles on polymers: **a**
interior; **b** interface;
c exterior

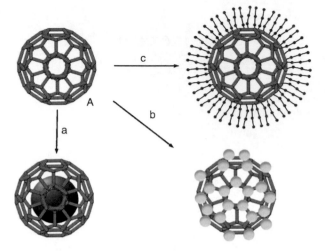

ing proteins. Study of the biomineralization in the last two decades and attempts to mimic the process have been highly successful although details are far from being completely understood [1, 2].

Use of biological nanosized biological systems as a template for selective synthesis of inorganic nanoparticles can therefore revolutionize the current technology. This chapter is devoted to a brief overview the synthesis of well defined inorganic nanoparticles using interior and exterior surface of most commonly biological templates. The chapter follows conventional classification of this reaction on the type of template with particular attention to proteins cages where a considerable array of information has been collected and comparison of different methods. Synthesis of nanoparticles on viral templates is highly promising due to diversity of shapes that can be potentially obtained by these methods but comparison of different studies that uses different templates is rather complicated.

This chapter will follow classification with regards to biological template used in the synthesis and subclassification with regards to the synthetic method applied for the irreversible step of nanoparticle formation (Fig. 7.1).

7.2   Protein Cages

Protein cages are essentially monodisperse nanosized molecular structures that self-assemble from individual protein subunit building blocks. Size and biological role of molecular cages are diverse. Smaller molecular cages such as ferritin, small heat shock proteins (HSP), DNA binding proteins from starved cells (DPS) serve as biomineralization sites for sequestration and storage of iron. Self-assembly of 12 (HSP) or 24 subunits produces a highly symmetric cage with diameter 9–12 nm that is isolated from the environment by a 6 nm protein shell. All these cages possess negatively charged amino acid residues that are capable of binding iron cations

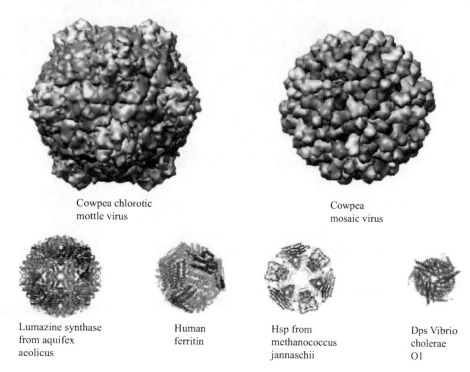

Cowpea chlorotic
mottle virus

Cowpea
mosaic virus

Lumazine synthase
from aquifex
aeolicus

Human
ferritin

Hsp from
methanococcus
jannaschii

Dps Vibrio
cholerae
O1

Fig. 7.2 Schematic representation of protein cages with different sizes

thus serving as nucleation sites for growth of iron hydroxide core Synthesis of nanoparticles in these proteins is essentially mimics of a natural process of biomineralization in these proteins allowing a large variety of spherical nanoparticles (Fig. 7.2).

Protein cages that are formed by viral proteins which serve for storage viral enzymes and nucleic acids are considerably larger (25–30 nm). Although these proteins does not serve in nature for biomineralization the presence of multiple charges groups on the internal or external surfaces of viral proteins can often served as nucleation sites allowing efficient synthesis of nanoparticles. Viral proteins possess a range of different shapes thus opening extensive possibilities of shape control of inorganic nanoparticles. Medium sized protein cages with sizes 15–20 nm are enzyme complexes such as lumazine synthase and pyruvate dehydrogenase are not used in nature for storage purposes yet possess well defined highly symmetric 3-D structure similar to other protein cages.

The resultant cages possess a defined interior cavity that can be used as template for growing inorganic nanoparticles. The protein shell of the resultant hybrid structures can be subsequently removed thus releasing inorganic nanoparticles with narrow distribution of sizes and molecular weight. Alternatively, the protein shell can be used for self assembly into hierarchic functional assemblies. Both the external and internal surfaces of protein shell can be modified using conventional genetic or chemical methodologies thus opening extensive possibilities to control nucleation

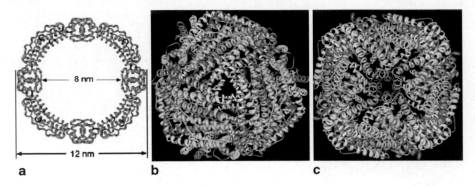

Fig. 7.3 a Inner cavity in apoferritin; **b** threefold channels in apoferritin; **c** fourfold channels in apoferritin

on the inner surface and molecular assembly sites on the external surface which are beyond the scope of the process.

7.2.1 Synthesis of Nanoparticles in the Inner Cavity of Natural Ferritin

7.2.1.1 Structural Features of Ferritins and Ferritins-Like Cages and Their Role in Biology

Historically, ferritins were the first protein cages to be used as templates for synthesis of inorganic nanoparticles. Ferritins are metalloproteins that serve for storage of iron in essentially all living organisms. Two types of natural ferritins of different size should be distinguished—maxi mammalian and bacterial ferritin comprised of 24 subunits and much less common mini ferritin of bacteria and archaea built of 12 subunits that are considered in a special section below. All ferritin molecules possess similar size and shape and consist of the protein apoferritin shell and the mineralized iron hydroxide core that in maxi ferritins.

The diameter of the protein shell in maxi ferritin is 12–13 nm diameter and of the inorganic core is 5–8 nm. The protein shell is composed from total 24 H and L-type protein subunits. Each H subunit possess a catalytic ferroxidase center capable of very rapid oxidation of iron(II) cation using dioxygen as an oxidant and producing hydrogen peroxide as a byproduct. The L-type subunit lack catalytic activity but possess extended stability. It coassembles with H-subunits in different ratios that are species- and tissue dependent. The HL ratio changes from zero (in bacterial and plant ferritins) to 5:1 in horse spleen ferritin. The protein shell of ferritin is unusually stable in pH range 2–10 and temperatures up to 70°C with HL ferritins are considerably more stable than H only ferritin [3]. No appreciable dissociation of the ferritin or apoferritin shell under a wide range of experimental conditions was observed (Fig. 7.3).

The cavity of maxi-ferritins is capable of accommodating up to 4500 iron atoms as a porous ferrihydrite iron(III) core with a diameter ~7–8 nm. Most data indicate that iron enters and exits the cavity of animal ferritins via the eight hydrophilic threefold channels. The reported diameter of these channels is 4 Å although a considerable flexibility has been reported [4, 5] and compounds with molecular diameter as large as 6 Å are capable to pass through these channels. In any case the process of iron deposition is diffusionally controlled. The hydrophobic channels are thought to be involved in protons and oxygen transport and diffusion although long range electron tunneling mediated by aromatic groups within the protein shell has been suggested. Small molecules such as reductants/oxidants and iron chelators can enter and leave the ferritin shell during the processes of iron deposition and release.

External/internal charges distribution on the ferritin surface and the number of nucleation centers are two crucial factors that has to be taken into account designing nanoparticle synthesis. The 24-subunits inner cavity of ferritin is connected through threefold and fourfold channels to the outside environment. The threefold channels are hydrophilic and leaned with negatively charged residues of aspartic and glutamic acids, this feature is common for ferritins of different type. The fourfold channels are hydrophobic. The recent crystal structure of the plant ferritin showed that they are leaned with several histidine residues. No metal ions have been observed around the channels in contrast to other ferritins. The role of the fourfold channels in metal ion deposition in ferritin cavity is less clear [6]. Several studies have confirmed that selective mineralization of inner cavity of ferritin with divalent metal cations over the monovalent cations similar to the calcium channels [7]. The electrostatic potential in human H-chain homopolymeric ferritin has been calculated showing that electrostatic potential of the threefold channels leads towards the inner cage of the ferritin, while fourfold channel leads to the ferritin outside [8].

The first step in the nanoparticles preparation is seeding the nucleation centers in inner cavity of ferritin. The initial centers of nucleation are residues of acidic amino acids such as glutamate and aspartate on the inner surface of the apoferritin shell. Both H- and L-subunits can serve as nucleation centers although L-subunits play the predominant role in nucleation. It has been suggested that the number of the primary nucleation centers influence the polymorphic properties of the resultant nanoparticle. The more primary nucleation centers are available within the inner core the wider will be polymorphism of the prepared nanoparticles. The nature of primary nucleation centers in ferritin is not exclusive for iron cations and a large variety of metal ions can be seeded inside the ferritin core [9]. However, as evidenced by Extended X-ray Absorption Fine Structure (EXAFS) experiments on Cu^{2+} deposition into ferritin metal cations can form an amorphous network of oxygen/hydroxide-coordinated cations that probably serve as the secondary nucleation sites [10].

Most commonly operations with ferritin are conducted in water solutions as solubility of ferritin in organic solvents is very low. However, conversion of carboxylic residues on the exterior surface of ferritin with aminoalkyl chain drastically improves solubility of ferritins in organic solvents [11].

7.2.1.2 Preparation of Nanoparticles in the Interior Cavity of Ferritin

Since the interior surface of the protein shell of ferritin contain acidic functions that serve as nucleation centers a wide range of inorganic nanoparticles can be synthesized within the apoferritin nanoreactor. Most of the research has been focused on preparation of polymorphic composites of metal oxides and chalcogenides as well as metal alloys, although more complex hybride organic–inorganic particles can be prepared. Transmission electron microscopy (TEM) of ferritin possessing a completed iron core revealed homogeneously sized nanoparticles commensurate in size with the interior diameter of ferritin (7.3 ± 1.4 nm). After assembly of the inorganic core the apoprotein shell can be removed by a number of methods including calcination, hydrolysis, or other methods [12–16].

A variety of reactions can be employed for the nanoparticles synthesis although the essential requirement for all these reactions will be irreversible formation of precipitates from soluble precursors. The process of precipitation must proceed slower then passage of ions through threefold channels of ferritin otherwise precipitation in solution and on the outside surface of ferritin will occur.

A variety of inorganic particles has been prepared by this approach and many more can be done. Because of the diversity of products it is more convenient to classify known processes by the type of reaction used for preparation of particles.

Most commonly used methods are:

a) Catalytic and non catalytic oxidation of transition metal cations with molecular oxygen
b) Non-catalytic oxidation with other oxidants
c) Prototropic reactions involving formation of metal hydroxides and oxides
d) Precipitation of insoluble salts

7.3 Oxidation with Dioxygen

Iron and other metal hydroxides–oxides. The predominant form of protein unbound iron cations in the reducing environment of the cell is iron(II). Iron(II) cations and dioxygen are native substrate for ferroxidase unit that is placed on the inner cavity of ferritin and the process of deposition of iron(III) hydroxide in the inner cavity of ferritin occur in vast majority of living organisms.

The first step in iron mineralization in such ferritin is the reaction of ferrous cations with free oxygen with formation diferric peroxo complex and hydrogen peroxide. In maxi-recombinant-L-exclusive ferritins lack of ferroxidase activity considerably slows the rate of iron accumulation. Negatively charged amino acids in the inner shell bind ferrous cations and serve as nucleation centers for non-catalytic oxidation of iron(II) cations with molecular oxygen.

$$2Fe^{2+} + O_2 \rightarrow [Fe(III)-O-O-Fe(III)] \tag{1}$$

$$[Fe(III)-O-O-Fe(III)] + H_2O \rightarrow [Fe(III)-O-Fe(III)] + H_2O_2 \quad (2)$$

$$[Fe(III)-O-Fe(III)] + xH_2O \rightarrow Fe_2O_3(H_2O)_{x-2} + 4H^+ \quad (3)$$

$$2Fe^{2+} + O_2 + xH_2O \rightarrow Fe_2O_3(H_2O)_{x-2} + 4H^+ + H_2O_2 \quad (4)$$

The general procedure for preparing iron oxide nanoparticles is the following: (1) the ferrous cation usually in a form of ferrous ammonium sulfate are added to a hollow apoferritin in buffered solution under anaerobic conditions; (2) the pH of the solution is adjusted to $8 \div 8.5$ with sodium hydroxide; (3) an oxidative agent which is added together with basification of solution or straight after this slowly completes the reaction. The protons generated during the reaction are neutralized with continuous titration with sodium hydroxide.

The composition of iron oxide core in natural ferritin is uneven. Depending on a number of factors the iron hydroxide inorganic core can be crystallized in the form of ferrihydrite, hematite, and magnetite [17]. The crystallization process can be attenuated by a number of factors including temperature. The initially formed ferrihydride form can be transformed into more thermodynamically stable hematite by heating at ca. 100°C [18–20]. This is possible using recombinant ferritin of hyperthermophilic archaeon *Pyrococcus furiosus* loaded with up to 1000 Fe which is stable up to 120°C [20].

Iron(II) cations can be prepared in situ through photochemical reduction of iron(III) citrate complexes by xenon-arc lamps. The labile iron(II) cations are capable of passing through hydrophilic threefold channels in ferritin and reoxidized by oxygen [21]. The photochemical method produces very low but stable concentration of labile cation and is useful for other metals possessing highly unstable lower oxidation state such as europium and titanium cations that were successfully deposited in ferritin with up to 1000 metal ions per ferritin.

A substantial component of the core in natural ferritins is phosphate ions that constitute from 10% molar percent in human ferritin to up to 50% in bacterial and plant ferritins [22]. Along with iron, ferritin also accumulates phosphate anions. Mammalian ferritin contains approximately 10 iron cations per phosphate group, 6 while plant and bacterial ferritin 8–12 have much higher iron-to-phosphate ratio approaching a 1:1 ratio in bacterial ferritin [23, 24]. In chemical experiments, conducting the deposition of iron into horse spleen ferritin in the presence of arsenate, vanadate, molybdate ions results in their incorporation into the inorganic core [25]. Although in all cases the final product contained approximately 2000 iron atom the stoichiometry of the anion incorporation was dramatically different. While phosphate as a hard anion was incorporated with 1:4.7 stoichiometry other anions presumably softer ones were incorporated in much larger amounts with arsenate 1:2.2, vanadate 1:1.3, and molybdate 1.5:1 ratio to iron.

7.4   Oxidation by Other Agents

Oxidation with molecular oxygen is not always the optimal way for synthesis due to relatively large reaction time and difficulty controlling the reaction stoichiometry. Other oxidizing agents have been successfully used not only to accelerate deposition of the inorganic core into the inner cavity of ferritin but also to change the composition of the inorganic core. Hydrogen peroxide is highly efficient oxidant that oxidizes a variety of metal cation in Fenton-type reactions and can be used for immobilization of metal(II) cations preconcentrated in ferritin.

$$2Co_2 + aq + H_2O_2 aq \rightarrow 2Co(O)OHs + 4H + aq + H_2O \qquad (7.1)$$

Since the oxidation is usually rapid special precautions are taken to avoid formation of polymeric metal hydroxides in the solution. Repetitive sequential addition of small amounts of metal salt and hydrogen peroxide was used for preparation of nanophase cobalt oxyhydroxide [26]. Demineralized ferritin (apo-ferritin) was treated with aliquots of Co(II) and H_2O_2 at pH 8.5 and rapidly formed a homogeneous olive-green solution, the result of a spatially constrained oxidative mineralization reaction. Theoretical loading of 2250 Co per ferritin molecule was achieved by 18 addition cycles.

Simultaneous slow addition of deaerated solutions of metal salts and hydrogen peroxide to solution of apoferritin at elevated temperature 50–65°C also produces selective deposition of metal oxides. Using this method Fe(II) and Co(II) cations were used for preparation of mixed oxide magnetic nanoparticles of the general formula $Fe_3-xCo_xO_4$ (x e 0.33) with theoretical loading of 1000 metal cations per ferritin [27]. Oxidation of 3 mM Co(II) ion loaded into 0.5 mg/mL apoferritin at pH 8.3 buffer solution with hydrogen peroxide at 50°C produced Co_3O_4 nanoparticles [28].

Growth of the inorganic core by this method is initiated at nucleation centers on the inner surface of the protein shell and continues toward the center. If the process is interrupted at 1000 Co/ferritin hollow nanoparticles of Co_3O_4 are produced [29]. The resultant spheres are highly porous as evidenced by filling the hollow interior in further deposition of cobalt oxide is done (Fig. 7.4).

Oxidation can also be done using trimethylamine N-oxide. This agent is considerably milder than hydrogen peroxide and allows avoiding Fenton process which produces hydroxyl radical capable of degrading the protein shell of ferritin [30]. Selective production of magnetite core has been achieved by oxidation of iron(II) using equimolar amount trimethylamine N-oxide [31]. The method allowed growth of magnetite cores up to 1000 Fe using sequential additions in portions not more than 140 Fe/protein. The obtained particles possess distinctive magnetic properties [32, 33].

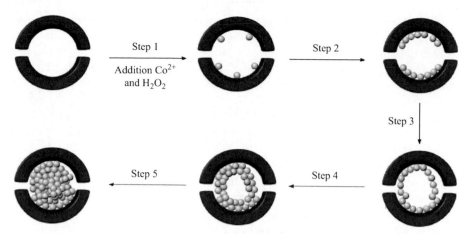

Fig. 7.4 Growth of hollow nanoparticles of Co_3O_4 in the interior cavity of apoferritin

7.5   Hydrolysis

The ability of the inner cavity of ferritin to serve as nucleation centers allows it accelerate the growth on metal hydroxide core. The supersaturation of metal hydroxide concentration can be induced by pH adjustment. The process conditions have to be very carefully controlled to avoid precipitation of metal hydroxide outside the apoferritin shell. Accurate control of pH without 0.01 pH unit was used for fabrication of nickel(II) and chromium(III) hydroxide nanoparticles. It was found the presence of carbon dioxide and ammonia considerably increases the yield of nanoparticles, most probably because of simplifying precision pH control in the resultant buffered solution [34].

Similar mechanism of pH adjustment was used for deposition of uranyl cations into ferritin [35, 36] through addition of pH 7.0 phosphate buffer to a mixture of uranyl acetate and apoferritin. In this case theoretical maximal loading of uranyl hydroxide (800/ferritin) can be achieved and even surpassed. In this case, the ferritin shell is disintegrated and polydisperse nanoparticles are formed.

7.6   Other Chemical Reactions

In addition to the previously described processes a large number of insoluble metal salts can be deposited in the inner cavity of ferritin. The mechanism of the process relies on selective accumulation of metal cations in the inner cavity of ferritin where the a reaction with a second component—anion source leading to formation of insoluble polymers takes place. A variety of semiconducting nanoparticles can be synthesized by this way.

This process requires diffusion of both components through hydrophilic channels in ferritin which bring certain limitation on the size of reactants. In addition the process of diffusion is relatively slow and an important requirement to the reaction is slow rate as otherwise a deposition of particles outside ferritin will take place. To conform the second requirement special modifications for slowing down the reaction are used.

An important group of reactions is used for preparation of semiconductor particles of metal chalcogenides. This reaction can be induces by different sources of chalcogenides such as hydrogen sulfide. Amorphous iron sulfide minerals containing either 500 or 3000 iron atoms in each cluster have been synthesized in situ within the inner cavity of horse spleen ferritin [37]. Iron-57 Mössbauer spectroscopy indicated that most of the iron atoms in the 3000-iron atom cores are trivalent, whereas in the 500-iron atom clusters, approximately 50% of the iron atoms are Fe(III), with the remaining atoms having an effective oxidation state of about $+2.5$.

Cations of iron, zinc, and cadmium can be also selectively nucleated by amino acid residues of ferritin and selective growth of nanoparticles inside ferritin is possible. The reported binding sites are not restricted to acidic side chains and involve Glu45, Cys48, His49, Arg52, and Glu53 residues located near the proposed ferrihydrate site. Since formation of these particles from metal cations and hydrosulfide or hydroselenide cation is very rapid special precautions are used to prevent precipitation outside. For example, selective deposition of zinc selenide [38] and cadmium selenide [39] inside the apoferritin is done utilizing selenourea for slow generation of selenide anion. Alternatively slowing down of the reactions can be achieved by decreasing concentration of a cation. The same CdSe in ferritin was synthesized using slow reaction of cadmium-EDTA complex with sodium hydroselenide [40].

Slow rate of diffusion through ferritin channels can be an important asset. Slowing down of the reaction is not sufficient for metal that are not well bound by ferritin, for example for calcium, barium, and strontium salts. However, even in this case if deposition of insoluble particles is prevented by addition of polyelectrolytes such as PMAA or sodium polyphosphate. Since polyelectrolytes are incapable of passing the interiors of ferritin the reaction inside result in controlled synthesis of Ba, Sr, Ca carbonate particles [41].

Lutetium phosphate nanoparticles have been deposited inside the cavity horse spleen ferritin. Selectivity in the deposition here was achieved by preferential concentration of Lu^{3+} cations in the negatively charged ferritin cavity. After diffusion of Lu^{3+}, the electrostatic potential was reversed and phosphate anions were able to enter the apoferritin cavity. Loading achieved by this method is lower than in the case of slow addition of reagent but due as high as 500 cations per ferritin due to high affinity of lutetium cations to ferritin [42]. Similar approach was use for synthesis of nanoparticles of zinc, cadmium and lead phosphate [43]. By using mixtures of metal cation it was possible with desired ratios of metal cations in these particles.

7.7 Preparation of Metal Nanoparticles in Ferritin by Reduction

Reduction of certain metal oxides can provide an access to metal nanoparticles. Most commonly, sodium borohydride used as a reductant. In most cases this approach is used for preparation of nanoclusters from relatively small number of metal cations that are more tightly bound to nucleation centers in ferritin. Although synthesis of metal nanoparticles is a mature field the use of ferritin provides several interesting advantages. At first, the ferritin shell protects growing nanoclusters from association. Furthermore, formation of hybrid alloy nanoparticles composed from two or more metals is possible in a wide range of ratios. Finally, catalytic activity of nanoclusters enclosed in ferritin shell is limited to substrates capable of diffusing through narrow three-ford channels in the shell, thus giving an opportunity for adjusting chemoselectivity.

Reduction of metal oxides nickel and cobalt alloy mixtures [44] prepared from metal sulfates by reduction with $NaBH_4$. The obtained mixed nanoclusters with compositions from $Co_{25}Ni_{75}$ to $Co_{75}Ni_{25}$ possess tunable magnetic properties.

Hydrophilic amino acid residues of the protein shell in ferritins serve as nucleation centers for hard and moderately hard metal cations. Growth of nanoparticles of noble metals cannot be done by these methods as carboxylic acids are not efficient nucleation sites for soft metal cations. In this case other amino acid residues such as cysteine and histidine can serve as nucleation centers. It was found that Pd^{2+} cations are capable of selective binding to horse spleen apoferritin. This process was studied by X-ray crystallography which indicated the presence of up to 216 Pd^{2+} cations at different binding sites on the inner surface of ferritin and threefold channels [45]. Even in the presence of a 500-fold excess of K_2PdCl_4 only 365 Pd^{2+} cations can be bound to ferritin (Fig. 7.5).

Subsequent treatment of ferritin bound Pd^{2+} cations with sodium borohydride resulted in catalytically active palladium metal nanoclusters [46]. Since the process leaved the protein shell intact the catalysis can be done only for reactants capable of diffusing though ferritin channels. Indeed only small olefins were hydrogenated.

Similar preparation of gold–palladium core–shell nanoparticles [47] was achieved through stepwise addition of chloroauric acid and dipotassium tetrachloropaladate to provide either alloy Au/Pd nanoparticles or shell–core Au/Pd nanoparticles after subsequent reduction with sodium borohydride. The investigation of nucleation centers showed that Au has an affinity to cysteine, methionine and histidine amino acid residues, presumably due to its reduced form Au(I) formation. Crystal structure analysis shows that divalent palladium is ligated in extend to the mentioned amino acids by carboxyl groups of aspartic and glutamic acids. This finding suggests that palladium cations are able to bind to the interior surface of proteins and form nucleation centers even after deposition of Au (Fig. 7.6).

An analogy to these results loading of horse spleen ferritin with silver cation followed by treatment with sodium borohydride resulted in formation of highly heterogeneous nanoparticles with diameter up to 4 nm [48]. Partial loading of

Fig. 7.5 Structure of
Cd^{2+}–Pd^{2+}–horse spleen
ferritin with 192 Pd binding
sites (*purple*) on the interior
surface. The exterior surface
possess additional 24 palladium and cardmium binding
sites [45]

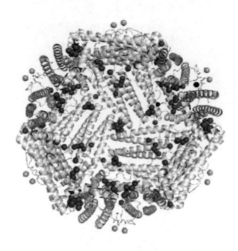

the inner cavity was observed for similar synthesis of gold nanoparticles [49]
Improvement of binding of these metal cations can be achieved using genetically
modified ferritins possessing increased amounts of cysteine and histidine [50].
However, the most promising were achieved by engineered ferritins that possess
an inner surface dodecapeptide on the inner cavity that was capable of reducing
silver ions to metallic silver [51]. Deposition of silver produced electron dense
particles with diameter 7 Å. An interesting development of this methodology is
in vivo deposition of silver into *E. Coli* that actively expressing the engineered
protein cages.

Higher loading can be achieved by simultaneous addition of metals in cation and
anionic form. Diffusion growth of mixed gold–silver nanoparticles was achieved
using treatment of horse spleen apoferritin with AgNO$_3$ and HAuCl$_4$ followed by
purification and reduction with sodium borohydride. The final composition reflects
the boundary conditions that have been applied for the syntheses. Gold containing
bimetallic nanoparticles are of interest as they show enhanced catalytic activity.
Homogeneous gold-silver alloy nanoparticles [52] have been prepared utilizing silver nitrate as a source for silver cations and chloroauric acid as a source for gold
containing anions. Relatively high concentration of ammonium hydroxide media
has been used for the reaction serving two purposes firstly to keep silver cations in
solution in the presence of chloride anions through their coordinative stabilization
with ammonia and secondly for basic pH adjustment. The obtained particles possessed relatively narrow distribution of size (Fig. 7.7).

7.7.1 Chemical Modifications of the Inorganic Core in Ferritin

The initially formed inorganic nanoparticles can be further transformed by either
by deposition/codeposition of additional compounds or chemical reactions of the
initial particles. The composition of the resultant hybrid particles can be adjusted
by simple change of the treatment time opening nearly endless possibilities.

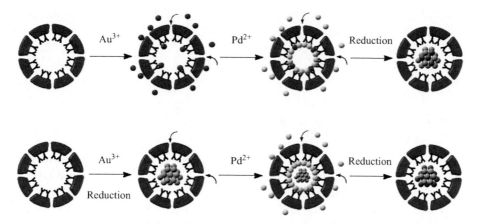

Fig. 7.6 Preparation of Au/Pd bimetallic nanoparticles in HL apoferritin. Au^{3+}, Pd^{2+}, Au^0, and Pd^0 are in *red*, *yellow*, *grey*, and *green*

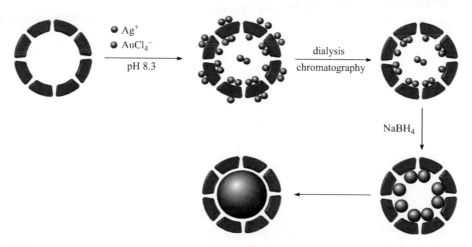

Fig. 7.7 Preparation of Ag–Au alloy nanoparticles through simultaneous treatment with cationic and anionic precursors

Photoreduction of Cu^{2+} cation by sacrificial electron donor such as citrate was catalyzed by ferritins partially loaded with ferryhydroxide core. The process results in formation of amorphous $Cu(0)$ cores inside the inner cavity of ferritin. Depending on $Cu(II)$/ferritin ratios can led to larger particle sizes, with loadings of 250, 500, 1000, and 2000 leading to average particle diameters of 4.5 ± 0.8, 9.7 ± 4.2, 12.7 ± 3.6, and 31.4 ± 10.1 nm, respectively. The process is likely to be accompanied by destruction of the ferritin shell as the size of these particles is considerably larger than the interior cavity of ferritin.

Reaction of copper ions deposited inside ferritin with ferricyanide anions produced Prussian blue particles [53]. The reaction proceeded only in the presence of
large concentrations urea since chaotropic agents are known to increase permeability of threefold channels in ferritin [54].

7.7.2   Removal of Ferritin Core

Removal of the protein shell from the resultant hybrid nanoparticles can be done
by a number of ways. High temperature oxidative calcination is a universal method
for removing any organics although high temperatures associated with these methods are not compatible with some nanoparticles. Room temperature ozonolysis can
be an alternative for metal oxide nanoparticles [55]. For example, iron and cobalt
nanoparticles were released from the protein shell through treatment of SiO_2 deposited particles with ozone and UV irradiation. Subsequent reduction with hydrogen
gas produced iron and cobalt metal nanoparticles with different sizes depending on
the initial loading of metal oxides into ferritin [56].

Enzymatic degradation of the protein shell of ferritin is possible by lysosomal
proteases [57]. Most of studies in the field are concentrated on *in vivo* systems
and most probably some commercially available proteases can degrade the ferritin
shell. This method can be useful for release of semiconducting nanoparticles such
as metal chalcogenides that are highly sensitive to oxidation.

7.7.3   Ferritin Like Listeria Dps Proteins

DPS proteins were originally discovered as stress proteins, which protect DNA
against oxidative stress during nutrient starvation. The Dps subfamily is diverse,
with many members promoting iron incorporation and others acting as immuno-
gens, neutrophile activators, cold-shock proteins, or constituents of fine-tangled
pili. Most commonly recombinant DPS protein from *Listeria Innocua* is used
for mineralization experiments. All DPS are considerably smaller than ferritin
and composed from 12 identical 18 kDa subunits. The resultant inner cavity is
only 5 nm in diameter and can hold up to 500 iron atoms. DNA binding by these
proteins was shown to suffice for protection against oxidative DNA damage and
might be mediated by magnesium ions, which bridge the protein surfaces with the
polyanionic DNA [58]. DPS possess ferroxidase catalytic center that has similar
functionally to ferritin despite substantial differences in amino acid sequence. Yet
oxidation of iron(II) cations by hydrogen peroxide proceeds considerably faster
that with oxygen, in line with the main role of DPS protein involving prevention
of generation of hydroxyl radicals through Fenton reaction. Iron(II) oxidation by
H_2O_2 occurs with a stoichiometry of 2 Fe^{2+}/H_2O_2 in both the protein-based fer-

roxidation and subsequent mineralization reactions, indicating complete reduction of H_2O_2 to H_2O [59].

$$2Fe^{2+} + H_2O_2 + (H_2O)_x \rightarrow Fe_2O_3(H_2O)_{x-1} + 4H^+ \tag{7.2}$$

Due to functional similarity between DPS from *Listeria inncua* and ferritins essentially all inorganic nanoparticles obtained in ferritin can be received inside DPS with corresponding decrease in size and molecular weight. The only exception of oxidation of M^{2+} cations with molecular oxygen which is usually too slow. The mineralized iron core in DPS obtained by stoichiometric oxidation with H_2O_2 is in the g-Fe_2O_3 form and is superparamagnetic [60]. Similar oxidation with H_2O_2 approach can be used for preparation of nanophase Co_3O_4 and $Co(O)OH$ [61, 62]. DPS form *Listariea inncua* was used for deposition of photo luminescent CdS particles with average diameter 4.2±0.4 nm. This was achieved by slowing the reaction down using tetraamminecadmium as Cd^{2+} source and thioacetic acid as the precursor of sulfide ions [63].

Replacing hydrophilic residues in a number of position on the inner sphere into hydrophobic does not affect function properties as most of resultant recombinant proteins retained 3D structure and the ability to mineralize iron in the presence of H_2O_2 [64].

7.7.4 HSP Proteins

Hsp16.5, isolated from the hyperthermophilic Archaeon *Methanococcus jannaschii*, is a member of the small heat shock protein (sHsp) family. Small Hsp have 12–42 kDa subunit sizes and have sequences that are conserved among all organisms. Hsp16.5 24 identical subunits assembled into protein cage that resemble the size and the appearance to ferritin with the assembled protein cage has an exterior diameter of 12 nm. HSP16.5 possesses very large pores at the threefold axes with ca. 3 nm diameter allowing facile diffusion on reagents with the inner cavity [65].

Since the biological role of HSPs does not involve mineralization they do not possess any ferroxidase activity. However hydrophilic amino acid residues on the inner surface of the protein shell can serve as nucleation centers for ion cation. As a result a non-catalyzed oxidation of Fe(II) by dioxygen takes place producing a poorly crystalline iron oxide particles in the interior cavity of HSPs [66].

Synthesis of platinum NP inside can be done through incubation with $PtCl_4^{2-}$ followed by reduction dimethylamine borane complex [67]. Size of the resultant particles depended on Pt loading ranging from 2.2 ± 0.7 nm for 1000 Pt/cage to 1 ± 0.2 nm for 250 Pt/cage indicating that formation of protein-associated nanoclusters rather than a single nanoparticles. The associated nanoclusters possessed a high catalytic activity in photoinduced generation of dihydrogen from water (Fig. 7.8).

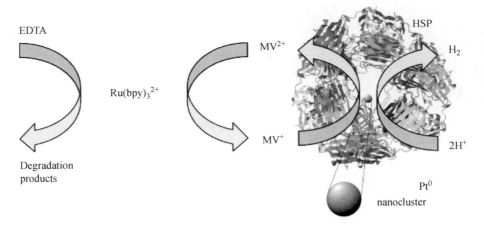

Fig. 7.8 Schematic representation of light mediated production of dihydrogen with platinum nanoclusters inside HSP. Methyl viologen (MV^{2+}) is used as an electron transfer mediator between the $Ru(bpy)_3^{2+}$ photocatalyst and Pt-HSP

Specific nucleation centers can be introduced into MjHSP through site directed mutagenesis. This process results in preparation of genetically modified subunits possessing artificially introduced cysteine or amine residues. Selective mineralization of CoPt nanoparticles was achieved through introduction of a short amino acid sequence KTHEIHSPLLHK that possess high specificity to Lo phase of CoPt. These nanoparticles were formed through treatment of MjHSP followed by reduction with sodium borohydride [68].

7.7.5   Enzymes as Templates for Mineralization

Subunits of a substantial number of enzymes such as lumazine synthase, pyruvate dehydrogenase, and others self assemble into highly symmetric structures resembling nanocages in the outer shape. The inner cavity of these enzymes is substantially larger than in ferritin thus potentially extending size of nanoparticles up to 15 nm. Growth of iron oxide particles inside lumazine synthase resulted in relatively low loading (2000 Fe/cage) [69]. Binding of iron were found to interfere with the self assembly by lumazine synthase subunits producing particles of larger size up to 30 nm were observed.

7.8   Growth of Nanoparticles on Viral Templates

Virus particles consists of nucleic acids surrounded by a protective coat of protein called a capsid that is formed from identical protein subunits called capsomers. Self assembly of viral capsid from protein subunits is many aspects resemble that of protein cages. In contrast to previously discussed nanocages viral coat proteins self assemble into nanoparticles of remarkable diversity in morphology. Most common viruses have size from 20 to 300 nm to produce shapes ranging from highly symmetric icosahedrons to rod like structures found in filoviruses.

Although viruses are not known to be involved in any natural biomineralization process they still can serve as templates for synthesis of nanoparticles. The viral capsid often bears a considerable percentage of ionized amino acids residues. The inner surface of capsids is positively charged to bind negatively charged viral genome and these charges can be used as nucleation centers of metal cations. The viral capsid after reaction serves as stabilizing capping agent preventing aggregation process in a way similar to nanocages and can serve for subsequent assembly of nanoparticles into functional devices. Exterior amino acids residues are predominantly negatively charged and also can serve as nucleation centers. The charge of the inner and/or the exterior surface of assembled viral capsids can be adjusted by controlling pH and/or ionic strength. Charge and nucleation sites of these proteins can also modified by covalent grafting with functionalized polymeric chains [70], and through genetic engineering of capsomer proteins to provide tailor-made amino acid residues on the interior surface [71–77]. Rather extensive modifications of amino acid residue are possible without appreciable degrading the self-assembly process and geometrical parameters of capsids (Fig. 7.9) [78].

There are several important challenges in application of viruses as protein templates. In contrast to other natural nanocages viral capsids do not possess appropriate channels for continuous influx of reagents for building the inner inorganic core. In some cases this can be compensated by swelling of the capsid by adjustment of pH and ionic strength. Another potential problem, directly related to morphological diversity of the obtained nanoparticles is their stability if the protein template is removed.

7.8.1 Icosahedral Viruses

Icosahedral viruses have 20 flat equilateral triangular faces arranged around a sphere. Each face can be composed from a number of individual capsomers example the protein shell of Cowpea chlorotic mottle virus is composed from 180 capsomers. Functional groups in capsomers that can serve as nucleation centers are evenly distributed on their surface so that viruses can direct mineralization uniformly. Icosahedral viruses with such as Cowpea chlorotic mottle virus (CCMV) [78–89] and Cowpea mosaic virus (CPMV) [14, 87, 90–94] possessing 28 nm outer size and the interior cavity about 18 nm are the most commonly used. The inner surface of the viruses regardless their shape and structure is hydrophobic and at neutral pH is positively charged due to inside extensive presence of lysine and arginine basic residues. For example, 6 arginines and 3 lysines from each out of 180 protomers are projecting towards interior of CCMV. It has been shown through genetic analysis that N-terminus is not required for in vitro assemblage of viral capsid [78].

tant for resorbable systems that the cross-linking does not interfere with degradation. In terms of composition, the primary linear (co-)polymers used in sub-micron and micron-sized particles are polyesters, for example poly(lactic acid) (PLA), poly(glycolic acid) (PGA) and copolymers of the two (PLGA). Polyethylene glycol (PEG) has been used as an outer surface or as part of a block copolymer to enhance hydrophilicity and compatibility [3].

Micelles formed from amphiphilic block copolymers typically provide a core-shell structure [6, 7]. In most cases, a hydrophobic core is able to act as the drug carrier while a hydrophilic shell, particularly using PEG, sterically stabilizes the micelles. Polymer vesicles can also be used. Micelles and vesicles have a variety of advantages over polymer particles. Their synthesis is relatively easy and reproducible, their circulation time is often long enough to allow drug accumulation in the target area, and their small size can allow them to reach areas within the body that larger (micron sized) particles cannot. However, because most micelles are made from the non-covalent association (self-assembly) of amphiphilic compounds or block copolymers, they are a dynamic system (i.e. micelles are forming and disintegrating on a constant basis). Furthermore, micelles may not form at all, or be particularly stable, as a result of even small environmental changes, such as dilution or changes in ionic strength. Therefore, the stability of micelles will alter greatly depending on the location in the body, and thus so will solubilization of the drug. As a result, delivery effectiveness may not be that high.

Polymer particles, whether based on linear polymers, cross-linked polymers or micelles, can also be made to passively or actively target specific cells [6, 8]. Passive targeting simply requires that the circulation time of the drug carrier be long enough to allow the particle and drug to reach the intended destination via random diffusion. However, active targeting requires specific recognition of a particular cell type by the carrier. The drug carrier, therefore, needs to be functionalized, typically on the surface, with an appropriate ligand that will recognize the target and preferentially bond to it. These active targeting drug delivery species show improved efficiency and specificity [8].

6.2 Molecular Structures of Degradable Polymers

Commonly degradable polymers include labile linkages within the backbone of material, which could be esters, orthoesters, anhydrides, carbonates, amides, etc. Over the past 40 years this area of polymers has been dominated by poly(α-hydroxy acids) (polyesters) such as PLA, PGA and PLGA copolymers [9]. These types of degradable polymers are seen on the medical market in the form of sutures, clips, staples and some drug delivery devices [2]. As medical technology advances, so do the requirements for the polymeric material. Some desired traits include: controllable erosion rates, surface degradation/erosion mechanism, and moldability with strong yet flexible mechanical properties.

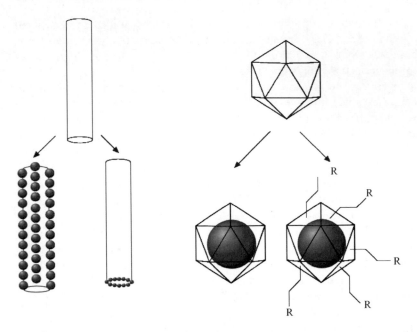

Fig. 7.9 Synthesis of nanoparticles on viral template. **a** Growth of nanowires on philoviruses; **b** growth of nanocrystals on the end of TMV nanorod; **c** growth of nanoparticles inside a viral capsid; **d** subsequent modification of viral capsid with nanparticle inside

The electrostatic interaction with these amino acid residues is the primary driving force in the encapsidation of nucleic acids. Cowpea chlorotic mottle virus capsid undergoes reversible swallowing of the capsid at pH > 6.5 by approximately 10%. This process opens 60 pores with diameter up to 2 nm resulting in increased permeability to molecules and ions. At this stage diffusion of precursors for synthesis of nanoparticles in the interior cavity of the virus is possible. Swallowing and shrinking are fully reversible and can be repeated several times to achieve the desired loading (Fig. 7.10) [81].

Selective deposition of inorganic material into the interior CCMV by pH control includes growth of polyoxometallates from tungstate and vanadate anions [81] as well as titanium oxide [83].

In contrast to pH dependent transition of Cowpea mosaic virus (CPMV) is highly stable in the pH range 3.5–9 without appreciable swelling. However, the protein capsid of CPMV is substantially more porous allowing diffusion of ions and neutral molecules into the capsid followed by nucleation and formation of nanoparticles (Fig. 7.11).

A number of studies was also performed on deposition of metal cations on the external surface of icosahedral viral templates. The process results in electrostatically driven deposition of metal cations and/or anion. Subsequent reduction with sodium borohydride results in formation of a rough metallic shells (Fig. 7.12) [87, 95].

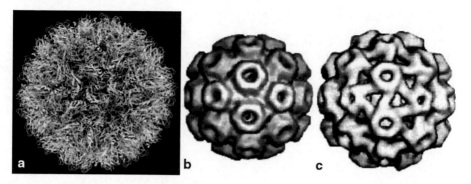

Fig. 7.10 a Structure of CCMV based on X-ray data. **b** Reconstruction of unswollen viral capsid of CCMV. **c** Swollen capsid with 2 nm pores [81]

Fig. 7.11 Cowpea mosaic virus (CPMV)

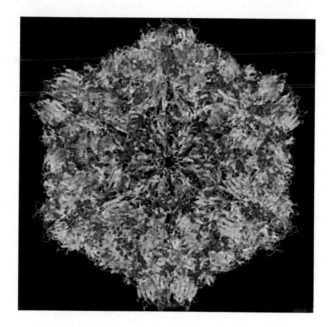

7.8.2   Filoviruses

Assembly of inorganic particles on viral templates of filoviruses can provide rode type shapes that cannot be prepared by conventional methods of growth. Classical experiments conducted by the group of Mann demonstrated versatility of this approach through synthesis of highly stable tobacco mosaic virus (TMV) [96]. Each TMV particle consists of 2130 identical capsomer arranged in a helical motif around a single strand of RNA. The viral capsid is a 300 nm × 18 nm hollow protein tube, possessing a 4 nm-wide central cavity. Both internal and external surfaces of the capsid are rich in both positively and negatively charged amino acid residues

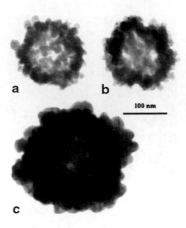

a   b

100 nm

c

of glutamate, aspartate, lysine, and agrinine that can serve as nucleation centers.
Controlling the electric charge of these residues c by pH it was possible to deposit
a layer of inorganic material using the same methods described for nanocages. Precipitation of CdS and PbS resulted in growth of disordered aggregates of crystalline
particles of 5 nm average size giving a rough appearance to the resultant hybrid
material. In contract, formation of largely amorphous SiO_2 and ferrihydrite layers
was largely homogeneous (Fig. 7.13).

Considerable attention is being attracted to synthesis of metallic nanowires on
filoviral templates. Rod like viruses such as tobacco and tomatoes mosaic virus with
length up to 1500 nm is and diameter 45–75 nm possess all necessary properties.
Deposition of metal such as silver or gold on the external surface of filamentous
viral capside followed by $NaBH^4$ hydrogenation can produce nanowires that are
needed for a number of applications such as advanced batteries. The important limitation here that in contrast to polymerization of amorphous materials [97, 98], is
synthesis of contiguous nanowires rather than deposition of isolated nanoparticles
on the capsid [99]. This problem may be treated by using more complex methods
of deposition, for example creation of mainly isolated small particles from gold
by $NaBH_3$ reduction followed by connecting these nanoparticles with Au^{3+} cations
that are reduced much more slow [100]. Slower reduction with methanol was used
for preparation of platinum nanowires [101]. Coverage can be improved by multiplication of sequential addition of $HAuCl_4$ and sodium borohydrides although the
surface of the resultant nanowires is rough in all cases [102].

Another approach toward formation of smooth nanowires involve slowing down
the process of deposition by using different reductants. In contrast to sodium borohydride reduction photochemical reduction of Ag(I) salts at pH 7 resulted in nucleation and constrained growth of discrete Ag nanoparticles aligned within the 4 nm-
wide internal channel of tobacco mosaic virus [103]. Although these nanoparticles
did not form a contiguous nanowire this method may be ised for 1D preparation
of assemblies of nanoparticles. Using another slow electroless deposition method
copper nanowires with diameter 3 nm and length up to 150 nm were produced from
copper [104], nickel, and cobalt (Fig. 7.14) [105].

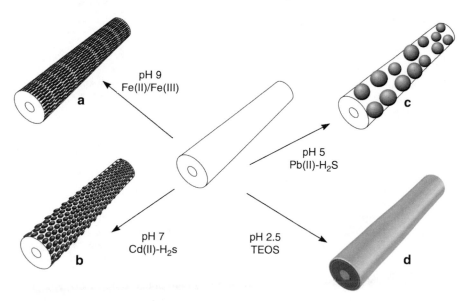

Fig. 7.13 Routes for the synthesis of nanotube composites using TMV templates using **a** oxidative hydrolysis; **b, c** coprecipitation; **d** sol–gel condensation of silica

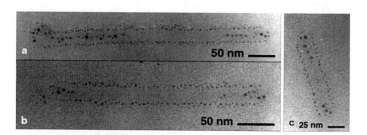

Fig. 7.14 a–c TEM microphotographs of mutant E95Q/D109N virus showing rows of 3±2 nm silver particles on the outer surface and in the inner channel of the virus [103]

Genetic engineering of recombinant rotavirus VP6 protein with cationic and anionic metal salts of silver, gold, palladium, and platinum allowed preparation of hybrid filaments with length as large as 1.5 μm [106]. In these cases the mineralization after sodium borohydride reduction was in the form of randomly distributed small nanoparticles with diameter 3–11 nm and no contiguous larger nanoparticles were formed. The small nanoparticles possessed considerable catalytic activity. Preparation of contiguous nanoparticles is less essential for synthesis of aligned magnetic nanoparticles, and a number of these syntheses have been reported including $CoPt_3$ and CoPt (up to 25% weight) [107] and Rh/Fe_3O_4 [108].

References

1. Sigel, A.; Sigel, H.; Sigel, R. K. O. *Biomineralization : from nature to application*; John Wiley & Sons: Chichester, 2008.
2. Bauerlein, E., Biomineralization of unicellular organisms: An unusual membrane biochemistry for the production of inorganic nano- and microstructures *Angewandte Chemie-International Edition* **2003**, *42*, 614–641.
3. Gallois, B.; dEstaintot, B. L.; Michaux, M. A.; Dautant, A.; Granier, T.; Precigoux, G.; Soruco, J. A.; Roland, F.; ChavasAlba, O.; Herbas, A.; Crichton, R. R., X-ray structure of recombinant horse L-chain apoferritin at 2.0 angstrom resolution: Implications for stability and function *Journal of Biological Inorganic Chemistry* **1997**, *2*, 360–367.
4. Liu, X. S.; Patterson, L. D.; Miller, M. J.; Theil, E. C., Peptides selected for the protein nanocage pores change the rate of iron recovery from the ferritin mineral *Journal of Biological Chemistry* **2007**, *282*, 31821-31825.
5. Galvez, N.; Ruiz, B.; Cuesta, R.; Colacio, E.; Dominguez-Vera, J. M., Release of iron from ferritin by aceto- and benzohydroxamic acids *Inorganic Chemistry* **2005**, *44*, 2706-2709.
6. Masuda, T.; Goto, F.; Yoshihara, T.; Mikami, B., Crystal structure of plant ferritin reveals a novel metal bindingsite that functions as a transit site for metal transfer in ferritin *The Journal of Biological Chemistry* **2010**, *285*, 4049-4059.
7. Takahashi, T.; Kuyucak, S., Functional properties of threefold and fourfold channels in ferritin deduced from electrostatic calculations *Biophysical Journal* **2003**, *84*, 2256-2263.
8. Douglas, T.; Ripoll, D. R., Calculated electrostatic gradients in recombinant human H-chain ferritin *Protein Science* **1998**, *7*, 1083-1091.
9. Pead, S.; Durrant, E.; Webb, B.; Larsen, C.; Heaton, D.; Johnson, J.; Watt, G. D., METAL-ION BINDING TO APO, HOLO, AND RECONSTITUTED HORSE SPLEEN FERRITIN *Journal of Inorganic Biochemistry* **1995**, *59*, 15-27.
10. Ceolin, M.; Galvez, N.; Sanchez, P.; Fernandez, B.; Dominguez-Vera, J. M., Structural aspects of the growth mechanism of copper nanoparticles inside apoferritin *European Journal of Inorganic Chemistry* **2008**, 795-801.
11. Wong, K. K. W.; Colfen, H.; Whilton, N. T.; Douglas, T.; Mann, S., Synthesis and characterization of hydrophobic ferritin proteins *Journal of Inorganic Biochemistry* **1999**, *76*, 187-195.
12. Clemente-Leon, M.; Coronado, E.; Primo, V.; Ribera, A.; Soriano-Portillo, A., Hybrid magnetic materials formed by ferritin intercalated into a layered double hydroxide *Solid State Sciences* **2008**, *10*, 1807-1813.
13. Hess, D. M.; Naik, R. R.; Rinaldi, C.; Tomczak, M. M.; Watkins, J. J., Fabrication of Ordered Mesoporous Silica Films with Encapsulated Iron Oxide Nanoparticles using Ferritin-Doped Block Copolymer Templates *Chemistry of Materials* **2009**, *21*, 2125-2129.
14. Niu, Z. W.; Kabisatpathy, S.; He, J. B.; Lee, L. A.; Rong, J. H.; Yang, L.; Sikha, G.; Popov, B. N.; Emrick, T. S.; Russell, T. P.; Wang, Q., Synthesis and Characterization of Bionanoparticle-Silica Composites and Mesoporous Silica with Large Pores *Nano Research* **2009**, *2*, 474-483.
15. Thota, S.; Kumar, J., Sol-gel synthesis and anomalous magnetic behaviour of NiO nanoparticles *Journal of Physics and Chemistry of Solids* **2007**, *68*, 1951-1964.
16. Tominaga, M.; Han, L.; Wang, L. Y.; Maye, M. M.; Luo, J.; Kariuki, N.; Zhong, C. J., Formation of water-soluble iron oxide nanoparticles derived from iron storage protein *Journal of Nanoscience and Nanotechnology* **2004**, *4*, 708-711.
17. Galvez, N.; Fernandez, B.; Sanchez, P.; Cuesta, R.; Ceolin, M.; Clemente-Leon, M.; Trasobares, S.; Lopez-Haro, M.; Calvino, J. J.; Stephan, O.; Dominguez-Vera, J. M., Comparative structural and chemical studies of ferritin cores with gradual removal of their iron contents *Journal of the American Chemical Society* **2008**, *130*, 8062-8068.
18. Pu, Z. F.; Cao, M. H.; Jing, Y.; Huang, K. L.; Hu, C. W., Controlled synthesis and growth mechanism of hematite nanorhombohedra, nanorods and nanocubes *Nanotechnology* **2006**, *17*, 799-804.

19. Parker, M. J.; Allen, M. A.; Ramsay, B.; Klem, M. T.; Young, M.; Douglas, T., Expanding the temperature range of biomimetic synthesis using a ferritin from the hyperthermophile Pyrococcus furiosus *Chemistry of Materials* **2008**, *20*, 1541-1547.
20. Klem, M. T.; Young, M.; Douglas, T., Biomimetic synthesis of photoactive alpha-Fe2O3 templated by the hyperthermophilic ferritin from Pyrococus furiosus *Journal of Materials Chemistry* **2009**, *20*, 65-67.
21. Klem, M. T.; Mosolf, J.; Young, M.; Douglas, T., Photochemical mineralization of europium, titanium, and iron oxyhydroxide nanoparticies in the ferritin protein cage *Inorganic Chemistry* **2008**, *47*, 2237-2239.
22. Li, C.; Qi, X.; Li, M.; Zhao, G.; Hu, X., Phosphate facilitates Fe(II) oxidative deposition in pea seed (Pisum sativum) ferritin *Biochimie* **2009**, *91*, 1475-1481.
23. Wade, V. J.; Treffry, A.; Laulhère, J. P.; Bauminger, E. R.; Cleton, M. I.; Mann, S.; Briat, J. F.; Harrison, P. M., Structure and composition of ferritin cores from pea seed (Pisum sativum) *Biochimica Et Biophysica Acta* **1993**, *1161*, 91-96.
24. Rohrer, J. S.; Islam, Q. T.; Watt, G. D.; Sayers, D. E.; Theil, E. C., Iron environment in ferritin with large amounts of phosphate, from Azotobacter vinelandii and horse spleen, analyzed using extended X-ray absorption fine structure (EXAFS) *Biochemistry* **1990**, *29*, 259-264.
25. Polanams, J.; Ray, A. D.; Watt, R. K., Nanophase iron phosphate, iron arsenate, iron vanadate, and iron molybdate minerals synthesized within the protein cage of ferritin *Inorganic Chemistry* **2005**, *44*, 3203-3209.
26. Douglas, T.; Stark, V. T. Nanophase cobalt oxyhydroxide mineral synthesized within the protein cage of ferritin *Inorganic Chemistry* **2000**, *39*, 1828-1830.
27. Klem, M. T.; Resnick, D. A.; Gilmore, K.; Young, M.; Idzerda, Y. U.; Douglas, T., Synthetic control over magnetic moment and exchange bias in all-oxide materials encapsulated within a spherical protein cage *Journal of the American Chemical Society* **2007**, *129*, 197-201.
28. Tsukamoto, R.; Iwahor, K.; Muraoka, M.; Yamashita, I., Synthesis of Co3O4 nanoparticles using the cage-shaped protein, apoferritin *Bulletin of the Chemical Society of Japan* **2005**, *78*, 2075-2081.
29. Kim, J. W.; Choi, S. H.; Lillehei, P. T.; Chu, S. H.; King, G. C.; Watt, G. D., Cobalt oxide hollow nanoparticles derived by bio-templating *Chemical Communications* **2005**, 4101-4103.
30. Arosio, P.; Levi, S., Ferritin, iron homeostasis, and oxidative damage *Free Radical Biology and Medicine* **2002**, *33*, 457-463.
31. Wong, K. K. W.; Douglas, T.; Gider, S.; Awschalom, D. D.; Mann, S., Biomimetic synthesis and characterization of magnetic proteins (magnetoferritin) *Chemistry of Materials* **1998**, *10*, 279-285.
32. Gider, S.; Awschalom, D. D.; Douglas, T.; Mann, S.; Chaparala, M., CLASSICAL AND QUANTUM MAGNETIC PHENOMENA IN NATURAL AND ARTIFICIAL FERRITIN PROTEINS *Science* **1995**, *268*, 77-80.
33. Gider, S.; Awschalom, D. D.; Douglas, T.; Wong, K.; Mann, S.; Cain, G., Classical and quantum magnetism in synthetic ferritin proteins *Journal of Applied Physics* **1996**, *79*, 5324-5326.
34. Okuda, M.; Iwahori, K.; Yamashita, I.; Yoshimura, H., Fabrication of nickel and chromium nanoparticles using the protein cage of apoferritin *Biotechnology and Bioengineering* **2003**, *84*, 187-194.
35. Hainfeld, J. F., Uranium-loaded apoferritin with antibodies attached: molecular design for uranium neutron-capture therapy *Proceedings of the National Academy of Sciences of the United States of America* **1992**, *89*, 11064-11068.
36. Meldrum, F. C.; Wade, V. J.; Nimmo, D. L.; Heywood, B. R.; Mann, S., Synthesis of inorganic nanophase materials in supramolecular protein cages *Nature* **1991**, *349*, 684-687.
37. Douglas, T.; Dickson, D. P. E.; Betteridge, S.; Charnock, J.; Garner, C. D.; Mann, S., Synthesis and structure of an iron(III) sulfide-ferritin bioinorganic nanocomposite *Science* **1995**, *269*, 54-57.
38. Iwahori, K.; Yoshizawa, K.; Muraoka, M.; Yamashita, I., Fabrication of ZnSe nanoparticles in the apoferritin cavity by designing a slow chemical reaction system *Inorganic Chemistry* **2005**, *44*, 6393-6400.

39. Yamashita, I.; Hayashi, J.; Hara, M., Bio-template synthesis of uniform CdSe nanoparticles using cage-shaped protein, apoferritin *Chemistry Letters* **2004**, *33*, 1158-1159.
40. Xing, R. M.; Wang, X. Y.; Yan, L. L.; Zhang, C. L.; Yang, Z.; Wang, X. H.; Guo, Z. J., Fabrication of water soluble and biocompatible CdSe nanoparticles in apoferritin with the aid of EDTA *Dalton Transactions* **2009**, 1710-1713.
41. Li, M.; Viravaidya, C.; Mann, S., Polymer-mediated synthesis of ferritin-encapsulated inorganic nanoparticles *Small* **2007**, *3*, 1477-1481.
42. Wu, H.; Engelhard, M. H.; Wang, J.; Fisher, D. R.; Lin, Y., Synthesis of lutetium phosphate-apoferritin core-shell nanoparticles for potential applications in radioimmunoimaging and radioimmunotherapy of cancers *Journal of Materials Chemistry* **2008**, *18*, 1779-1783.
43. Liu, G.; Wu, H.; Dohnalkova, A.; Lin, Y., Apoferritin-templeated synthesis of encoded metallic phosphate nanoparticle tags *Analitical Chemistry* **2007**, *79*, 5614-5619.
44. Galvez, N.; Valero, E.; Ceolin, M.; Trasobares, S.; Lopez-Haro, M.; Calvino, J. J.; Dominguez-Vera, J. M., A bioinspired approach to the synthesis of bimetallic CoNi nanoparticles *Inorganic Chemistry* **2010**, *49*, 1705-1711.
45. Ueno, T.; Abe, M.; Hirata, K.; Abe, S.; Suzuki, M.; Shimizu, N.; Yamamoto, M.; Takata, M.; Watanabe, Y., Process of Accumulation of Metal Ions on the Interior Surface of apo-Ferritin: Crystal Structures of a Series of apo-Ferritins Containing Variable Quantities of Pd(II) Ions *Journal of the American Chemical Society* **2009**, *131*, 5094-5100.
46. Ueno, T.; Suzuki, M.; Goto, T.; Matsumoto, T.; Nagayama, K.; Watanabe, Y., Size-selective olefin hydrogenation by a Pd nanocluster provided in an apo-ferritin cage *Angewandte Chemie-International Edition* **2004**, *43*, 2527-2530.
47. Suzuki, M.; Abe, M.; Ueno, T.; Abe, S.; Goto, T.; Toda, Y.; Akita, T.; Yamadae, Y.; Watanabe, Y., Preparation and catalytic reaction of Au/Pd bimetallic nanoparticles in Apo-ferritin *Chemical Communications* **2009**, 4871-4873.
48. Dominguez-Vera, J. M.; Galvez, N.; Sanchez, P.; Mota, A. J.; Trasobares, S.; Hernandez, J. C.; Calvino, J. J., Size-controlled water-soluble Ag nanoparticles *European Journal of Inorganic Chemistry* **2007**, 4823-4826.
49. Zhang, L.; Swift, J.; Butts, C. A.; Yerubandi, V.; Dmochowski, I. J., Structure and activity of apoferritin-stabilized gold nanoparticles *Journal of Inorganic Biochemistry* **2007**, *101*, 1719-1729.
50. Butts, C. A.; Swift, J.; Kang, S. G.; Di Costanzo, L.; Christianson, D. W.; Saven, J. G.; Dmochowski, I. J., Directing Noble Metal Ion Chemistry within a Designed Ferritin Protein *Biochemistry* **2008**, *47*, 12729-12739.
51. Kramer, R. M.; Li, C.; Carter, D. C.; Stone, M. O.; Naik, R. R., Engineered protein cages for nanomaterial synthesis *Journal of the American Chemical Society* **2004**, *126*, 13282-13286.
52. Shin, Y.; Dohnalkova, A.; Lin, Y., Preparation of homogeneous gold-silver alloy nanoparticles using the apoferritin cavity as a nanoreactor *Journal of Physical Chemistry C* **2010**, *114*, 5985-5989.
53. Galvez, N.; Sanchez, P.; Dominguez-Vera, J. M., Preparation of Cu and CuFe Prussian Blue derivative nanoparticles using the apoferritin cavity as nanoreactor *Dalton Transactions* **2005**, 2492-2494.
54. Liu, X. F.; Jin, W. L.; Theil, E. C., Opening protein pores with chaotropes enhances Fe reduction and chelation of Fe from the ferritin biomineral *Proceedings of the National Academy of Sciences of the United States of America* **2003**, *100*, 3653-3658.
55. Liu, G.; Debnath, S.; Paul, K. W.; Han, W. Q.; Hausner, D. B.; Hosein, H. A.; Michel, F. M.; Parise, J. B.; Sparks, D. L.; Strongin, D. R., Characterization and surface reactivity of ferrihydrite nanoparticles assembled in ferritin *Langmuir* **2006**, *22*, 9313-9321.
56. Hosein, H. A.; Strongin, D. R.; Allen, M.; Douglas, T., Iron and Cobalt oxide and metallic nanoparticles prepared from ferritin *Langmuir* **2004**, *20*, 10283-10287.
57. Radisky, D. C.; Kaplan, J., Iron in cytosolic ferritin can be recycled through lysosomal degradation in human fibroblasts *Biochemical Journal* **1998**, *336*, 201-205.
58. Bellapadrona, G.; Stefanini, S.; Zamparelli, C.; Theil, E. C.; Chiancone, E., Iron Translocation into and out of Listeria innocua Dps and Size Distribution of the Protein-enclosed

Nanomineral Are Modulated by the Electrostatic Gradient at the 3-fold "Ferritin-like" Pores *Journal of Biological Chemistry* **2009**, *284*, 19101-19109.

59. Su, M. H.; Cavallo, S.; Stefanini, S.; Chiancone, E.; Chasteen, N. D., The so-called Listeria innocua ferritin is a Dps protein. Iron incorporation, detoxification, and DNA protection properties *Biochemistry* **2005**, *44*, 5572-5578.

60. Allen, M.; Willits, D.; Mosolf, J.; Young, M.; Douglas, T., Protein cage constrained synthesis of ferrimagnetic iron oxide nanoparticles *Advanced Materials* **2002**, *14*, 1562-+.

61. Allen, M.; Willits, D.; Young, M.; Douglas, T., Constrained synthesis of cobalt oxide nanomaterials in the 12-subunit protein cage from Listeria innocua *Inorganic Chemistry* **2003**, *42*, 6300-6305.

62. Resnick, D. A.; Gilmore, K.; Idzerda, Y. U.; Klem, M. T.; Allen, M.; Douglas, T.; Arenholz, E.; Young, M., Magnetic properties of Co3O4 nanoparticles mineralized in Listeria innocua Dps *Journal of Applied Physics* **2006**, *99*.

63. Iwahori, K.; Enomoto, T.; Furusho, H.; Miura, A.; Nishio, K.; Mishima, Y.; Yamashita, I., Cadmium sulfide nanoparticle synthesis in Dps protein from Listeria innocua *Chemistry of Materials* **2007**, *19*, 3105-3111.

64. Swift, J.; Wehbi, W. A.; Kelly, B. D.; Stowell, X. F.; Saven, J. G.; Dmochowski, I. J., Design of functional ferritin-like proteins with hydrophobic cavities *Journal of the American Chemical Society* **2006**, *128*, 6611-6619.

65. Bova, M. P.; Huang, Q. L.; Ding, L. L.; Horwitz, J., Subunit exchange, conformational stability, and chaperone-like function of the small heat shock protein 16.5 from Methanococcus jannaschii *Journal of Biological Chemistry* **2002**, *277*, 38468-38475.

66. Flenniken, M. L.; Willits, D. A.; Brumfield, S.; Young, M. J.; Douglas, T., The small heat shock protein cage from Methanococcus jannaschii is a versatile nanoscale platform for genetic and chemical modification *Nano Letters* **2003**, *3*, 1573-1576.

67. Varpness, Z.; Peters, J. W.; Young, M.; Douglas, T., Biomimetic synthesis of a H-2 catalyst using a protein cage architecture *Nano Letters* **2005**, *5*, 2306-2309.

68. Klem, M. T.; Willits, D.; Solis, D. J.; Belcher, A. M.; Young, M.; Douglas, T., Bio-inspired synthesis of protein-encapsulated CoPt nanoparticles *Advanced Functional Materials* **2005**, *15*, 1489-1494.

69. Shenton, W.; Mann, S.; Colfen, H.; Bacher, A.; Fischer, M., Synthesis of nanophase iron oxide in lumazine synthase capsids *Angewandte Chemie-International Edition* **2001**, *40*, 442-445.

70. Zhang, Z.; Buitenhuis, J.; Cukkemane, A.; Brocker, M.; Bott, M.; Dhont, J. K. G., Charge Reversal of the Rodlike Colloidal fd Virus through Surface Chemical Modification *Langmuir*, *26*, 10593-10599.

71. Brumfield, S.; Willits, D.; Tang, L.; Johnson, J. E.; Douglas, T.; Young, M., Heterologous expression of the modified coat protein of Cowpea chlorotic mottle bromovirus results in the assembly of protein cages with altered architectures and function *Journal of General Virology* **2004**, *85*, 1049-1053.

72. Chatterji, A.; Ochoa, W.; Shamieh, L.; Salakian, S. P.; Wong, S. M.; Clinton, G.; Ghosh, P.; Lin, T. W.; Johnson, J. E., Chemical conjugation of heterologous proteins on the surface of cowpea mosaic virus *Bioconjugate Chemistry* **2004**, *15*, 807-813.

73. Nam, K. T.; Peelle, B. R.; Lee, S. W.; Belcher, A. M., Genetically driven assembly of nanorings based on the M13 virus *Nano Letters* **2004**, *4*, 23-27.

74. Ochoa, W. F.; Chatterji, A.; Lin, T. W.; Johnson, J. E., Generation and structural analysis of reactive empty particles derived from an icosahedral virus *Chemistry & Biology* **2006**, *13*, 771-778.

75. Steinmetz, N. F.; Lin, T.; Lomonossoff, G. P.; Johnson, J. E. In *Viruses and Nanotechnology* 2009; Vol. 327, p 23-58.

76. Uchida, M.; Flenniken, M. L.; Allen, M.; Willits, D. A.; Crowley, B. E.; Brumfield, S.; Willis, A. F.; Jackiw, L.; Jutila, M.; Young, M. J.; Douglas, T., Targeting of cancer cells with ferrimagnetic ferritin cage nanoparticles *Journal of the American Chemical Society* **2006**, *128*, 16626-16633.

77. Young, M.; Willits, D.; Uchida, M.; Douglas, T., Plant viruses as biotemplates for materials and their use in nanotechnology *Annual Review of Phytopathology* **2008**, *46*, 361-384.
78. Douglas, T.; Strable, E.; Willits, D.; Aitouchen, A.; Libera, M.; Young, M., Protein engineering of a viral cage for constrained nanomaterials synthesis *Advanced Materials* **2002**, *14*, 415-+.
79. Allen, M.; Bulte, J. W. M.; Liepold, L.; Basu, G.; Zywicke, H. A.; Frank, J. A.; Young, M.; Douglas, T., Paramagnetic viral nanoparticles as potential high-relaxivity magnetic resonance contrast agents *Magnetic Resonance in Medicine* **2005**, *54*, 807-812.
80. Aniagyei, S. E.; Kennedy, C. J.; Stein, B.; Willits, D. A.; Douglas, T.; Young, M. J.; De, M.; Rotello, V. M.; Srisathiyanarayanan, D.; Kao, C. C.; Dragnea, B., Synergistic Effects of Mutations and Nanoparticle Templating in the Self-Assembly of Cowpea Chlorotic Mottle Virus Capsids *Nano Letters* **2009**, *9*, 393-398.
81. Douglas, T.; Young, M., Host-guest encapsulation of materials by assembled virus protein cages *Nature* **1998**, *393*, 152-155.
82. Hu, Y. F.; Zandi, R.; Anavitarte, A.; Knobler, C. M.; Gelbart, W. M., Packaging of a polymer by a viral capsid: The interplay between polymer length and capsid size *Biophysical Journal* **2008**, *94*, 1428-1436.
83. Klem, M. T.; Young, M.; Douglas, T., Biomimetic synthesis of beta-TiO2 inside a viral capsid *Journal of Materials Chemistry* **2008**, *18*, 3821-3823.
84. Li, H. Y.; Klem, M. T.; Sebby, K. B.; Singel, D. J.; Young, M.; Douglas, T.; Idzerda, Y. U., Determination of anisotropy constants of protein encapsulated iron oxide nanoparticles by electron magnetic resonance *Journal of Magnetism and Magnetic Materials* **2009**, *321*, 175-180.
85. Liepold, L.; Anderson, S.; Willits, D.; Oltrogge, L.; Frank, J. A.; Douglas, T.; Young, M., Viral capsids as MRI contrast agents *Magnetic Resonance in Medicine* **2007**, *58*, 871-879.
86. Sikkema, F. D.; Comellas-Aragones, M.; Fokkink, R. G.; Verduin, B. J. M.; Cornelissen, J.; Nolte, R. J. M., Monodisperse polymer-virus hybrid nanoparticles *Organic & Biomolecular Chemistry* **2007**, *5*, 54-57.
87. Slocik, J. M.; Naik, R. R.; Stone, M. O.; Wright, D. W., Viral templates for gold nanoparticle synthesis *Journal of Materials Chemistry* **2005**, *15*, 749-753.
88. Suci, P. A.; Klem, M. T.; Arce, F. T.; Douglas, T.; Young, M., Assembly of multilayer films incorporating a viral protein cage architecture *Langmuir* **2006**, *22*, 8891-8896.
89. Chen, C.; Daniel, M. C.; Quinkert, Z. T.; De, M.; Stein, B.; Bowman, V. D.; Chipman, P. R.; Rotello, V. M.; Kao, C. C.; Dragnea, B., Nanoparticle-templated assembly of viral protein cages *Nano Letters* **2006**, *6*, 611-615.
90. Wang, Q.; Lin, T. W.; Tang, L.; Johnson, J. E.; Finn, M. G., Icosahedral virus particles as addressable nanoscale building blocks *Angewandte Chemie-International Edition* **2002**, *41*, 459-462.
91. Wang, Q.; Kaltgrad, E.; Lin, T. W.; Johnson, J. E.; Finn, M. G., Natural supramolecular building blocks: Wild-type cowpea mosaic virus *Chemistry & Biology* **2002**, *9*, 805-811.
92. Martinez-Morales, A. A.; Portney, N. G.; Zhang, Y.; Destito, G.; Budak, G.; Ozbay, E.; Manchester, M.; Ozkan, C. S.; Ozkan, M., Synthesis and Characterization of Iron Oxide Derivatized Mutant Cowpea Mosaic Virus Hybrid Nanoparticles *Advanced Materials* **2008**, *20*, 4816-+.
93. Kang, S.; Suci, P. A.; Broomell, C. C.; Iwahori, K.; Kobayashi, M.; Yamashita, I.; Young, M.; Douglas, T., Janus-like Protein Cages. Spatially Controlled Dual-Functional Surface Modifications of Protein Cages *Nano Letters* **2009**, *9*, 2360-2366.
94. Flenniken, M. L.; Uchida, M.; Liepold, L. O.; Kang, S.; Young, M. J.; Douglas, T., A Library of Protein Cage Architectures as Nanomaterials *Viruses and Nanotechnology* **2009**, *327*, 71-93.
95. Radloff, C.; Vaia, R. A.; Brunton, J.; Bouwer, G. T.; Ward, V. K., Metal nanoshell assembly on a virus bioscaffold *Nano Letters* **2005**, *5*, 1187-1191.

96. Shenton, W.; Douglas, T.; Young, M.; Stubbs, G.; Mann, S., Inorganic-organic nanotube composites from template mineralization of tobacco mosaic virus *Advanced Materials* **1999**, *11*, 253-+.
97. Niu, Z.; Liu, J.; Lee, L. A.; Bruckman, M. A.; Zhao, D.; Koley, G.; Wang, Q., Biological templated synthesis of water-soluble conductive polymeric nanowires *Nano Letters* **2007**, *7*, 3729-3733.
98. Niu, Z. W.; Bruckman, M.; Kotakadi, V. S.; He, J. B.; Emrick, T.; Russell, T. P.; Yang, L.; Wang, Q., Study and characterization of tobacco mosaic virus head-to-tail assembly assisted by aniline polymerization *Chemical Communications* **2006**, 3019-3021.
99. Lim, J. S.; Kim, S. M.; Lee, S. Y.; Stach, E. A.; Culver, J. N.; Harris, M. T., Formation of Au/Pd Alloy Nanoparticles on TMV *Journal of Nanomaterials* **2010**.
100. Kumagai, S.; Yoshii, S.; Matsukawa, N.; Nishio, K.; Tsukamoto, R.; Yamashita, I., Self-aligned placement of biologically synthesized Coulomb islands within nanogap electrodes for single electron transistor *Applied Physics Letters* **2009**, *94*.
101. Gorzny, M. L.; Walton, A. S.; Evans, S. D., Synthesis of High-Surface-Area Platinum Nanotubes Using a Viral Template *Advanced Functional Materials* **2010**, *20*, 1295-1300.
102. Bromley, K. M.; Patil, A. J.; Perriman, A. W.; Stubbs, G.; Mann, S., Preparation of high quality nanowires by tobacco mosaic virus templating of gold nanoparticles *Journal of Materials Chemistry* **2008**, *18*, 4796-4801.
103. Dujardin, E.; Peet, C.; Stubbs, G.; Culver, J. N.; Mann, S., Organization of metallic nanoparticles using tobacco mosaic virus templates *Nano Letters* **2003**, *3*, 413-417.
104. Balci, S.; Bittner, A. M.; Hahn, K.; Scheu, C.; Knez, M.; Kadri, A.; Wege, C.; Jeske, H.; Kern, K., Copper nanowires within the central channel of tobacco mosaic virus particles *Electrochimica Acta* **2006**, *51*, 6251-6257.
105. Knez, M.; Bittner, A. M.; Boes, F.; Wege, C.; Jeske, H.; Maiss, E.; Kern, K., Biotemplate synthesis of 3-nm nickel and cobalt nanowires *Nano Letters* **2003**, *3*, 1079-1082.
106. Plascencia-Villa, G.; Saniger, J. M.; Ascencio, J. A.; Palomares, L. A.; Ramirez, O. T., Use of Recombinant Rotavirus VP6 Nanotubes as a Multifunctional Template for the Synthesis of Nanobiomaterials Functionalized With Metals *Biotechnology and Bioengineering* **2009**, *104*, 871-881.
107. Kobayashi, M.; Seki, M.; Tabata, H.; Watanabe, Y.; Yamashita, I., Fabrication of Aligned Magnetic Nanoparticles Using Tobamoviruses *Nano Letters* **2010**, *10*, 773-776.
108. Avery, K. N.; Schaak, J. E.; Schaak, R. E., M13 Bacteriophage as a Biological Scaffold for Magnetically-Recoverable Metal Nanowire Catalysts: Combining Specific and Nonspecific Interactions To Design Multifunctional Nanocomposites *Chemistry of Materials* **2009**, *21*, 2176-2178.

Chapter 8
Particle Deposition in the Human Respiratory Tract

Philip K. Hopke and Zuocheng Wang

Abstract Ambient airborne particles are now strongly linked with adverse health effects with stringent standards being set to protect public health particularly for smaller sized particles (particles with aerodynamic diameters less than 2.5 μm, $PM_{2.5}$). However, as noted elsewhere in this volume, there are a variety of medicines and therapeutic agents that can be effectively delivered through aerosolization and deposition in the lungs. Thus, the details of particle deposition in the human respiratory tract are important to human health for both good or ill. Airborne particles can penetrate into various portions of the respiratory tract during inhalation with a small fraction depositing on the airway surfaces. In this way, various hazardous or beneficial materials can be introduced into the body. Particle deposition in human airways takes exposure to particles to deposited dose in specific locations in the body. It is vital to the effective administration of pharmaceutical aerosols by inhalation to be able to deliver the particles to targeted regions of the respiratory tract.

Keywords Deposition • Extra-thorasic • Thorasic • Alveolar

8.1 Introduction

Ambient airborne particles are now strongly linked with adverse health effects with stringent standards being set to protect public health particularly for smaller sized particles (particles with aerodynamic diameters less than 2.5 μm, $PM_{2.5}$). However, as noted elsewhere in this volume, there are a variety of medicines and therapeutic agents that can be effectively delivered through aerosolization and deposition in the lungs. Thus, the details of particle deposition in the human respiratory tract are important to human health for both good or ill. Airborne particles can penetrate into various portions of the respiratory tract during inhalation with a small fraction depositing on the airway surfaces. In this way, various hazardous or beneficial materials can be introduced into the body. Particle deposition in human airways takes

P. K. Hopke (✉)
Department of Chemical & Biomolecular Engineering, Clarkson University,
8 Clarkson Avenue, CAMP Annex, Box 5708, Potsdam, NY 13699, USA
e-mail: phopke@clarkson.edu

Center for Air Resources Engineering and Science, Clarkson University,
Box 5708, Potsdam, NY 13699, USA

E. Matijević (ed.), *Fine Particles in Medicine and Pharmacy,*
DOI 10.1007/978-1-4614-0379-1_8, © Springer Science+Business Media, LLC 2012

exposure to particles to deposited dose in specific locations in the body. It is vital to the effective administration of pharmaceutical aerosols by inhalation to be able to deliver the particles to targeted regions of the respiratory tract.

Respiratory deposition of particles occurs in a system of changing geometry with a flow that changes with time and cycles in direction. This variability presents difficulties in obtaining a full understanding of particle deposition. Thus, experimental data and empirically derived equations resulting from those data have been the primary basis for estimating deposition. Deposition in the different regions of the respiratory system depends on many factors including: Breathing rate, mouth or nose breathing, lung volume, respiration volume, existing airway disease, the particle size/shape, airflow direction, and specific location in the respiratory system.

In this chapter, the structure of human respiratory tract is introduced. Then, some basics of particle deposition in human airway are presented. Finally, experimental and model deposition estimates are provided showing that deposition in the lungs can be predicted with a reasonable degree of accuracy.

8.2   Respiratory Anatomy

In order to understand the deposition of aerosols in the respiratory tract, it is useful to understand aspects of human respiratory system. Figure 8.1 provides a schematic representation of the human respiratory system. The system is typically divided into three broad regions: the extrathoracic region, the tracheo-bronchial region, and the alveolar region.

The extrathoracic region also referred to as the 'upper airways', mainly includes the nose, mouth and throat. It is the region that is proximal to the trachea. The extrathoracic region includes the following components:

1. The oral cavity (i.e. the mouth);
2. The nasal cavity (i.e. the nose);
3. The larynx, which is the constriction at the entrance to the trachea that contains the vocal cords.
4. The pharynx, which is the region between the larynx and either the mouth or nose.

The pharynx itself can be divided into parts that include the pathway from the larynx to the mouth (oropharynx) and the nose (nasopharynx). The term 'throat' usually refers to the pharynx and larynx.

Immediately distal to the extrathoracic region is the tracheo-bronchial region, sometimes also called the 'lower airways'. This region consists of the airways that conduct air from the larynx to the gas exchange regions of the lung, starting with the trachea, passing through the bronchi, bronchioles, and stopping at the end of the so-called 'terminal bronchioles'. The bronchi are the first six or so generations of branched airways after the trachea. The two airways branching off the trachea are called the main bronchi. These two bronchi branch into the lobar bronchi (of which there are two in the left lung and three in the right lung) that subsequently

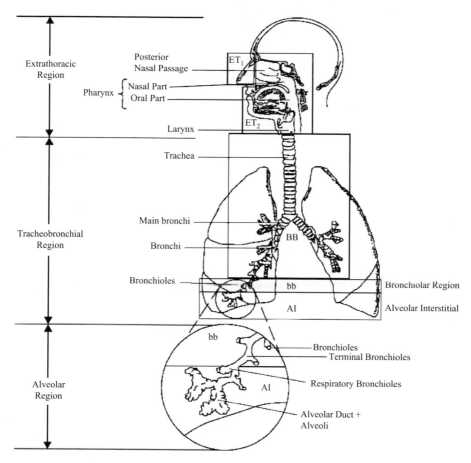

Fig. 8.1 The human respiratory system based on International Commission on Radiological Protection [1] and the U.S. Environmental Protection Agency. [2]

branch into the segmental bronchi. Taken together, the extrathoracic and tracheo-bronchial airways are called the 'conducting airways' since they conduct air to the gas-exchange regions of the lung.

The tracheo-bronchial airways are covered in a mucus layer that overlays fine hairs (cilia) that are attached to the airway walls. In healthy lungs, the cilia continually wave in synchronized motion to cause the mucus to move up the airway resulting in clearance of particles depositing in the bronchial region (generally within 24 h after inhalation).

Distal to the tracheo-bronchial airways is the alveolar region that is sometimes called the pulmonary region. Taken together, the tracheo-bronchial and alveolar regions are called the lung. The pulmonary or alveolar region consists of the alveoli and capillaries that serve them. It is in the alveolar region that gas exchange takes place. The total surface area of the alveoli in a typical adult is about 75 m^2 and this area is perfused with more than 2000 km of capillaries.

8.3   Basics of Particle Deposition

During inhalation, particles are transported with the inspired air through the extra-thoracic airways and the bifurcating tracheo-bronchiolar system to the gas exchange region of the lung. A certain number of these particles deposit in the respiratory system by touching the wet surfaces. Particle deposition in the human respiratory tract is the link between exposure and dose. The physical and chemical characteristics of inhaled particles can have a major effect on the nature of the effects produced in humans. Not all particles will be equivalent from a biological standpoint. A major determinant of how a particle is deposited, retained in the body, and produces effects depends on the particle size and its chemical composition.

8.4   Mechanisms of Deposition

The human airway has a complex geometry and results in a constantly changing hydrodynamic flow field. The particles deposit from the flow via several mechanisms including: inertial impaction, gravitational sedimentation, diffusion, interception, sand electrostatic attraction.

8.4.1   *Impaction*

Impaction occurs when a particle has sufficient inertia so that its trajectory deviates from the fluid streamline originally carrying it and contacts an airway surface. This mechanism is highly dependent on aerodynamic diameter or stopping distance (the distance a particle with a given initial velocity will travel in still air in the absence of external forces). Consequently, this mechanism is limited to the deposition of large particles that happen to be close to the airway walls. For large particles where inertia is the primary deposition mechanism and molecular slip can be assumed to be insignificant, then the total radial displacement of the particle from its original streamline can be expressed approximately as [3]:

$$\Delta = \frac{\rho_p d_p^2}{18\eta} \cdot \frac{\pi U}{2} = \frac{\rho_0 d_a^2}{18\eta} \cdot \frac{\pi U}{2}$$

Where Δ is total radial displacement of the particle from its original streamline, ρ_p is the particle density, d_p is the physical diameter of the particle, ρ_0 is particle density of 1, d_a is the *aerodynamic* diameter of the particle, U is the streamline velocity, and η is the viscosity of the air. This equation implies that the product of squared aerodynamic diameter and flow rate can be used in the analysis of particle impaction deposition in human airways.

Fig. 8.2 Physical model of
a human nose showing the
deposition of spherical particles of sodium fluorescein

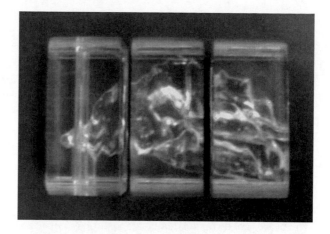

In lung bifurcations, for example, the greatest deposition occurs at or near the
carina, the dividing point at the tracheal bifurcation, and to lesser degree at other bifurcations. This increased deposition is caused by the sharp bending of the streamlines at the bifurcations as they pass close to the carina. In the human nose, an
experimental study using a physical model [4] has shown that for micrometer sized
particles, the deposition "hot spots" are in the vestibule where the air streamlines
bend most sharply and the nasal valve where the inhaled air has the highest velocity
(Fig. 8.2).

8.4.2   Sedimentation

Sedimentation represents deposition under the action of gravity. During sedimentation, a particle acquires its terminal settling velocity v when gravitational forces are
balanced by viscous resistive forces of the gas. The terminal settling velocity for
spherical particles, v, is then

$$v = \frac{\rho_p d_p^2 g}{18\eta}$$

where g is the gravitational constant. For sizes bigger than 1 μm diameter, the Cun-
ningham slip correction factor has not to be applied to take the discontinuity of the
surrounding medium into consideration.

Respirable particles acquire this terminal settling velocity in less than 0.1 ms [3],
a fraction of less than 1% of the typical transit time of the airflow in any airway
generation of the bronchial tree. For all practical purposes, the particles may be considered to reach their terminal settling velocity instantaneously. The probability of a
particle for depositing on airway walls by gravitational settling is proportional to the
distance a particle will cover within the airways and is thus proportional to the square
of the particle diameter. Considering the respiratory flow rate Q, gravitational depo-

sition is proportional to the parameter d^2/Q. Gravitational deposition mostly happens as the settling out of particles in the smaller airways of the bronchioles and alveoli, where the air flow is low and airway dimensions are small. Sedimentation has its maximum removal effect in horizontally oriented airways. Hygroscopic particles may grow in size as they pass through the warm, humid air passages, thus increasing the probability of deposition by sedimentation. Recent studies [5] have shown that the sedimentation deposition of micron sized particles in human nasal airway is not critical due to the relatively high flow rates and the shortness of the nasal airways.

8.5   Diffusion

The process of diffusion depicts deposition occurring via random Brownian motion. Brownian motion is the random wiggling motion of a particle due to the constant bombardment of air molecules. Diffusion is the primary mechanism of deposition for particles less than 0.1 μm in diameter and is governed by geometric rather than aerodynamic size. Diffusion happens as particles transport from a region of high concentration to a region of lower concentration. Diffusional deposition occurs mostly when the particles have just entered the nasopharynx, and is also most likely to occur in the smaller airways of the pulmonary (alveolar) region, where air flow is low. Ingham [6] developed an analytical, matched-asymptote solution for uniform flow penetration through a circular tube, P_t, given by

$$P_t = \frac{4}{\alpha_1^2} e^{-\alpha_1^2 \mu} + \frac{4}{\alpha_2^2} e^{-\alpha_2^2 \mu} + \frac{4}{\alpha_2^2} e^{-\alpha_3^2 \mu}$$

$$+ \left\{ 1 - 4(1/\alpha_1^2 + 1/\alpha_2^2 + 1/\alpha_3^2) \right\} e^{-\frac{4\mu^{1/2}}{\sqrt{\pi} \left[1 - 4(1/\alpha_1^2 + 1/\alpha_2^2 + 1/\alpha_3^2) \right]}}$$

Where $\mu = DL/R^2U$, D is the diffusion coefficient, L is the length of the tube, and R is the radius of the tube. The parameters α_1^2 (= 5.783186), α_2^2 (= 30.471262), α_3^2 (= 74.887007) are the zeros of the zero order Bessel function of the first kind, $J_o(\alpha_n) = 0$ [7]. This equation is valid for tube Reynolds number, $Re_t = 2U\rho R/\eta < 1200$ and tube Peclet number, $Pe_t = RU/D > 100$.

8.6   Interception

For irregular particle shapes such as fibrous aerosols, another deposition process becomes relevant; namely, interception. It occurs when a particle contacts an airway surface due to its physical size or shape while its center-of-mass remains on a fluid streamline. Unlike impaction, particles that are deposited by interception do not deviate from their air streamlines. Interception passes most likely to occur in small airways or when the air streamline passes close to an airway wall. Interception is most significant for fibers, which easily contact airway surfaces due to their length.

8.7 Electrostatic Forces

Most ambient particles become neutralized by the background concentrations of air ions to a Boltzmann distribution of low charge state, but many freshly generated particles can be highly electrically charged. The level of charge may vary widely, depending on the nature of the material, the mode of generation, and the age of the particles. Charged particles may have an enhanced deposition. It has been observed experimentally that charged particles cause a greater particle deposition in the lung than uncharged spherical particles [8, 9] and fibers [10, 11]. A threshold level of charge is required to enhance deposition. Yu [12] found this critical charge to be about 50 elementary charges for 1 μm spherical particles. Electrostatic forces may increase deposition up to a factor of 2, but generally deposition is influenced to a lesser extent, by only about 10% [12].

8.8 Experimental Studies of Particle Deposition

8.8.1 Lung Deposition Studies

Most of the deposition studies performed directly on people generally measured total deposition by measuring the concentration prior to inhalation and the concentration remaining on exhalation. These measurements were generally made many years ago. The resulting limited data were summarized by Schlesinger [13] and are presented in Fig. 8.3.

However, it is more important to understand the deposition of the particles in the various regions of the lung since that is where their therapeutic or deleterious actions will take place.

8.8.2 Particle Deposition in Human Nasal Airway

The human nose filters, warms, and humidifies inspired air. It is the important first line of protection that captures harmful particles and vapor pollutants, preventing them from reaching more delicate structure of the deeper lung. In the process, however, the nasal cavity places itself at risk from these insults. Researchers have been trying to understand local transport and deposition in the nasal cavity and to determine the pollutant concentration introduced to the lungs. Studies of spherical particle deposition in human airway with human volunteers (*in vivo*), physical casts (*in vitro*) and computational fluid dynamic simulation have been performed intensively

Fig. 8.3 Total particle deposition in the human respiratory tract. (Data taken from [14] based on Schlesinger [13])

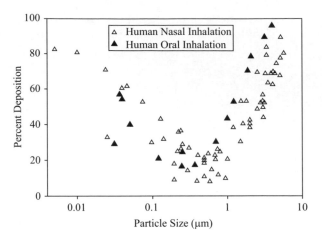

during past three decades to understand deposition mechanism and deposition patterns. Data obtained from human volunteers were initially used for estimates of regional deposition in the thoracic region. The Task Group on Lung Dynamics [15] adopted an empirical relationship for nasal deposition of particles based on *in vivo* data of Pattle [16] for particles > 1 μm in diameter.

Yu et al. [17] derived similar deposition equations in nasal and oral airways by summarizing *in vivo* deposition data from several studies with two logarithmic functions of the inertial parameter. Their analysis showed great variability of deposition efficiency in both nasal and oral regions under similar experimental conditions. Because of the variability of the data sets, it is not possible to confirm the assumption that the head airway deposition is a unique function of the inertial parameter.

Kelly et al. [18] used two replicas of the same nasal airways manufactured by different stereolithography machines to determine the inertial particle deposition efficiency. One replica called the SLA replica such that thicker build layers generally resulted in greater airway surface roughness in the build direction. The other was called Viper replica. They compared results with deposition efficiencies reported for models manufactured by other techniques from the same magnetic resonance imaging scans. Deposition in the replicas was measured for particles of aerodynamic diameter between 1 and 10 μm and constant inspiratory flow rates ranging from 20–40 L/min.

In a study based on data of four human subjects [19], it was found that, while the intersubject variability and the flowrate-induced intrasubject variability could not be eliminated by plotting particle deposition efficiency, PDE, against d_a^2Q, both could be accounted for when PDE was plotted against a parameter containing the pressure drop across the nasal passage, $d_a^2P^{2/3}$. The authors concluded that $d_a^2P^{2/3}$ was a more universal parameter than d_a^2Q.

Swift [20] obtained deposition data by using two nasal airway replicas made from MRI scans with detailed airway dimensions. This work was then extended to sizes of the order 1 nm [21]. Cheng [22] suggested that the difference in deposition between the infant and adult was due to nasal airway geometry. Swift's group [23, 24] used more than 10 human subjects to investigate the effects of particle size, flow rate, nostril shape, and nasal passage geometry on human nasal particle deposition with aerodynamic regime particles. They first used 40 healthy, nonsmoking adults (24 male, 16 female) to determine the effect of nostril dimensions, nasal passage geometry, and nasal resistance on particle deposition efficiency at a constant flow rate. Nasal resistance is defined as the ratio of the trans-nasal pressure drop to the flow rate. Polydisperse 1–10 μm diameter particles entering the nose and leaving the mouth were measured and the nasal particle deposition efficiency was calculated for bilateral and unilateral flow. Bilateral flow means both nasal passages were open and unilateral flow means left nasal passage open and right nostril occluded, or right nasal passage open and left nostril occluded when the experiments were performed. The bilateral flow rate was 30 L/min and right or left unilateral flow was 20 L/min.

Their research showed that, besides particle diameter and air flow rates, nostril dimensions and minimum nasal cross-sectional area (A_{min}) also significantly affected the particle deposition efficiency (PDE). However, even after inclusion of ellipticity of nostril (E, nostril length to width ratio) and A_{min} in PDE equations, there still remained a large intersubject variation in deposition. They concluded that aerosol deposition in the nasal passage has been difficult to predict and suggested further studies should evaluate the effect of other parts of the nose in deposition.

Particle deposition in human respiratory airway is strongly influenced by three major factors: physical (particle characteristics, flow rates); physiological (respiratory ventilation and pattern), and morphological (airway size and shape). The nasal valve and the main nasal airway (turbinate region) have very special significance to nasal functions as well as to air flow dynamics. The nasal valve is slit-shaped and is located at the narrow end of the funnel-shaped vestibule of the nose. After the nasal valve, the flow expands into the much wider main nasal airway. The expansion disturbs the flow and the resulting decrease in velocity, change in direction, and formation of eddies in the olfactory region tend to help air conditioning and sensing. While the main nasal passage is relatively fixed in its dimensions, the nasal valve is collapsible. Its shape and size are modified by the activity of alar muscles and by the pressure gradient between the ambient air and the inspired air. Relaxation, weakness or paralysis of the voluntary dilatory muscles allows the nasal valve to collapse during inspiration. Collapse may also occur during a sniff and in some cases with greatly increased inspiration. In the main nasal airway, the passages are narrow (1~3 mm) and ribbon-like structures. The irregular lateral walls which line the main nasal airway are clothed by ciliated respiratory mucosa which has abundant capacitance vessels—venous sinusoids whose blood content and volume determine the lumen of the nasal airway. Nasal cycle, exercise, posture, ambient temperature, medication, irritants, even pain and emotion can induce vascular changes of lumen. All result in the change of the nasal cavity with consequent airflow resistance.

Wang et al. [4] also used MRI-image based replicate human noses to measure the deposition of supermicron fibers and spherical particles. Three different nose models were employed. They found that fibrous particle deposition could be calculated from spherical particle results using equations for impaction equivalent diameters of the fibers (cylinder shape) as given in that study. An empirical model developed using a Stokes number and a non-dimensional relaxation time was used to fit the deposition data for both the fibrous and spherical particles resulting in a good fit. In addition, a method to estimate the pressure drop across the nasal models was also suggested. Thus, particle deposition in the human nose can be predicted if the nasal cavity dimensions are known such as could be obtained from MRI images.

8.8.3   Regional Lung Deposition

There have been a limited number of studies of regional deposition of particles within the tracheobronchial and alveolar regions. These studies have primarily been made with poorly soluble particles labeled with radioactive material [25–32]. These measurements permit the localization of the deposited particles within a general region of the lungs (bronchus, bronchioles, alveolar).

These measurements have been supplemented by a number of measurements of deposition in physical models of the human tracheobronchial tree [33–40]. Most of these measurements are made under a steady flow rather than a realistic cyclical flow. There have been some cyclical flow measurements (e.g., [37]) where they found somewhat lower deposition, but the results from the steady flow can be scaled to provide good agreement with cyclical flow in models and deposition in human subjects. These results have provided the basis for the development of semi-empirical models that are generally used to predict particle deposition over a range of flow rates, lung geometries and size, gender, age, and other population characteristics.

8.9   Empirical Deposition Models

8.9.1   ICRP Model

The deposition models were initially developed to provide estimates of the radiation dose produced by the inhalation and deposition of radioactive particles. The International Commission on Radiological Protection (ICRP) has taken the lead in model development dating back to the 1960s [15]. This early work was then updated to take into account all of the research conducted in the intervening period [41]. This dosimetric model for the respiratory tract included a revised regional particle deposition model for individuals as a function of age, gender, breathing rate, and clearance [42]. The respiratory tract is separated into anatomical regions using ICRP's Publication 66 notation: extrathoracic (ET), bronchial (BB), bronchiolar (bb) and alveolar-interstitial (AI). Each region of the respiratory tract is considered to be an equivalent particle filter that act in series. The detailed equations are given

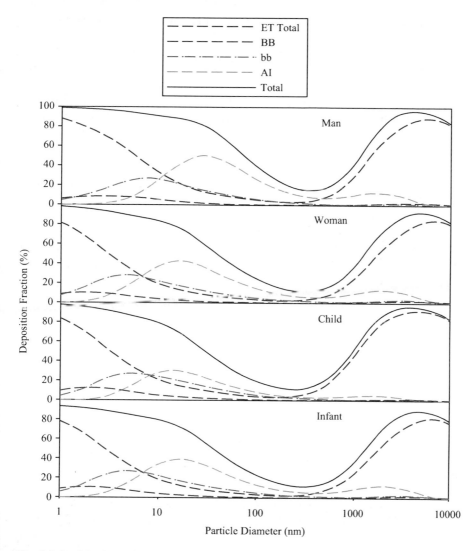

Fig. 8.4 Particle deposition as a function of particle size, age and gender based on the ICRP deposition model

by James et al. [43]. The model has been implemented in the computer model, LUDEP [44], to make the model easy to use. Figure 8.4 shows the mean respiratory deposition pattern for men, women, children (age 10) and infants (age 1) as a function of particle diameter from 1 to 10 μm for typical scenarios of activity levels and their related breathing rates. Figure 8.4 shows that there are differences in total as well as regional deposition among the major categories of individuals. This model was developed primarily to provide estimates of the dose from deposited radioactivity and has been widely used for the estimation of the dose from radon decay products (e.g., [45]).

Particle deposition models for inhaled particulate matter such as the ICRP model use morphologic data from symmetric idealizations of the lung airway tree structure such as that of Weibel [46] or Yeh [47, 48]. This approach leads to a typical-path description of the lung [49] that provides reasonably good average deposition agreement with experimental compartmental deposition data. However, typical-path models lack the resolution to examine deposition in individual airways and to thereby address critical heterogeneities [50]. Although the lower airways may be reasonably characterized in a symmetric fashion, there are major asymmetries in the upper airways of the tracheobronchial tree. These asymmetries may lead to different deposition patterns in these airways as well as in the apportionment of air flow, and thus in particle transport, to the different lung lobes. Alternate models have been developed to deliver more detailed results that could pertain to specific individuals and help define conditions that would permit targeting the deposition to specific locations in the respiratory tract.

8.10   Multiple-Path Model

Because of the need to have a more complete and accurate depiction of the conducting airways in a particle deposition model, but the inability to measure the entire human airway tree, the available measurements and associated statistical analyses were used to develop a stochastic model from which 10 five-lobe, asymmetric, structurally different, human airway trees were derived [50, 51]. This model is similar in structure to that originally developed for rats [52]. The MPPD model calculates the deposition and clearance of monodisperse and polydisperse aerosols in the respiratory tracts of rats and human adults and children (deposition only) for particles ranging in size from ultrafine (0.001 μm) to coarse (20 μm). Using more recent data, children's lungs have been redone to more effectively reflect the fact that children's lungs are not simply a smaller version of an adult lung [53].

The models are based on single-path and multiple-path methods for tracking air flow and calculating aerosol deposition in the lung. The single-path method calculates deposition in a typical path per airway generation, while the multiple-path method calculates particle deposition in all airways of the lung and provides lobar-specific and airway-specific information. Within each airway, deposition is calculated using theoretically derived efficiencies for deposition by diffusion, sedimentation, and impaction within the airway or airway bifurcation. Deposition in the nose and mouth is determined using empirical efficiency functions. The MPPD model includes the calculation of particle clearance in adults following deposition.

Figure 8.5 shows a comparison of regional deposition as estimated by the ICRP and MPPD models for resting conditions. In this case, there are 12 breaths per minute producing a breathing rate of 7.5 L/m. It can be seen that in general there is reasonable agreement between these estimates. This result should be expected since

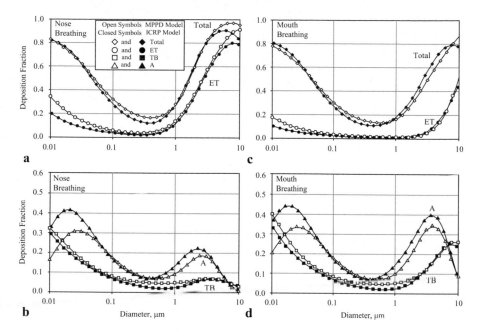

Fig. 8.5 Comparison of regional deposition results from the ICRP (LUDEP) and the MPPD models for a resting breathing pattern **a** and **b** are for nose breathing; **c** and **d** are for mouth breathing. (Taken from [14])

the empirical models are based on the same base of experimental data. However, more details can be obtained from the MPPD output and there is greater flexibility in the choice of input parameters. There are four options for the model used to depict the human lung morphology. The user has the choice of functional residual capacity (FRC) and the Upper Respiratory Tract (URT) volume. FRC is defined as the volume of the lung at end of a normal expiration. URT is the volume of the respiratory tract from the nostril or mouth down to the pharynx. In humans, the oral cavity and nasal passages are assumed to occupy the same volume. Particle input data include density and their size distributions can be monodisperse or polydisperse with specified values. There are two available exposure scenarios: constant and variable exposure. In the constant exposure scenario, inhalation occurs at a fixed tidal volume and breathing frequency. Steady breathing is assumed separately at the inhaled and exhaled flow rates at a specified breathing frequency. Uniform flow rates are assumed to be the average parabolic velocity at an airway inlet. It is possible to vary body orientation [54] to reflect if the subject is standing or lying down. The breathing frequency and tidal volume are input variables so subject specific values could be used. The breathing mode is also specified as nasal, oral, oronasal, and endotracheal. All of these options may be important in determining the deposition in patients undergoing respiratory therapy. The aerosol concentration is also specified. The clearance mechanism can be turned off if desired.

The MPPD model (v 2.0) is available for downloading from the internet at http://www.ara.com/products/mppd.htm. It represents an excellent compilation of the current state of knowledge in respiratory deposition of stable particles in a easy to use program that could be used to tailor particle characteristics to target deposition in specific areas of the respiratory tract.

8.10.1   Hygroscopic Growth

Particles entering the respiratory tract are immediately exposed to a temperature of 37°C and 99.5% relative humidity. Particles containing water soluble or partially soluble material will then deliquesce and grow in size in a time frame of the order of 100 ms [55]. Several models have been developed to account for the hygroscopic growth of particles [56–60]. However, these approaches all pertain to water soluble materials. Li and Hopke [61] developed a semi-empirical model that assumes that the particle contains both water-soluble and insoluble materials. Upon exposure to high humidity air, the soluble compounds will dissolve into the solution and form a liquid shell, while the insoluble compounds form a solid core. A large number of typical particle types have been characterized with respect to its hygroscopicity [61–66].

Since the solid core is surrounded by liquid shell, the liquid–gas interface remains the same as in a fully soluble material. By introducing a parameter, Z, defined as mass ratio of insoluble materials to soluble materials, the hygroscopic growth ratio can be derived as

$$\frac{d_1}{d_0} = \left(\frac{\rho_0}{\rho_1} \left[1 + \frac{i}{(1+\chi)M_s} \frac{M_w H}{R_1 - H} \right] \right)^{1/3} = \left(\frac{\rho_0}{\rho_1} \left[1 + \beta \frac{M_w H}{R_1 - H} \right] \right)^{1/3}$$

$$\beta = \frac{i}{(1+\chi)M_s}$$

$$\chi = \frac{m_{insoluble}}{m_{soluble}}$$

where d_1 and d_0 are the diameter of grown and dry particle, respectively; Px and Po are the respective densities; Mw is the molecular weight of water while Ms is the apparent average molecular weight of soluble components; H is the relative humidity in the system expressed as a fraction; i is the degree of dissociation of soluble components in water, which equals the number of ions or molecules formed upon dissolution. R_1 reflects the Kelvin effect, which is defined by Kelvin equation.

$$R_1 = \frac{p}{p_0} = \exp\left(\frac{4\sigma M_w}{RT\rho d} \right)$$

where p is the actual partial pressure of water, Po is the vapor pressure of water at temperature T, tr is surface tension of water, R is the gas constant, T is the absolute temperature and p is the density of the droplet liquid. Under the assumptions that $p_1 = p_o = 1$ and $i = 1$, all quantities in the equation except β are either measurable or known. Therefore, by employing a least squares fit of experimental data to this equation, the value of β can be estimated. For any drug formulation, it is then possible to generate an aerosol and make measurements as outlined by Li et al. [62], it is then possible to estimate the change in size resulting from hygroscopic growth. The modified size can then be used in the particle deposition model to determine the likely deposition and dose of that agent to various portions of the respiratory tract.

8.11   Conclusions

Over the past half century, a considerable amount of work has been devoted to the understanding of particle deposition in the human respiratory tract. Although individuals have a considerable degree of variability in the morphology of their respiratory system and breathing patterns, it is possible though measurements including sophisticated imaging to understand the parameters that would characterize any specific person. From the values of these parameters, an understanding of the chemistry of the particulate material, control of the particle size and shape, and exposure conditions can be established to maximize the deposition of particles in specific areas of the respiratory tract where they will be most efficacious.

References

1. International Commission on Radiological Protection. 1994. Human respiratory tract model for radiological protection: a report of a task group of the International Commission on Radiological Protection. Oxford, United Kingdom: Elsevier Science Ltd. (ICRP publication 66; Annals of the ICRP: v. 24, pp. 1-482).
2. U.S. Environmental Protection Agency. 1996. Air quality criteria for particulate matter. Research Triangle Park, NC: National Center for Environmental Assessment-RTP Office; report nos. EPA/600/P-95/001aF-cF. 3v.
3. Hinds, W.C., *Aerosol technology: Properties, behavior, and measurement of airborne particles*. (2nd ed.), New York: Wiley, 1999.
4. Wang, Z., Hopke, P.K., Ahmadi, G., Cheng, Y.S., Baron, P.A.. J. Aerosol Sci. 2008, 39, 1040–1054.
5. Su, W.C., Cheng, Y.S.,2005. Deposition of fiber in the human nasal airway. *Aerosol Science and Technology, 39*, 888–901.
6. Ingham, D.B.,1975. Diffusion of Aerosols from a Stream Flowing Through a Cylindrical Tube, *J. Aerosol Sci.* 6, 125-132.
7. Tan, C. W.,1969. Diffusion Of Disintegration Products of Inert Gases In Cylindrical Tubes, *Int. J. Heat Mass Transfer* 12, 471-478.
8. Prodi, V., Mularoni, A.,1985. Electrostatic lung deposition experiments with human and animals. *Ann. Occup. Hyg.* 29: 229-240

9. Ferin, J., Mercer, T.T., Leach, L.J.,1983. The effect of aerosol charge on the deposition and clearance of titanium dioxide particles in rats. *Environ. Res.* 31: 145-151.
10. Vincent, J.H., Johnston, W.B., Jones, A.D., Johnston, A.M.,1981. Static electrification of airborne asbestos: A study of its causes, assessment and effects on deposition in the lungs of rats. *Am. Ind. Hyg. Asssoc. J.* 42:711-721.
11. Vincent, J.H.,1985. On the practical significance of electrostatic lung deposition of isometric and fibrous aerosols. *J Aerosol Sci.* 16: 511-519.
12. Yu, C.P.,1985. Theories of electrostatic lung deposition of inhaled aerosols. *Ann. Occup. Hyg.* 29: 219-227.
13. Schlesinger, R. B. (1989) Deposition and clearance of inhaled particles. In: McClellan, R. O.; Henderson, R. F., eds. *Concepts in inhalation toxicology.* New York, NY: Hemisphere Publishing Corp.; pp. 163-192.
14. U.S. Environmental Protection Agency. 2004. Air Quality Criteria for Particulate Matter, Volume II of II, Environmental Protection Agency Report No. EPA/600/P-99/002bF, October 2004.
15. International Commission on Radiation Protection, 1966. Task Group on Lung Dynamics, *Health Phys.* **12**, 173-207.
16. Pattle, R.E.,1961. The Retention of Gases and Particles in the Human Nose. In: *Inhaled Particles and Vapours*, C.N.Davies, ed., Pergamon Press, Oxford, UK.
17. Yu, C.P., Diu, C.K., Soong, T.T.,1981. Statistical Analysis of Aerosol Deposition in Nose and Mouth. *Am. Ind. Hyg. Assoc. J.* 42:726–733.
18. Kelly, J.T., Asgharian, B., Kimbell, J., Wong, B.A.,2004. Particle deposition in human nasal airway replicas manufactured by different methods, Part I: Inertial regime particles. *Aerosol Science and Technology, 38*, 1036–1071.
19. Stahlhofen, W., Rudolf, G., James, A.C.,1989. Intercomparison of experimental regional aerosol deposition data. *Journal of Aerosol Medicine, 2*, 285–308.
20. Swift, D.L.,1991. Inspiratory Inertial Deposition of Aerosols in Human Airway Replicate Casts: Implication for the Proposed NCRP Lung Model. *Radiat. Prot. Dosim.* 38:29–34.
21. Swift, D.L., Montassier, N., Hopke, P.K., Karpen-Hayes, K., Cheng, Y.S., Su, Y-F., Yeh, H.C., Strong, J.C.,1992. Inspiratory Deposition of Ultrafine Particles in Human Nasal Replicate Casts, J. Aerosol Sci. 23, 65-72.
22. Cheng, Y.S.,2003. Aerosol deposition in the extrathoracic region. *Aerosol Science and Technology, 37*, 659–671.
23. Kesavanathan, J., Swift, D.L.,1998. Human nasal passage particle deposition: The effect of particle size, flow rate, and anatomical factors. *Aerosol Science and Technology, 28*, 457–463.
24. Kesavanathan, J., Bascom, R., Swift, D.L.,1998. The effect of nasal passage characteristics on particle deposition. *Journal of Aerosol Medicine*, 11, 27–39.
25. Stahlhofen, W., Gebhart, J., Heyder, J.,1980. Experimental determination of the regional deposition of aerosol particles in the human respiratory tract, Am. Ind. Hyg. Assoc. J. 41, 385-398.
26. Heyder, J., Gebhart, J., Rudolf, G., Schiller, C.F., Stahlhofen, W.,1986. Deposition of particles in the human respiratory tract in the size range 0.005–15 µm, J. Aerosol Science, 17, 811-825.
27. Wilson, F.J., Jr, Hiller, F.C., Wilson, J.D., Bone, R.C.,1985. Quantitative deposition of ultrafine stable particles in the human respiratory tract. *J Appl Physiol* 58,223–229.
28. Schiller, C.F., Gebhart, J., Heyder, J., Rudolf, G., Stahlhofen, W.,1988. Deposition of monodisperse insoluble aerosol particles in the 0.005 to 0.2 µm size. *Ann Occup Hyg* 32 (Suppl. 1), 41–49.
29. Anderson, P.J., Wilson, J.D., Hiller, F.C.,1990. Respiratory tract deposition of ultrafine particles in subjects with obstructive or restrictive lung disease *Chest* 97, 1115–1120.
30. Jaques, P.A., Kim, C.S.,2000. Measurement of total lung deposition of inhaled ultrafine particles in healthy men and women. *Inhal Toxicol* 12, 715-731.

31. Kim, C.S., Jaques, P.A.,2000. Respiratory dose of inhaled ultrafine particles in healthy adults. *Philos Trans R Soc Lond A* 358, 2693–2705.

32. Brown, J.S., Zeman, K.L., Bennett, W.D.,2002. *American Journal of Respiratory and Critical Care Medicine* 166, 1240-1247.

33. Schlesinger, R.B., Lippmann, M.,1972. Particle Deposition in Casts of the Human Upper Tracheobronchial Tree, *Am. Ind. Hyg. Assoc. J.* 33, 237–251.

34. Schlesinger, R.B., Bohning, D.E., Chan, T.L., Lippmann, M.,1977. Particle Deposition in a Hollow Cast of Human Tracheobronchial Tree, *J. Aerosol Sci.* 8:429–445.

35. Chan, T.L., Lippmann, M., Cohen, V.R., and Schlesinger, R.B.,1978. Effect of Electrostatic Charges on Particle Deposition in a Hollow Cast of the Human Larynx-tracheobronchial Tree, *J. Aerosol Sci.* 9, 463–468.

36. Chan, T.L., Lippmann, M.,1980. Experimental Measurements and Empirical Modeling of the Regional Deposition of Inhaled Particles in Humans, *Am. Ind. Hyg. Assoc. J.* 41, 399–409.

37. Gurman, J.L., Lippmann, M., Schlesinger, R B.,1984. Particle Deposition in Replicate Casts of Human Upper Tracheobronchial Tree Under Constant and Cyclic Inspiratory Flow: I. Experimental, *Aerosol Sci. Technol.* 3, 245–252.

38. Cohen, B.S., Susman, R.G., Lippmann, M.,1990. Ultrafine Particle Deposition in a Human Tracheobronchial Cast, *Aerosol Sci. Technol.* 12, 1082–1091.

39. Sussman, R.G., Cohen, B. S., Lippmann, M.,1991. Asbestos Fiber Deposition in Human Tracheobronchial Cast. I. Experimental, *Inhal. Toxicol.* 3, 145–160.

40. Zhou, Y., Cheng, Y.-S.,2005. Particle Deposition in a Cast of Human Tracheobronchial Airways, *Aerosol Sci. Technol.* 39, 492-500.

41. International Commission on Radiological Protection (ICRP). 1994. *Human Respiratory Tract Model for Radiological Protection.* ICRP publication 66. Ann. ICRP 24, Nos 1–3.

42. Frappe, F.N., Rannou, A.,1998. General Solution of the ICRP 66 Model and Its Application to Given Radionuclides, *Rad. Protect. Dosim.* 79, 29-32.

43. James, A.C., Stahlhofen, W., Rudolf, G., Egan, M.J., Nixon, W., Gehr, P., Briant, J.K.,1991. The Respiratory Tract Deposition Model Proposed by the ICRP Task Group, *Rad Protect. Dosim.* 38, 159-165.

44. Birchall, A., Bailey, M. R., James, A. C.,1991. LUDEP: A Lung Dose Evaluation Program, *Radiat. Prot. Dosim.* 38, 167–174.

45. Hopke, P.K., Jensen, B., Li, C.S., Montassier, N., Wasiolek, P., Cavallo, A.J., Gatsby, K., Sokolow, R.H.,1995. Assessment of the Exposure to and Dose from Radon Decay Products in Normally Occupied Homes, *Environ. Sci. Technol.* 29:1359-1364.

46. Weibel, E. R. 1963. *Morphometry of the human lung.* Berlin: Springer-Verlag.

47. Yeh, H.-C. 1980. *Respiratory tract deposition models.* LF-72, UC-48. Inhalation Toxicology Research Institute, Albuquerque, NM.

48. Yeh, H. C., Schum, G. M. 1980. Models of human lung airways and their application to inhaled particle deposition. *Bul!. Math. Biol.* 42, 461-480.

49. Yu, C. P. 1978. Exact analysis of aerosol deposition during steady breathing. *Powder Technol.* 21, 55-62.

50. Asgharian, B., Hofmann, W., Bergmann, R. 2001. Particle deposition in a multiple-path model of the human lung. *Aerosol Sci. Technol.* 34:332-339.

51. Winter-Sorkina, R. de; Cassee, F.R. (2002) From concentration to dose: factors influencing airborne particulate matter deposition in humans and rats. Bilthoven, The Netherlands: National Institute of Public Health and the Environment (RIVM); report no. 650010031/2002. Available: http://www.rivm.nl/bibliotheek/rapporten/650010031.html (13 June 2003).

52. Anjilvel, S., Asgharian, B.,1995. A multiple-path model of particle deposition in the rat lung. Fundam. Appl. Toxicol. 28, 41- 50.

53. Asgharian, B., Ménache, M.G., Miller, F.J.,2004. Modeling age-related particle deposition in humans. *J Aerosol Med* 17:213-224.

54. Asgharian, B., Price, O.T., Oberdörster, G.,2006. A modeling study of the effect of gravity on airflow distribution and particle deposition in the lung. Inhalation Toxicology 18, 473-481.

55. Ferron, G.A., Oberdörster, G., Henneberg, R.. 1989. Estimation of the Deposition of Aerosolized Drugs in the Human Respiratory Tract Due to Hygroscopic Growth, J. Aerosol Medicine 2, 271-284.

56. Ferron, G.A.,1977. The size of soluble aerosol particles as a function of the humidity of the air: application to the human respiratory tract, *J. Aerosol Sci.* 8, 251-267.

57. Martonen, T.B.,1982. Analytical Model of Hygroscopic Particle Behavior in Human Airways, *Bull. Math. Bio.* 44, 425-442.

58. Martonen, T.B., Bell, K.A., Phalen, R.F., Wilson, A.F., Ho, A.,1982. Growth Rate Measurements and Deposition Modeling of Hygroscopic Aerosols in Human Tracheobronchial Models, *Ann. Occup. Hyg.* 26, 93-108.

59. Egan, M.J., Nixon, W.,1989. On the Relationship between Experimental Data for Total Deposition and Model Calculations - Part II: Application to Fine Particle Deposition in the Respiratory Tract, *J. Aerosol Sci.* 20, 149-156.

60. Ferron, G.A., Hayder, B., Kreyling, W.G.,1985. A Method for Approximation of the Relative Humidity in the Upper Human Airways, *Bull. Math. Bio.* 47, 565-589.

61. Li, W., Hopke, P. K.,1993. Initial size distributions and hygroscopicity of indoor combustion aerosol particles. Aerosol Sci. Technol. 19, 305-316.

62. Li, W., Montassier, N., Hopke, P.K.,1992. A System to Measure the Hygroscopicity of Aerosol Particles, *Aerosol Sci. Technol.* 17, 25-35.

63. Li, W., Hopke, P.K.,1994. Hygroscopicity of Consumer Aerosol Products, *J. Aerosol Sci.* 25, 1342-1351.

64. Dua, S.K., Hopke, P.K., Raunemaa, T.,1995. Hygroscopic Growth of Consumer Spray Products, *Aerosol Sci. Technol*, 23, 331-340

65. Dua, S. K., Hopke, P.K.,1996. Hygroscopic Growth of Assorted Indoor Aerosols, *Aerosol Sci. Technol.* 24, 151-160.

66. Dua, S.K., Hopke, P.K., Raunemaa, T.,1999. Hygroscopicity of Diesel Aerosols, *Water, Air, Soil Pollution* 112, 247-257,

Chapter 9
Detection of Oxidative Stress Biomarkers Using Functional Gold Nanoparticles

Maria Hepel and Magdalena Stobiecka

Abstract Biomarkers of oxidative stress are biomolecules that can be utilized in the diagnosis of diminished capacity of a biological system to counteract an over-production or invasion of reactive oxygen species and other radicals. In this Chapter, the detection methods of the oxidative stress biomarkers, such as glutathione (GSH), homocysteine (Hcys) and cysteine (Cys), based on their interactions with monolayer-protected gold nanoparticles are described. The nanoparticle utilization in a solution-phase analysis as well as in a multifunctional sensory film preparation is presented. The interactions of AuNP with GSH, a tripeptide maintaining the redox potential level in eukaryotic cells, have been investigated using resonance elastic light scattering (RELS) and plasmonic UV–vis spectroscopy. The high sensitivity of the RELS measurements enables monitoring of ligand exchanges and the bio-marker-induced AuNP assembly. The viability of designing simple and rapid assays for the detection of GSH and Hcys is discussed. The surface plasmon band broadening and bathochromic shift are consistent with the biomarker-induced AuNP assembly and corroborate the RELS measurements and HR-TEM imaging. The results of molecular dynamics and quantum mechanical calculations support the mechanism of the formation of GSH- and Hcys-linkages in the interparticle interactions and show that multiple H-bonding can occur. The functionalized gold nanoparticles have also been shown to enhance the design of molecularly-templated conductive polymer films for the detection of biomolecules and to amplify the analytical signal by AuNP labeling. Novel designs of molecularly imprinted orthophenylenediamine (oPD) sensor films are presented. The biolyte-induced assembly of monolayer protected gold nanoparticles is evaluated in view of bioanalytical applications and the design of novel sensory films for molecularly-templated conductive polymers for microsensor arrays.

Keywords Biomarkers of oxidative stress • Gold nanoparticles • RELS • Surface plasmon • Molecularly-templated polymer films • Glutathione • Homocysteine • Cysteine

M. Hepel (✉)
Department of Chemistry, State University of New York, Potsdam, NY 13676, USA
e-mail: hepelmr@potsdam.edu

E. Matijević (ed.), *Fine Particles in Medicine and Pharmacy*,
DOI 10.1007/978-1-4614-0379-1_9, © Springer Science+Business Media, LLC 2012

9.1   Introduction

Oxidative stress is a condition of cells and organisms characterized by the decreased capacity to counteract the invasion or overproduction of reactive oxygen species. The balance, normally existing between the production of the latter and a biosystem's ability to readily detoxify the damaging radicals, is disturbed under the oxidative stress, leading to the development of many diseases. Therefore, it is desirable to diagnose and control the oxidative stress as early as possible. The oxidative stress can be diagnosed by analyzing the oxidative stress biomarkers.

The biomarkers of oxidative and nitrosative stress have recently been the subject of extensive studies [1, 2] as the new evidence demonstrates ever increasing number of related diseases. The oxidative stress has been suggested as the causative factor in aging [3] and many diseases [4–7] such as cadiovascular, diabetes, cancer, autism spectrum disorders (ASD), and others. Among the biomarkers of oxidative stress are small biomolecules such as: ubiquinol [8] which is very labile in the oxidation of low-density lipoprotein (LDL), glutathione (GSH) which is depleted in the presence of organic radicals and peroxides [9], homocysteine [10, 11] which has been found at elevated levels in atherosclerosis [12–17], Alzheimer disease [5, 6], dementia [5], and poses an increased risk of birth defects [7]. Some biomarkers of oxidative stress are necessary to maintain healthy homeostasis (e.g. glutathione), while others participate in the development of diseases (e.g. homocysteine, which is a risk factor for cardiovascular disease). For instance, decreased levels of glutathione and increased levels of oxidized glutathione (GSSG) have been observed in plasma, serum and urine samples from individuals diagnosed with ASD [4, 18–20]. Homocysteine (Hcys), which is a sulfur-containing amino acid, is formed during a metabolism of methionine to cysteine (Scheme 9.1) but the increased concentration of Hcys in plasma ($C_{Hcys} > 15$ μM) is a risk factor for many disorders, including cardiovascular [12–15], renal [21], Alzheimer's [5, 6], and other diseases [22]. Redox-related alterations, measured usually as the change in the concentration ratio of GSH/GSSG which is the main redox level maintaining couple in organisms, may also be heritable. Deviations from healthy biomarker concentration levels may result from deficiency of certain vitamins, e.g. B12 and folic acid (in hyperhomocysteinemia). The investigations of oxidative stress biomarkers are important to understand their behavior and role in organisms and to develop sensors and assays for their rapid detection and diagnosis of stress-related disorders.

The GSH/GSSG couple is the main redox regulation system in living organism's homeostasis (Scheme 9.2). It protects cells against organic peroxides and damaging radicals, and is involved in signaling processes associated with cell apoptosis. The diminished active GSH levels in cells and body fluids and the reduced antioxidation capacity [1] to protecting against radicals have been found to increase susceptibility to autism [19, 20], diabetes [18], and other diseases [1, 18, 20, 24–28]. The low GSH levels have been found to be caused by oxidative stress and exposure to toxic heavy metals (Hg, Cd, Pb). GSH and phytochelatines with general structure (γ-Glu-Cys)$_n$Gly have been found to participate as the capping agents [29, 30] in

Metabolism of Homocysteine

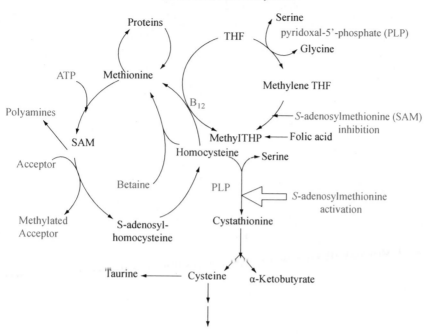

Scheme 9.1 Metabolism of the oxidative stress biomarker homocysteine. (Adapted from Selhub [23], with permission from Annual Reviews of Nutrition)

heavy-metal sulfide nanoparticles formed in living organisms in natural detoxification processes [31–34].

The reactivities and interactions of the oxidative stress biomarkers have been investigated in conjunction with the development of molecularly-templated polymer films with biorecognition capabilities designed for biomarkers detection [35], fluorimetric assays based on specific reactions [36–38], electrochemical sensors [39, 40], colorimetric assays based on nanoparticle assembly [41–44], and the design of immunosensors [45] and other sensors for the analysis of biomarkers or utilizing biomarkers in the sensory film design [46–48]. In particular, in studies of biomolecule-induced gold nanoparticle assembly, the kind of interparticle interactions is the key element of the functionalized nanoparticle self-affinity [49–51]. The interparticle forces include electrostatic [52], zwitterionic [41, 52], van der Waals forces [53], as well as hydrogen bonding forces [53–55]. The investigations of functionalized spherical gold nanoparticles and gold nanorods for application in novel assays for GSH [55], cysteine [44, 56–60] and homocysteine [41, 42, 57, 61] have been reported. The gold nanoparticle cores with protective shells of self-assembled monolayers (SAM) of thiolates [62, 63], surfactants [59, 64, 65], citrate ions [60], and others can be utilized in the analysis. A difference in the sensitivity of the gold nanoparticle assembly process to structurally similar cysteine and homocysteine molecules, which differ only by one CH_2 group, has been found [44, 58, 59]. Prob-

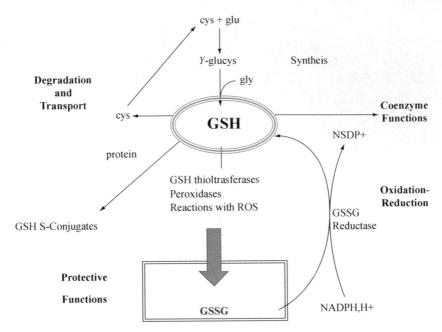

Scheme 9.2 Metabolism of glutathione (GSH), according to Anderson [66] (with permission from Elsevier)

ing the interactions of biomolecules with gold nanoparticles and their influence on surface plasmon resonance and the elastic light scattering cross-section has potential applications in the development of novel assays for these molecules.

Recently, GSH has been utilized as the active component of various sensory films on electrodes. GSH is a tripeptide (glutamate-cysteine-glycine) with sulfhydryl group able to form a strong Au-S bond and therefore, it adsorbs readily on gold surfaces [46–48, 67, 68] and forms protective shells on AuNP [43, 55, 69–71]. We have investigated the interactions and electrochemical reactivity of Hg(II) on GSH-modified gold piezoelectrodes [46, 67, 72]. The GSH-SAM permeability to ionic species has been demonstrated for Hg(II) [46, 67], Pb(II) [47], Ni(II) [47], and Cu(II) [48] using metal adatom probe, nanogravimetry and chronoamperometry. Gooding et al. [73] studied GSH bonded to mercaptopropionic acid as the sensor for Cd(II). The interactions of adsorbed GSH with Cu^{2+} have also been studied [48, 68]. It has been shown that GSH-SAM's formed on Au piezoelectrodes can act as chemically controlled ion gates [47, 74–77] and as templates for metal depositions [48].

Thin self-assembled monolayers (SAM) on a gold electrode surface [47, 48, 78–82] have been successfully utilized in designing sensors [46, 67, 83–85], biorecognition systems [86–89], molecular electronic devices [83, 90], nanoimprints [91, 92], catalysts [93–95], etc. Much attention has recently been focused on SAM-protected inorganic nanoparticles [96–103], including gold [50, 54, 79, 80, 104–109], silver [106], CdS/CdSe quantum dots [110, 111], n-TiO_2, silica nanoparticles [112] and others, in view of their novel applications in nanomedicine [113–116], molecular biology [117–127], and biotechnology [104, 112]. Thus far, SAM-protected gold nanoparticles (AuNP) have been most widely investigated due to their unique optical and electronic properties associated with surface plasmon resonance [128–131] and quantum size confinement [132, 133].

The high sensitivity of surface plasmon (SP) resonance of AuNP to modifications in protective shells, as well as to the interparticle interactions and molecularly-mediated assembly make the gold nanoparticle based designs [50, 134–136] highly responsive to external stimuli [137]. The response to small interacting molecules [41] and to complexing metal cations have already been reported and utilized in analytical procedures. Furthermore, biorecognition layers on AuNP, designed to detect single mismatches in DNA strands, have been advocated for rapid and reliable bioanalytical applications [104, 109, 118, 125, 126, 135–138]. The compatibility of AuNP with biorecognition structures based on biotin-streptavidin interactions, antigen-antibody affinity, and enzyme-based immunosensors has become apparent as novel architectures are reported in growing numbers [121, 122, 124, 139]. In view of these interests, the effect of small biomolecules on SP frequency has also been investigated [58]. Such molecules as aminoacids [41, 58, 140–142] and small peptides [43, 55, 58] have been found to influence the spectral SP characteristics under favorable experimental conditions. For instance, by properly designing the protecting SAM shell of AuNP, an SP-based colorimetric assay with enhanced selectivity toward homocysteine in a matrix of aminoacids has been developed [41, 58].

The assembly of gold nanoparticles mediated by GSH has been studied by Sudeep et al. [43] and by Zhong et al. [55] and strong effects of pH and electrolyte concentration have been found. A model based on two-point zwitterionic interparticle interactions has been proposed [69]. Directional growth of GSH-linked gold nanorod assemblies has been observed by Kou et al. [70] whereas the competitive adsorption of GSH and thiolated oligonucleotides has been investigated by Ackerson et al. [71]. Recently, GSH has been found to participate in the degradation of Pt(II)-based DNA intercalators utilized in chemotherapy. GSH adsorbs also on Ag nanoparticles [143] and on Ag_2S semiconductor nanoparticles [144] owing to the strong binding through its cysteine moiety S-atom to a Ag atom of the substrate. Different kinds of interparticle interactions and bonding have been recently discussed by Maye et al. [145].

In this chapter, recent progress of investigations of oxidative stress biomarkers, and in particular their interactions with gold nanoparticles, is presented. The main focus is on the biophysical implications of these interactions on the development of novel nanoparticle-based assays and new sensory films for bioanalytical applications. The emphasis is placed on the utilization of the plasmonic absorbance spectroscopy and the resonance elastic light scattering (RELS) spectroscopy. The latter technique has been recently developed [146–152] as a very sensitive technique for the detection and analysis of bioorganic complexes. The strong resonant light scattering from molecules and nanoparticles in solution results from the absorption of photons followed by an immediate coherent re-emission of light without any energy loss. We have found the RELS spectroscopy to be extremely useful [152] in studies of the interactions of small biomolecules with metal nanoparticles.

9.2 Monolayer-Protected Gold Nanoparticles in the Analysis of Biomarkers of Oxidative Stress

The high stability of gold colloids, on one hand, and their sensitivity to ligand-exchangeable biomolecules able to induce assembly, on the other hand, provide ample opportunities to devise simple assays for a variety of biolytes. In this section, analytical aspects of the interactions of a biomarker of oxidative stress, GSH, with gold nanoparticles in solution are evaluated. Particular attention is paid to GSH-mediated AuNP assembly processes and the use of resonance elastic light scattering and surface plasmon absorbance spectroscopy to monitor these processes. The propensity of Hcys for AuNP cross-linking is discussed in Sect. 9.3.

9.2.1 Determination of GSH Using Resonance Elastic Light Scattering (RELS)

We have explored the utility of RELS spectroscopy to study small biomolecule-induced interparticle cross-linking. The RELS technique has been developed recently [146–151] as a very sensitive technique for the detection and analysis of bioorganic complexes and assembly processes. The strong resonant light scattering from molecules and nanoparticles in solution results from the absorption of photons followed by an immediate coherent re-emission of light without any energy loss. The gold nanoparticles are known to be very good scatterers of light. The high cross-section of AuNP for light scattering stems from the efficient reflectivity of conduction electrons oscillating at frequencies of visible light in the form of a surface plasmon. We have found the RELS spectroscopy to be extremely useful in studies of the interactions of small biomolecules with metal nanoparticles [152]. In this section, we describe the interactions of GSH with AuNP and elucidate the mechanism of the multi-step process leading to the assembly of GSH-capped AuNP networks.

Typical RELS spectrum for a citrate-capped $AuNP_{5\ nm}$ solution is presented in Fig. 9.1a, curve a1, for a constant excitation wavelength $\lambda_{ex} = 640$ nm (1.94 eV). The scattering intensity peak with a Gaussian peak shape centered at $\lambda_{em} = \lambda_{ex} = 640$ nm and with a narrow linewidth of $\Delta\lambda = 14$ nm confirms that the effects due to radiation broadening, density fluctuation, fluorescence, and inelastic Raman scattering are negligible. The background intensity is very low (virtually zero) providing excellent conditions for a sensitive analysis with well defined RELS peaks.

In the presence of GSH, the light scattering from $AuNP_{5\ nm}$ nanoparticles is strongly enhanced (Fig. 9.1a, curve a2). The solution pH is 3.21–3.27 which is within the range of predominantly neutral (zwitterionic) form of GSH (pH = 2.04–3.53). The enhancement of RELS from $AuNP_{5\ nm}$ by GSH molecules is attributed to the size increase of AuNP due to the ligand exchange (i.e. replacing short-chain citrate molecules in the nanoparticle shell with longer-chain GSH molecules) and/or interparticle bridging interactions leading to AuNP assembly.

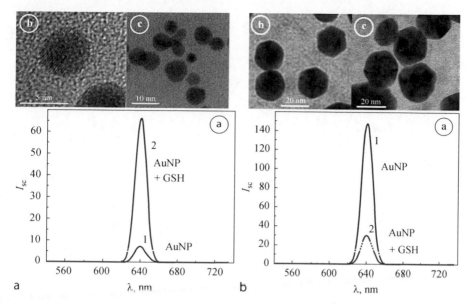

Fig. 9.1 Gold nanoparticle size-effect on RELS spectra of GSH-capped AuNP with diameter: **a** 5 nm, **b** 22.5 nm. **a**: (*a*) Light scattering spectra for 10.1 nM AuNP$_{5\,nm}$ in the absence (1) and in the presence (2) of 5 mM glutathione, recorded for the incident beam wavelength $\lambda_{ex} = 640$ nm (spectra recorded within 1 min of glutathione injection); (*b*) high-resolution TEM image of AuNP; atomic rows with distance 0.24 nm are seen; (*c*) HR-TEM image of small GSH-linked AuNP$_{5\,nm}$ assemblies. **b**: (*a*) Light scattering spectra for 1.42 nM AuNP$_{22\,nm}$, in the absence (1) and in the presence (2) of 5 mM glutathione; (*b, c*) HR-TEM images of AuNP before (*b*) and after (*c*) addition of 5 mM GSH. Spectra recorded within 1 min of glutathione injection; incident beam wavelength: $\lambda_{ex} = 640$ nm (from [156], with permission from Elsevier)

The analysis of the particle-size dependence for RELS signals has recently been performed [152]. The strong sixth-power dependence of elastic scattering intensity I_{sc} on the nanoparticle diameter a follows from the Rayleigh equation for light scattering from small particles:

$$I_{sc} = I_0 N \frac{(1 + \cos^2 \theta)}{2R^2} \left(\frac{2\pi}{\lambda}\right)^4 \frac{\left((n_p - n_s)^2 - 1\right)}{\left((n_p - n_s)^2 + 2\right)} \left(\frac{a}{2}\right)^6 \qquad (9.1)$$

where n_p and n_s are the refractive indices for the particles and the solution, respectively, λ is the wavelength of incident light beam, θ is the scattering angle, N is the number of particles, and I_0 is the constant. Taking into account the decrease in particle concentration due to the assembly, as well as $\lambda = $ const and other experimental conditions (θ, R, I_0) unchanged, the increase of the effective particle diameter a_{rel} can be estimated using the formula [152]:

$$a_{rel} = \frac{a_1}{a_0} = \sqrt[3]{\frac{I_{sc,1}}{I_{sc,0}}} \qquad (9.2)$$

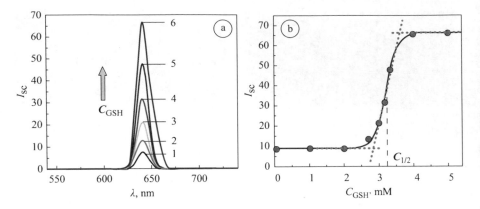

Fig. 9.2 a Elastic light scattering spectra for 10.1 nM $AuNP_{5\,nm}$ for different concentrations of GSH, recorded within 1 min of GSH injection, CGSH [mM]: (1) 0, (2) 2.67, (3) 3.0, (4) 3.17, (5) 3.33, (6) 5; **b** Dependence of Isc vs. CGSH; incident beam wavelength: $\lambda_{ex} = 640$ nm (from [156], with permission from Elsevier)

where indices 0 and 1 stand for the particles before and after GSH addition, respectively. From the data of Fig. 9.1a, the scattering intensity increase is: $I_{sc,1}/I_{sc,0} = 9.09$ and, hence, $a_{rel} = 2.09$. This means that most likely small aggregates composed of only few nanoparticles (e.g 2–6) are formed.

The RELS signal, measured for small Au nanoparticles, $AuNP_{5\,nm}$, within 1 min of reaction time, depends strongly on GSH concentration and increases monotonously with C_{GSH}. This is illustrated in RELS spectra presented in Fig. 9.2. The dependence of $I_{sc,max}$ on C_{GSH} is sigmoidal, as shown in the inset indicating on a kinetic threshold taking place near the inflection point (Fig. 9.2).

It is imperative that the analytical determinations of GSH utilizing the AuNP based assays are performed with well-defined small AuNP (e.g. 5 nm diameter) since there is a strong AuNP-size dependence of RELS sensitivity for GSH analysis. We have found that for larger AuNP, the GSH concentration dependence of RELS signal, characterized by the high slope ($\partial I_{sc}/\partial C_{GSH}$), may change considerably and this slope may even become negative, as shown in the example for $AuNP_{22\,nm}$ in Fig. 9.1b. While the analysis using large $AuNP_{22\,nm}$ can still be performed, it is highly recommended to utilize 5 nm AuNP because of better system stability. To elucidate these extraordinary differences in the behavior of $AuNP_{5\,nm}$ and $AuNP_{22\,nm}$, further investigations of full-scan RELS spectroscopy, long-term scattering evolution, and UV–vis plasmonic spectroscopy have been carried out.

The full-scan RELS spectra for small and large AuNP's are presented in Fig. 9.3. For small $AuNP_{5\,nm}$ nanoparticles, a dramatic increase of the resonant scattering intensity upon addition of GSH is observed in the entire photon energy range scanned. The maximum of RELS signal for $AuNP_{5\,nm}$ alone is found at $\lambda_{max} = 625$ nm and it is red-shifted from the absorption maximum $\lambda_{max,A} = 516$ nm by $\Delta\lambda = 109$ nm. The

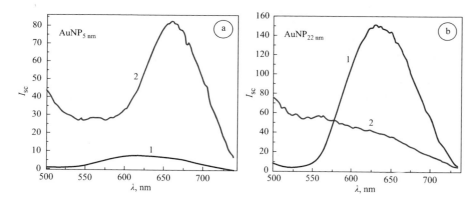

Fig. 9.3 RELS spectra I_{sc} vs. λ for: **a** small nanoparticles (AuNP$_{5\,nm}$) and **b** larger nanoparticles (AuNP$_{22\,nm}$) in the absence of GSH (1) and in the presence of 5 mM GSH (2), recorded within 1 min of GSH injection (from [156], with permission from Elsevier)

large RELS peak generated by the interactions of GSH with AuNP$_{5\,nnm}$ is shifted even farther to the longer wavelengths and appears at $\lambda_{max} = 655$ nm.

Similar RELS spectra for larger nanoparticles (AuNP$_{22\,nm}$), presented in Fig. 9.3b, show that the RELS maximum for AuNP$_{22\,nm}$ alone is much larger than that for AuNP$_{5\,nm}$ alone, consistent with stronger scattering for larger particles. However, the injection of GSH causes an increase in scattering only in the shorter wavelength region, $\lambda < 580$ nm, whereas a decrease in scattering intensity is observed in the wavelength range above $\lambda = 580$ nm. Interestingly, the wavelength of the RELS maximum for AuNP$_{22\,nm}$ in the absence of GSH ($\lambda_{max} = 628$) is virtually unchanged from that for AuNP$_{5\,nm}$, despite of the remarkable differences in other characteristics and in their interactions with GSH.

In RELS spectra for both AuNP$_{5\,nm}$ and AuNP$_{22\,nm}$, in the absence of GSH, we observe a strong scattering in the photon energy range below that of the surface plasmon oscillation energies. For AuNP$_{5\,nm}$, I_{sc} begins to increase at $\lambda > 527$ nm and for AuNP$_{22\,nm}$, I_{sc} increases at $\lambda > 524$ nm. These values are quite close to the surface plasmon absorption maxima $\lambda_{max,A} > 516$ nm (for AuNP$_{5\,nm}$) and $\lambda_{max,A} > 528$ nm (for AuNP$_{22\,nm}$). The origin of the broad scattering peaks with $\lambda_{max} > \lambda_{SP,max}$ nm is ascribed to the increasing reflectivity of free electrons in Au at wavelengths longer than the Frohlich wavelength: $\lambda > \lambda_{Froh} > \lambda_{SP,max}$ where the electrical conductance of nanoparticles becomes a true metallic conductance [153]. At shorter wavelengths, the free conduction electrons in Au are unable to follow fast changing electromagnetic field imposed by the incident light beam and the mechanism of light scattering changes to involve electronic transitions of Au atoms.

The scattering intensity increases with nanoparticle concentration as illustrated in Fig. 9.4 where the dependencies of I_{sc} on C_{AuNP} for AuNP$_{5\,nm}$ and AuNP$_{22\,nm}$ are presented. A higher slope $\partial I_{sc}/\partial C_{AuNP}$ is encountered for AuNP$_{22\,nm}$ than for AuNP$_{5\,nm}$ consistent with the enhanced scattering by larger particles.

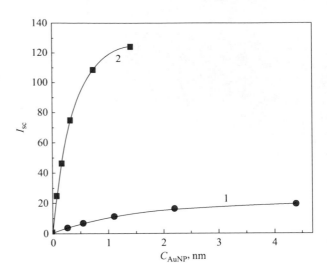

Fig. 9.4 Resonance elastic light scattering intensity I_{sc} on concentration of gold nanoparticles C_{AuNP} for: (1) $AuNP_{5\ nm}$ and (2) $AuNP_{22\ nm}$; incident beam wavelength: $\lambda_{ex} = 640$ nm (from [156], with permission from Elsevier)

9.2.2 Gold Nanoparticle-Based Colorimetric Analysis of GSH

The colorimetric analysis of GSH is based on the evolution of surface plasmon absorbance of AuNP in response to the interactions with GSH molecules. Extensive studies of the surface plasmon absorbance for various AuNP systems have been carried out by the Zhong group [41, 58, 145, 154]. In particular, from studies of the GSH-mediated assembly of AuNP [55, 69] it follows that the presence of NaCl (10 mM) stimulates GSH-mediated assembly of $AuNP_{11\ nm}$ while the addition of NaOH (1.6–3.3 mM) reverses it. In this section, we discuss the variation in spectral characteristics of AuNP upon addition of GSH and their analytical implications for colorimetric determination of GSH.

The GSH concentration dependence of UV–vis absorption spectra for $AuNP_{5\ nm}$ is complex. Here, we present and analyze spectra recorded within 1 min of GSH injection. The color and intensity evolution for GSH-mediated assemblies of $AuNP_{5\ nm}$ is illustrated in Fig. 9.5a for C_{GSH} from 0.1 to 3.33 mM, at constant $AuNP_{5\ nm}$ concentration, $C_{AuNP} = 10.1$ nM. From the analysis of absorption maximum wavelength λ_{max} vs. C_{GSH} presented in Fig. 9.5b, one can see that λ_{max} increases from the initial value $\lambda_{max,ini} = 522$ nm to the final value $\lambda_{max,fin} = 576$ nm at the plateau established for $C_{GSH} \geq 2.6$ mM. The dependence of λ_{max} vs. C_{GSH} is determined by both the kinetics and thermodynamics of GSH-mediated $AuNP_{5\ nm}$ assembly. The critical concentration of GSH, leading in 1 min reaction time to the midway characterized by the $\lambda_{max,1/2} = 548$ nm, is $C_{GSH,1/2} = 2.0$ mM. Close to this concentration, at $C_{max} = 2.28$ mM, the absorbance A_{max} is largest. The decrease in A_{max} observed at $C_{GSH} > 2.28$ mM is due to $1/\lambda$-dependence of absorbance and to the rapidly progressing assembly.

The general trend of the evolution of A_{max}, i.e. an initial increase followed by a decrease, upon the addition of GSH, is not well understood. Extensive efforts have

Fig. 9.5 Effect of glutathione concentration on absorbance of 10.1 nM $AuNP_{5\,nm}$ nanoparticle solution: **a** absorbance spectra for C_{GSH} (mM): (1) 0.1, (2) 0.67, (3) 2.0, (4) 2.33, (5) 2.5, (6) 2.67, (7) 2.83, (8) 3.0, (9) 3.33; above: color change in response to GSH injections (1–9); **b** dependence of the surface plasmon band wavelength λ_{max} on C_{GSH} (from [156], with permission from Elsevier)

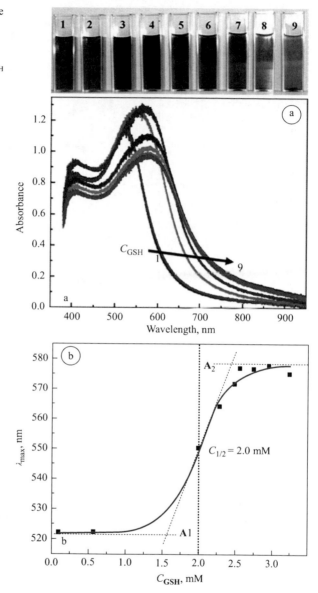

been devoted to calculate the SP band shift caused by the assembly [128–133] with immense success in accounting for the size and shape variability of nanoparticles, nanorods, and nanoplates. The broadening of the SP band for smaller and larger AuNP have been described by El-Sayed et al. [129, 130]. In the case of Fig. 9.5, the broadening is clearly observed concomitant with bathochromic band shift, both corroborating the assembly process.

Fig. 9.6 a Comparison of heights of AuNP capping molecules: citrate and glutathionate (atoms: *yellow*—sulfur, *red*—oxygen, *blue*—nitrogen, *grey*—carbon, *light grey*—hydrogen); **b** small assemblies of n GSH-linked AuNP's with diameter 2a0 < a < 3a0 (n = 3 to 6, marked at the assemblies), where a0 is the diameter of single AuNP (n = 1) (from [156], with permission from Elsevier)

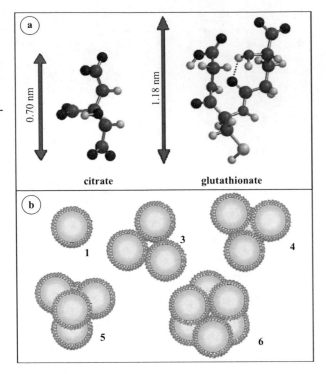

In view of the complex dependence of A_{max} on C_{GSH}, it is recommended to utilize longer and constant wavelength for analytical determination of GSH, *e.g.* $\lambda = 700$ nm to obtain linear calibration curves. The use of gold nanoparticles enables employing a simple colorimetric method for GSH analysis owing to the large extinction coefficient of AuNP in the range of SP band.

9.2.3 Mechanistic Aspects of the Interactions of GSH with AuNP

On the basis of measurements described in previous sections, we have evaluated the pathways of GSH interactions and reactivity with core–shell gold nanoparticles. Before presenting the mechanism, it is necessary to assess changes in the thickness of AuNP protecting monolayers. The thickness of a citrate shell around a AuNP is from 0.38 (flat orientation) to 0.70 nm (vertical, fully extended orientation) and the height of a GSH molecule adsorbed on a Au is on the order of 1.18 nm based on EQCN measurements and quantum chemical evaluation [47] for GSH adsorbed on solid QC/Au piezoelectrodes. The structure and dimensions of Cit and GSH molecules are shown in Fig. 9.6.

During the ligand exchange (AuNP@Cit$^+$GSH = AuNP@GSH$^+$Cit), the diameter of a single AuNP, with core of 5 nm dia., would increase from ca. 6.4 to 7.4 nm, or by 15.6%. This is rather a small change of the diameter and so it can not account for the observed ninefold scattering intensity increase observed upon injection of GSH to $AuNP_{5\text{ nm}}$ solution. The change in the refractive index on ligand replacement is small due to the similarity of organic shells and is thus neglected.

Therefore, upon addition of GSH to a solution of citrate-capped AuNP's, a GSH-mediated assembly of AuNP@Cit occurs. According to our estimates based on RELS measurements, the diameter of the assemblies is approximately doubled (the effective diameter: $a_1 = 2.1 \, a_0$, where a_0 is the diameter of the original citrate-capped AuNP). The GSH-capped AuNP assemblies are depicted in Fig. 9.6.

9.2.4 Significance of Interparticle Hydrogen Bonding and Electrostatic Interactions in Nanoparticle Assembly

In order to evaluate the viability of the formation of hydrogen bonding and their role in the interparticle interactions, molecular dynamics simulations and quantum mechanical calculations of structural interdependencies of interacting molecular capping agents and cross-linkers were conducted. It follows from the experimental observation that after several hours of relative stability of the GSH-capped AuNP assemblies, an extensive GSH-mediated AuNP network formation begins to settle down and results in the sedimentation of large ensembles and decrease in light scattering associated with the colloidal solution depletion. While the electrostatic forces, including the zwitterionic interactions, are established immediately after the ligand-exchange is completed, the formation of hydrogen bonds is not. The reason is that H-bonding is very sensitive to the spatial arrangement of atoms while the electrostatic forces act with spherical symmetry around the charge center. Therefore, it takes time to conformationally rearrange external atoms to positions of lowest energy constituting the H-bonding.

The interparticle interactions of GSH-capped AuNP have been analyzed by considering the formation of multiple hydrogen bonds as depicted in Fig. 9.7. It is seen that there are several possibilities for the formation of hydrogen bonds, from a single H-bond up to a triple H-bond. In Fig. 9.7a, a single H-bond COOH'–COOH" formed between COOH groups from two GSH molecules is presented. In Fig. 9.7b, another single H-bond COOH-NH_2 is shown which forms between COOH group from one GSH molecule and NH_2 group from another GSH molecule. Then, in Fig. 9.7c–e, double H-bonds are presented.

They can be formed as follows: (c) two H-bonds on the same couple COOH'–COOH" where one carboxyl is from one GSH molecule and one from the second GSH molecule; (d) COOH'–COOH" and COOH'–COOH''', where the same COOH' from one GSH molecule forms two H-bonds with two different carboxyl groups COOH" and COOH''' from the second GSH molecule; (e) one H-bond COOH'–NH_2 and one COOH"–COOH where two different carboxyl groups (COOH' and COOH") in one GSH molecule are involved. A triple H-bond is shown in Fig. 9.10f: two H-bonds in one carboxylate couple COOH'–COOH" plus one H-bond COOH'''–NH_2. This

Fig. 9.7 Hydrogen bonding between two glutathione molecules from shells of two interacting AuNP's; H-bonds marked with a dotted line; atoms: *yellow*—sulfur, *red*—oxygen, *blue*—nitrogen, *grey*—carbon, *light grey*—hydrogen); GSH molecules are separated with a *red dashed line* (from [156], with permission from Elsevier)

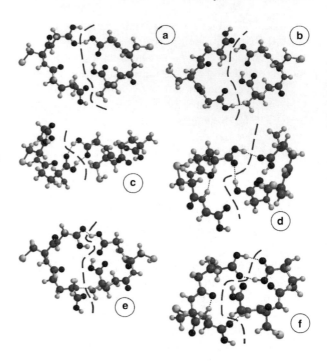

means that there is a lot of configurational flexibility in GSH molecules to achieve one- to triple-H-bonding and thus GSH-induced assembly of AuNP is feasible.

The formation of hydrogen bonds is augmenting strong zwitterionic interactions due to COO–NH_3^+ operating in the pH range from 2.04 to 3.53. At pH < 2.04 and pH > 3.53, electrostatic repulsions become the dominant force in the interparticle interactions between GSH–capped gold nanoparticles.

The results presented above corroborate the mechanism of interactions of glutathione molecules with citrate-capped gold nanoparticles and the interparticle interactions of AuNP. The prevailing role of multiple H-bonds formed in these interactions is consistent with experimental spectroscopic results for other AuNP systems with functionalized-thiol capping agents.

9.3 Detection of Homocysteine and Cysteine Based on Induced Gold Nanoparticle Assembly

The concentrations of the oxidative stress biomarkers, homocysteine (Hcys) and cysteine (Cys), in body fluids and cytosol are much lower that those of GSH and are typically in the low micromolar range. The concentration of Hcys higher than 15 µM is considered dangers to the human health and constitutes a condition called hyperhomocysteinemia. Despite of lower concentration, the interactions of Hcys with gold nanoparticles are much stronger that those of GSH and therefore, the ligand-exchange processes of Hcys and Cys with citrate-capped AuNP are observed at generally lower concentrations of these biomarkers than it is seen for GSH. The Hcys- and Cys-induced assembly of gold nanoparticles has been investigated for non-ionic fluorosurfactant-capped gold nanoparticles as well as for negatively-charged citrate-capped gold nanoparticles. Due to the system complexity, more detailed knowledge of the interactions and reactivities is necessary to understand better the system behavior. Despite of the fact that Cys molecules are smaller than Hcys by only one CH_2 group, they show clearly less activity than Hcys which can be utilized in their analytical determinations. The reasons leading to the differences in activities of these two thioaminoacids have been analyzed. For citrate-capped gold nanoparticles, the formation of surface complexes facilitated by electrostatic attractions and formation of double hydrogen bonds for both Hcys and Cys have been postulated. The conformational differences between these two kinds of complexes result in marked differences in the distance between–SH groups of the biomarkers to the gold surface and different abilities to induce nanoparticle assembly. Analytical implications of these mechanistic differences are discussed in the next sections.

9.3.1   Plasmonic Spectroscopy of Homocysteine- and Cysteine-Mediated Assembly of ZONYL-Capped Gold Nanoparticles

Fluorosurfactants provide similar advantages to other surfactants but, in addition, show high degree of chemical inertness. For these reasons they have recently been applied in chemical analysis [64]. The ZONYL fluorosurfactant is known to form self-assembled monolayers on gold surfaces rendering the surface more hydrophobic and significantly retarding the gold oxide formation processes [155]. In the case of AuNP, the fluorosurfactant stabilizes gold colloids by forming tight shells around nanoparticle cores with hydrophilic heads oriented toward Au surface and fluorocarbon tails forming hydrophobic non-interacting external surface. Although the surfactants of this type form water-tight shells, their bonding to a gold surface is not as strong as that of thiolates. Therefore, in the presence of thiols, ZONYL is replaced in a ligand exchange process. The thioaminoacids, such as Hcys and Cys, also replace ZONYL from a AuNP shell, provided that sufficiently high concentration of these agents is used and long enough time is allowed. The HR-TEM images of fluorosurfactant-capped AuNP's are presented in Fig. 9.8 before (a) and after (b–d) homocysteine-induced nanoparticle framework assembly.

The ligand exchange process taking place upon addition of homocysteine to ZONYL-capped AuNP can be monitored using SP-band absorbance of AuNP, as illustrated in Fig. 9.9. It is seen that the SP band shifts toward longer wavelengths and the maximum absorbance increases with increasing C_{Hcys}. These observations are consistent with ligand exchange process:

$$AuNP/FES_x + yHcys = AuNP/Hcys_y + xFES \qquad (9.3)$$

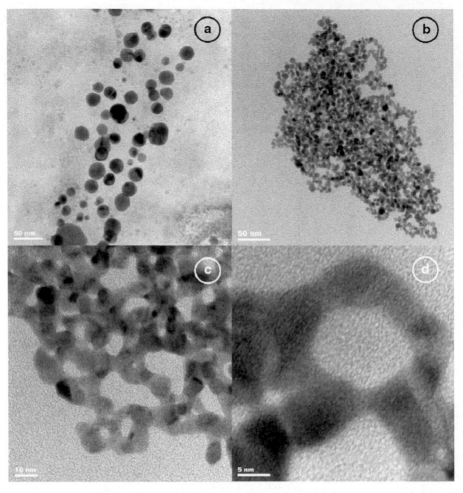

Fig. 9.8 HR-TEM images of ZONYL-capped gold nsanoparticles before **a** and after assembly with 15 μM homocysteine **b–d**; C_{AuNP} = 6 nM, C_{ZONYL} = 0.22%, pH = 6; bar size: **a** 50 nm, **b** 50 nm, **c** 10 nm, **d** 5 nm (from [152], with permission from Elsevier)

where $x \approx y$, followed by interparticle molecular linking of AuNP/Hcys through direct Hcys-Hcys interactions. At the pH of these experiments (pH = 6.0), homocysteine exists as a zwitterion with α-amino group protonated ($-NH_3^+$) and carboxylic group dissociated (COO^-). Therefore, the zwitterionic interparticle binding between Hcys-capped AuNP is playing a predominant role as recently discussed by Zhong et al. [41].

The bathochromic shift of the surface plasmon peak ($\Delta\lambda_{max}$ = 36 nm, for 16 μM Hcys) corresponds to the formation of small Hcys-linked AuNP ensembles. The increase of SP absorbance by 21% indicates on the collective oscillations of local surface plasmons in AuNP that form these ensembles. The collective oscillation

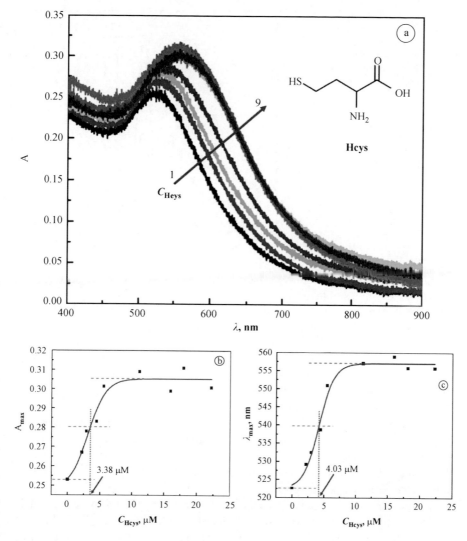

Fig. 9.9 a Absorbance spectra for ZONYL-capped AuNP for different concentrations of homo-cysteine, C_{Hcys} [μM]: (1) 0, (2) 2.22, (3) 3, (4) 4.44, (5) 5.56, (6) 11.11, (7) 16, (8) 18, (9) 22.22. $C_{AuNP} = 6$ nM, $C_{ZONYL} = 0.22\%$, pH = 6; dependence of **b** λ_{max} and **c** A_{max} vs. C_{Hcys} (from [152], with permission from Elsevier)

of local surface plasmons is excited when the distance d between AuNP is: $d < 5r$, where r is the AuNP radius.

The absorbance maximum increases with C_{Hcys} and reaches the saturation value at $C_{Hcys} > 7$ μM, with the half-absorbance change appearing at $C_{Hcys} = 3.38$ μM. The value of λ_{max} also reaches saturation at $C_{Hcys} > 7$ μM (Fig. 9.9c). Therefore, we can assume that above 7 μM Hcys concentration the ligand exchange process has been completed and nanoparticle shells are saturated with Hcys.

Fig. 9.10 Resonance elastic light scattering spectra for ZONYL-capped $AuNP_{5\,nm}$ for different concentrations of homocysteine, C_{Hcys} [µM]: (*1*) 0, (*2*) 2.22, (*3*) 3, (*4*) 4.44, (*5*) 5.56, (*6*) 14, (*7*) 16, (*8*) 18, (*9*) 22.22. $C_{AuNP} = 6$ nM, $C_{ZONYL} = 0.22\%$, pH = 6, $\lambda_{ex} = 550$ nm (from [152], with permission from Elsevier)

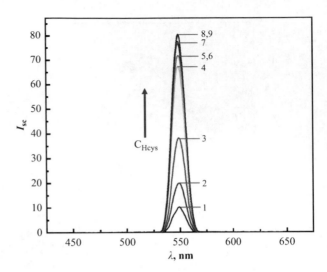

Extensive studies of the surface plasmon absorbance for various AuNP systems have been carried out by several groups [49, 50, 54, 62, 63, 104, 105, 117, 118, 125, 128, 135–137, 145]. In particular, it follows from studies of the homocysteine-mediated assembly of AuNP that the interparticle zwitterion interaction of the Hcys-Au system is particularly strong [41] and that the Hcys-mediated assembly of AuNP can be accelerated by an increased temperature and ionic strength of the solution thus reducing the barrier for Hcys attachment to gold nanoparticle surface [41]. Also, the assembly can be reversed by the pH change [41, 42].

Similar experiments performed with cysteine indicate that at higher concentrations ($C > 15$ µM) the kinetics of ligand exchange for both Hcys and Cys is very fast and the exchange is completed within 1 min of mixing AuNP solution with the thioaminoacids. However, at lower concentrations, the ligand exchange is considerably faster for Hcys than for Cys.

In summary, the colorimetric analysis of homocysteine and cysteine utilizing gold nanoparticles as active reagent is viable. Moreover, the sensitivity of the analysis is much higher than that for GSH due to the strong interactions of Hcys and Cys with AuNP. A simple and inexpensive detection of color changes with naked eye can be readily performed in the concentration range of these oxidative stress biomarkers from 0.5 to 5 µM.

9.3.2 RELS Monitoring of Homocysteine and Cysteine Interactions with ZONYL-Capped Gold Nanoparticles

The resonance elastic light scattering spectra for a ZONYL FSN surfactant-capped $AuNP_{5\,nm}$ in Hcys solutions are presented in Fig. 9.10. The strong RELS from

Fig. 9.11 Dependence of elastic light scattering intensity maximum $I_{sc,max}$ for ZONYL-capped AuNP on concentration of analytes: (*1*) homocysteine, (*2*) methionine, (*3*) alanine, (*4*) histidine, (*5*) glutathione, $C_{AuNP} = 6$ nM, $C_{ZONYL} = 0.22\%$, pH = 6, $\lambda_{ex} = 550$ nm (from [152], with permission from Elsevier)

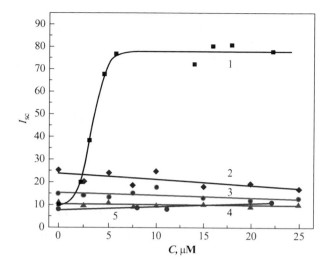

AuNP$_{5\ nm}$ in solution results from the absorption of photons at $\lambda_{ex} = 550$ nm, in the range of the SP absorption band. The addition of homocysteine to the ZONYL-capped AuNP$_{5\ nm}$ results in strong scattering enhancement. The solution pH maintained at a constant value, pH = 6.0, guarantees that Hcys is within the range of predominantly neutral (zwitterionic) form (pH = 2.22 to 8.87; pK$_{a,1}$ = 2.22 (–COOH), pK$_{a,2}$ = 8.87 (–NH$_2$), pK$_{a,3}$ = 10.86 (–SH)). The strong enhancement of RELS from AuNP$_{5\ nm}$ by Hcys molecules is then due to the aggregate formation associated with interparticle interactions with zwitterionic Hcy-Hcys cross-linking leading the scattering increase. From the data of Fig. 9.10, the scattering intensity increase is: $I_{sc,2}/I_{sc,1} = 7.8$ and, hence, $a_{rel} = 1.99$. This means that most likely small aggregates composed of 2–6 particles are formed.

The RELS intensity differs considerably for ZONYL-capped AuNP interacting with different aminoacids [36, 64, 152]. The plots of RELS intensity vs. aminoacid concentration measured at $\lambda_{ex} = 550$ nm for Hcys, methionine, alanine, histidine, and glutathione, are presented in Fig. 9.11. They show a strong increase of I_{sc} with C for homocysteine and apparent no response for other aminoacids and glutathione. The I_{sc} vs. C_{Hcys} dependence is sigmoidal with an inflection point at low Hcys concentration indicating a high affinity of Hcys for a Au surface. From a Boltzmann function fitted to the experimental data for Hcys and ZONYL, we obtain $I_{sc} = A_2 + (A_1 - A_2)/(1 + \exp[(C - C_{1/2})/s])$, where A_1, A_2—are the lower and higher I_{sc} plateaus, $C_{1/2}$ is the concentration at the inflection point, and s is the slope parameter. The value of $C_{1/2} = 3$ µM.

The longer elution time for Hcys than for Cys observed in C18 column chromatography experiments [36, 64] is consistent with higher affinity of Hcys than Cys to hydrophobic chains. In the setting of a ZONYL-capped AuNP, this would translate to a slower transfer of Hcys through a ZONYL shell. Since the opposite is observed, this means that other factors must also play a role.

Fig. 9.12 Dependence
of elastic light scattering
intensity maximum $I_{sc,max}$ for
citrate-capped $AuNP_{5\ nm}$ on
concentration of analytes:
(*1*) homocysteine, (*2*)
methionine, (*3*) alanine, (*4*)
histidine, (*5*) glutathione, (*6*)
cysteine, $C_{AuNP} = 3.8$ nM,
pH = 5, $\lambda_{ex} = 560$ nm (from
[152], with permission from
Elsevier)

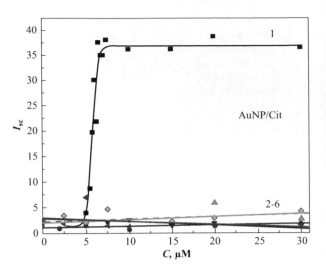

9.3.3   Interactions of Homocysteine and Cysteine with Citrate-Capped Core–Shell Gold Nanoparticles

Upon addition of homocysteine to citrate-capped AuNP, an increase in resonance elastic light scattering, similar to the one described for ZONYL-capped AuNP, is also observed, provided that the solution pH is carefully controlled. The increase in scattering intensity upon addition of 15 μM Hcys is $I_{sc,2}/I_{sc,1} = 19$ (Fig. 9.12). This large increase in scattering intensity clearly indicates on the homocysteine-induced assembly of AuNP. Utilizing Eq. (9.2), we obtain for the increase of particle diameter: $a_{rel} = 2.7$. Similar RELS experiments carried out for other aminoacid ligands and glutathione, presented in Fig. 9.12, show that the RELS response is highly selective to Hcys, consistent with recent findings [41, 55, 58] showing that thiol-containing aminoacids adsorb preferentially on a gold surface while glutathione (at neutral pH) is repelled from the citrate shell of nanoparticles. The mechanisms leading to this high selectivity are not well understood, though the importance of this selectivity for analytical determinations of homocysteine in a matrix of aminoacids and glutathione is high.

The protonation equilibria for species in solution and in the protective SAM environment of gold nanoparticle shells influence the interparticle interactions and thus the analytical determinations based on AuNP assembly. The following example illustrates these phenomena. The RELS measurements for Hcys and citrate-capped AuNP were carried out in solutions with three different pH: 2.0, 5.0, and 9.0. The plot of scattering intensity I_{sc} vs. C_{Hcys} for these three media is presented in Fig. 9.13. The three dependencies of I_{sc} vs. C_{Hcys} for different pH values show completely different behaviors. The curve 1 for pH = 2.0 shows a scattering intensity decrease with increasing C_{Hcys} and establishment of a plateau for $C_{Hcys} > 4$ μM. Curve 2 shows a sigmoidal shape with the onset of scattering at $C_{Hcys} = 5$ μM and

Fig. 9.13 Dependence of elastic light scattering intensity maximum $I_{sc,max}$ on concentration of homocysteine C_{Hcys} for citrate-capped $AuNP_{5 nm}$ for different solution pH: (*1*) pH = 5 and (*2*) pH = 2 (from [152], with permission from Elsevier)

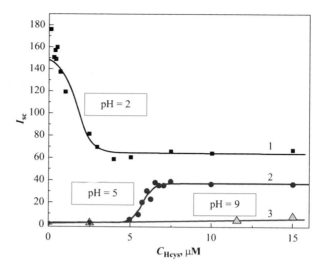

establishment of a new level of scattering intensity plateau for $C_{Hcys} > 7$ μM. In the case of the third curve, for pH = 9.0, there is virtually no scattering change seen for the entire concentration range of Hcys examined and the level of scattering is very low ($I_{sc} \approx 8$, for 20 μM Hcys). Note that the scattering intensity levels established for pH = 2.0 and pH = 5.0 at higher concentrations of Hcys, are different.

The elucidation of the mechanism of processes leading to the complex behavior of the citrate-capped AuNP—homocysteine system is the key element to understanding the reactivity and assembling properties of functionalized AuNP and their interactions with small biomolecules. The three $I_{sc} = f(C_{Hcys})$ characteristics presented in Fig. 9.13 represent three different regime as follows:

1. At pH = 9 (curve 3), the RELS intensity observed for all Hcys concentrations examined is due to the high gold colloid stability associated with strong electrostatic interparticle repulsions between deprotonated carboxyl groups that exist in the Hcys-shell after the ligand exchange.
2. The sigmoidal switching characteristics is observed at pH = 5 (curve 2) where citrates are predominantly deprotonated ($pK_{a,1} = 3.09$, $pK_{a,2} = 4.75$, $pK_{a,3} = 5.41$) but homocysteine exists as a zwitterion with protonated–NH_3^+ group and dissociated COO^- group ($pK_{a,1} = 2.22$ (COOH), $pK_{a,2} = 8.87$ (NH_2)). Thus, at low Hcys concentrations ($C_{Hcys} < 5$ μM), scattering is low since it is dominated by interparticle repulsions of negatively-charged citrate shells. As the ligand exchange process progresses, the citrate ions are being replaced by the neutral Hcys molecules. The progression is accelerated at higher Hcys concentrations. The switch from low elastic light scattering intensity to high intensity is observed in the concentration range: 5 μM$<C_{Hcys}<7$ μM. This saturation level attained at $C_{Hcys} = 7$ μM can is associated with small ensembles of Hcys-linked AuNP where the interparticle attractions are attributed to strong Hcys-Hcys zwitterionic interactions.

3. The strong scattering intensity observed in acidic environment (curve 1, pH = 2)
   and in the absence of homocysteine is due to the extensive interparticle hydrogen
   bonding between predominantly undissociated citric acid ligands in the AuNP
   shells ($pK_{a'1} = 3.09$, $pK_{a,2} = 4.75$, $pK_{a'3} = 5.41$, for citric acid). The H-bonding
   results in the formation of gold nanoparticle ensembles and a high RELS intensity. Upon the addition of homocysteine, the light scattering intensity unexpectedly decreases to a new level, approximately at 50% of the initial scattering
   intensity value. This is attributed to the dismantling of the initial citrate-linked
   gold nanoparticle ensembles and replenishing the nanoparticle shells with homocysteine in a ligand exchange process. While the newly formed shells are more
   strongly bound to the gold cores than citric acid based shells do, the Hcys molecules at pH = 2 are partially positively charged and cannot form as large the
   nanoparticle aggregates as citrate-capped AuNP do. However, for Hcys one
   should expect interparticle repulsions of Hcys-capped AuNP at pH = 2, since
   $pK_{a'1} = 2.22$ (COOH), $pK_{a,2} = 8.87$ (NH_2). Also, the RELS saturation level is
   higher that that at pH = 5. This means that despite of the ligand-exchange, still
   citric acid molecules participate in the interparticle bonding by cross-linking (or:
   bridging) through hydrogen bonding.

In summary, the scattering spectra present evidence for different types of interparticle interactions including the de-aggregation of citrate-capped gold nanoparticle
ensembles followed by their conversion to citrate-linked Hcys-capped nanoparticle
assemblies recently described [152].

9.3.4   Molecular Dynamics and Quantum Mechanical Calculations of Molecular Cross-Linking of Core–Shell Gold Nanoparticles

The two main monolayer-protective types of shells for AuNP described in previous sections differ considerably in their composition and properties, yet they both
provide selectivity toward homocysteine versus glutathione or cysteine in the
nanoparticle assembly process. The MD simulations have been carried out to evaluate the interactions of Hcys and Cys with a non-ionic fluorosurfactant-capped gold
nanoparticles [152] and to characterize the kind of intermediate structures that form
on approach of Cys and Hcys molecules to a charged citrate-capped gold nanoparticle [152, 156]. Here, we will concentrate only on the latter.

The interactions of cysteine and homocysteine with citrate-capping film have
been considered for a pH range corresponding to the experimental work (pH 5–6),
where the citrate shell is charged negatively providing a long-term stability for the
gold colloid, whereas both cysteine and homocysteine are in the form of zwitterions
with protonated–NH_3^+ group and dissociated–COO^- group. The main interaction of
the electrostatic nature between–COO^- group of the nanoparticle shell and–NH_3^+
group of the approaching thioaminoacid is expected with strong repulsions between

Fig. 9.14 Interactions of cysteine and homocysteine with citrate ions in a ligand exchange process: surface complex formation through hydrogen bonding calculated for **a** Cit-Cys and **b** Cit-Hcys using molecular dynamics, and electron density surfaces for $d = 0.08$ au^{-3}, with electrostatic potential map for **c** Cit-Cys and **d** Cit-Hcys; electrostatic potential: color coded from negative—*red* to positive—*blue* (from [152], with permission from Elsevier)

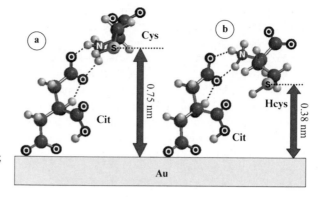

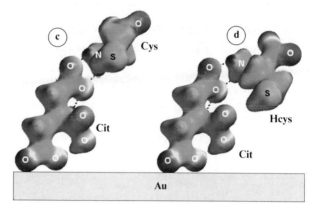

dissociated carboxylate groups of the citrate and Cys or Hcys molecules. The results of molecular dynamics simulations and quantum mechanical calculations obtained are presented below.

In Fig. 9.14, the interactions of cysteine and homocysteine with citrate ions in a ligand exchange process are analyzed. It is seen that both Cys and Hcys form intermediate surface complexes on approaching to a citrate-capped gold nanoparticle. Within the framework of electrostatic attractions between COO$^-$ group of the nanoparticle shell and NH$_3$$^+$group of the thioaminoacid, a double hydrogen bond is formed for both the Cit-Cys and Cit-Hcys complexes. Immediately seen is, however, a completely different configuration of the thioaminoacid in the surface complex formed. Whereas a cysteine molecule forms a kind of axial (linear) configuration extending out of the citrate protective SAM, the homocysteine tends to bend out of the axial conformation and toward the citrate side-chain and the electrode surface.

The lack of flexibility of the cysteine molecule has already been pointed out when comparing ring-forming abilities of these two molecules [44]. Here, the bending toward the citrates side chain results in the substantial difference in the distance of the thiol group to the gold surface. This distance is 0.75 nm for Cit-Cys surface complex and only 0.38 nm for Cit-Hcys complex. This difference can be translated to classifying the thioaminoacid position as being outside of the shell (in the case of cysteine) or inside the shell (in the case of homocysteine). The large difference in the observed light scattering intensity between Cys and Hcys can be explained by easier and faster penetration of Hcys into the citrate-dominated gold nanoparticle shell followed by citrate ligand replacement.

After the ligand exchange has been completed, the zwitterion-type interactions begin to operate leading to the nanoparticle assembly and manifested by the sharp increase in the resonance elastic light scattering, as observed experimentally. On the other hand, in the case of cysteine, the ligand exchange process is strongly hindered by cysteine inability to enter the citrate protective shell due to the axial conformation of the surface complex Cit-Cys.

9.4   Overview of Molecularly-Templated Biorecognition Polymer Films for Biosensing Applications

Extensive applications of biorecognition phenomena in chemical analytical procedures [157–167] and in a variety of human designed sensors are based on specificity and strong affinity interactions of the antibody-antigen system [158, 168–171]. The recent introduction of engineered oligonucleotide- or polypeptide-based aptamers [159, 160] as ligands mimicking antibodies indicates that they can also be used in sensors for various target molecules and provide additional advantages of higher packing density and improved structural flexibility. Similar in principle is the use of molecularly imprinted polymer films [172–178] whereby the polymerization of the polymer is carried out in situ in the presence of the target molecule which is then released from the template. Such a templated polymer film can exhibit specificity toward the target molecule and offers enhanced processibility and scalability. Therefore, this approach is promising for miniaturization and the development of microsensor arrays with multiple functionalities. Molecularly imprinted polymers show recognition properties similar to biological receptors but they are more stable and less expensive than biological systems [179]. This is in tune with the increasing need for sensitive techniques to analyze rapidly many components of living organisms.

Various transduction techniques have been investigated for application with molecularly imprinted polymers including: potentiometric [180], capacitive [181], conductimetric [182, 183], voltammetric [184], acoustic wave [185–191], colorimetric [192], surface plasmon spectroscopy [193], and fluorescence [194–197] detection. An enhancement of the analytical signal has been achieved by utilizing gold, silver or semiconducting nanoparticle labeling of target molecules [35]. Molecularly imprinted polymers represent very promising materials not only for sensory films but also for the selective solid-phase separation techniques [198, 199], such as the electrophoresis and chromatography.

The synthesis of templated polymer films plays the key role in accomplishing the desired target recognition level. Since molecular imprinting leads to the formation of binding sites for target analyte molecules in the supramolecular architecture of a polymer, the target molecules should interact with monomers during the polymerization stage and act as a template around which the polymer grows. After the removal of templating molecules, high affinity sites are left in the polymer matrix. Two distinct methodologies have been employed to synthesize molecularly imprinted polymers [200] depending on the nature of interactions between the template and monomer: covalent and non-covalent imprinting. The non-covalent imprinting is more versatile and easier than the former and it is most often utilized. It is based on hydrogen bonding, Van der Waals forces, electrostatic or hydrophobic interactions.

In this section, novel designs of molecularly imprinted sensor films are presented. The molecular imprinting, carried out in-situ by electropolymerization of orthophenylenediamine (oPD) in the presence of target molecules, is described for GSH molecules and GSH-capped gold nanoparticles (AuNP) as the targets. The monitoring of the formation of GSH-doped poly(oPD) and other templates using the electrochemical quartz crystal nanogravimetry (EQCN) is discussed in detail.

9.4.1   Template Design for Molecularly-Imprinted Conductive Polymer Films

The deposition of GSH-templated films has been carried out in-situ by electropolymerization of oPD on the surface of a quartz crystal piezoelectrode, either directly on a bare Au electrode sputtered on a QC resonator, or on a layer of AuNP network assembled on a QC/Au surface. Three types of the design of templated polymer films, employed in this work, are presented in Fig. 9.15.

The first type of sensors, QC/Au/PoPD(GSH), labeled T1, shown in Fig. 9.15a, was synthesized by direct electropolymerization of oPD in the presence of GSH. During the polymerization process, GSH molecules are trapped inside the polymer layer and become permanently embedded in the polymer. At the end of the polymerization stage, the GSH molecules present on the polymer surface make impressions in the film which, after dis-association of the templating molecules by hydrolysis, can be utilized in GSH recognition schemes.

The second type of sensors, QC/Au/PoPD(AuNP-SG), labeled T2, shown in Fig. 9.15b, was synthesized by electropolymerization of oPD in the presence of GSH-capped AuNP nanoparticles. Since the GSH-capped AuNP's tend to assemble due to hydrogen bonding, the solutions of oPD and GSH were mixed with AuNP solution right before the experiment and immediately transferred to the EQCN cell. During the polymerization process, the AuNP-SG nanoparticles are trapped inside the polymer layer and become permanently embedded in the polymer. At the end of the polymerization stage, the AuNP-SG nanoparticles present on the polymer surface leave impressions in the film which are then utilized in the recognition schemes of GSH and GSH-capped AuNP's.

The third type of sensors, QC/Au/AuNP/PoPD(GSH), labeled T3, shown in Fig. 9.15c, was synthesized by direct electropolymerization of oPD in the presence

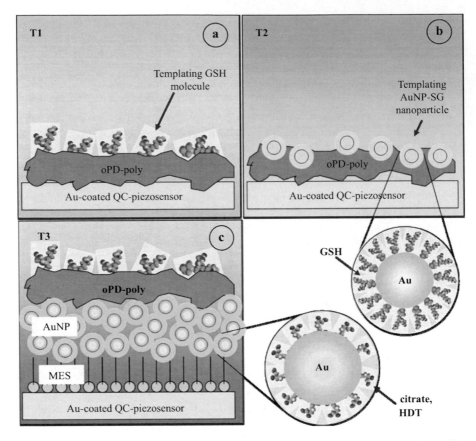

Fig. 9.15 Schematic of GSH-templated sensor designs: **a** GSH embedded in a PoPD polymer film (oPD-poly) on a Au piezoelectrode (sensor T1), **b** AuNP-SG nanoparticle-templated PoPD film on a Au piezoelectrode (sensor T2), and **c** GSH embedded in a PoPD film electrodeposited on a layer of AuNP network assembled on a SAM of MES on a Au piezoelectrode (sensor T3) (from [35], with permission from the Electrochemical Society)

of GSH, on a QC/Au substrate that was coated with a SAM of MES and a layer of AuNP network assembled on top of the SAM. The GSH molecules attached to the polymer surface at the end of the oPD polymerization stage leave impressions in the film which can be utilized for GSH detection.

The polymerization of GSH-templated films was investigated for different types of templates. Examples of typical polymerization characteristics are provided in Figs. 9.2, 9.3, 9.4, 9.5, 9.6. For each polymerization, the mass-to-charge ratio $p_{exp} = \partial m/\partial Q$, was analyzed. The electropolymerization was carried out either by successive potential scans from $E_1 = 0$ to $E_2 = +0.8$ V and back to E_1, or by potential pulses with $E_1 = 0$ to $E_2 = +0.8$ V and $E_3 = 0$, with pulse widths $\tau_1 = 1$ s, $\tau_2 = 300$ s.

In Fig. 9.16, the polymerization of oPD in the presence of GSH by potential scanning is illustrated. The simultaneous voltammetric (LSV) and nanogravimetric

Fig. 9.16 LSV and EQCN characteristics (first cycle) for a QC/Au electrode in 5 mM oPD + 10 mM GSH in 10 mM phosphate buffer solution: **a** current-potential (*1*) and mass-potential (*2*); **b** mass-charge (*1*) and potential-charge (*2*); $v = 100$ mV/s (from [35], with permission from the Electrochemical Society)

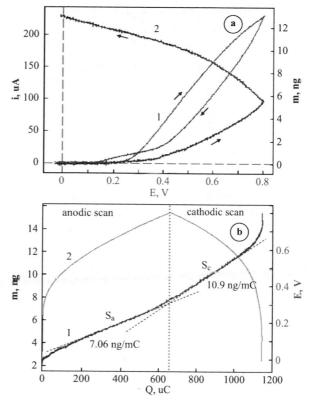

characteristics indicate an instant of the oPD oxidation at $E = 0.25$ V vs. Ag/AgCl and almost linear current increase in the potential range from $E = +0.3$ to $+0.6$ V, followed by a decrease of the current growth rate at higher potentials which may be due to transport limitations. It is seen that the apparent mass increases during the anodic potential scan. Further mass gain is also noted after the potential scan reversal. Moreover, the mass kept increasing even after current cessation at the end of the cathodic-going potential scan, at potentials $E < +0.15$ V. This type of a behavior is indicative of the formation of quasi-stable oPD radicals which are able to attach to the PoPD film after the oPD oxidation has ended. If this mechanism is correct then the Faradaic efficiency of the polymer formation must be impeded due to the diffusion of the oPD intermediates and oligomeric radicals out of the electrode surface. The Faradaic efficiency can conveniently be investigated using the mass-to-charge analysis. The experimental slope, $p_{exp} = \partial m / \partial Q$, can be compared to the theoretical slope calculated for a given reaction:

$$m = M \frac{Q}{nF} \tag{9.4}$$

$$p_{th} = \frac{\partial m}{\partial Q} = \frac{M}{nF} \tag{9.5}$$

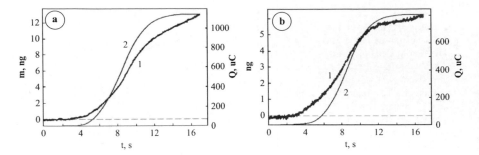

Fig. 9.17 Apparent mass (*1*) and charge (*2*) consumed in electropolymerization of a PoPD template from 5 mM oPD + 10 mM GSH in 10 mM phosphate buffer by potential scanning between $E_1 = 0$ and $E_2 = +0.8$ V at a scan rate of $v = 100$ mV/s: a 1-st cycle, b 17-th cycle (from [35], with permission from the Electrochemical Society)

where F is the Faraday constant ($F = 96{,}485$ C/equiv) and n is the number of electrons transferred. Here, for the electro-oxidation of oPD, we have:

$$C_6H_4(NH_2)_2 - 2e^- = -|NH - C_6H_4 - NH| - +2H^+ \qquad (9.6)$$
$$oPD \qquad\qquad\qquad PoPD\ unit$$

so that $n = 2$ and the molmass $M = 106$ g/mol (i.e. the molar mass of species deposited on the electrode minus molar mass of species detached from the electrode surface), assuming that the oxidized PoPD unit is cross-linked to the electrodic polymer film. Hence the theoretical value of p_{th} is: $p_{th} = 53 \times 10^{-5}$ g/C = 530 ng/mC.

We can see that the experimental value of p_{exp} is much lower: $p_{exp} = 7.06$ ng/mC. This means that a large majority of the oxidized oPD radicals escape to the solution before they are able to bind to the electrode surface and become part of it. The slope p_{exp} increases during the cathodic-going potential scan, $p_{exp} = 10.9$ ng/mC, but this is still too low a value in comparison with the theoretical expectation for an efficient Faradaic process.

The polymerization efficiency does not increase in subsequent potential cycles. For instance, after 16 cycles, the mass-to-charge is: $p_{exp} = 6.25$ ng/mC. Generally, smaller oPD oxidation currents and smaller mass gains are observed due to the increased polymer film resistance attributable to the enlarged film thickness. The decrease in the oxidation charge and mass gain between cycle 1 and 17 is illustrated in Fig. 9.17 on the m–t and Q–t plots.

The number of PoPD monolayers deposited in potential scanning polymerization procedure can be estimated by calculating an equivalent monolayer mass of PoPD. On the basis of quantum mechanical calculation of the electronic structure of the polymer, the definition of the equivalent monolayer is rather ambiguous because the benzene rings of oPD are not in plane or stacked parallel to each other in the PoPD. Therefore, we define here the equivalent PoPD monolayer as a densely packed layer of flat oPD molecules. The calculated surface area for a unit oPD $A = 27.1$ Å^2 is assumed. Then the maximum surface coverage is: $\Gamma = 3.69$ molc/cm^2

and $\gamma = 0.61$ nmol/cm^2. The monolayer mass is then: $m_{mono} = 65.0$ ng/cm^2 and for our quartz resonator: $m_{mono,QC} = 16.6$ ng/QC. It is seen then that in a single potential scan experiment only a fraction of the equivalent PoPD monolayer is being formed.

The electropolymerization of oPD to form GSH-templated films was also carried out using the potential step technique. Here, we present an example of the synthesis of a film of GSH-templated PoPD which was deposited on a QC/Au substrate coated with a SAM of MES with a layer of AuNP network assembled on top of the SAM.

In the potential step experiments, the potential program included three stages: $E_1 = 0$, $E_2 = +0.8$ V, and $E_3 = 0$, with pulse widths $\tau_1 = 1$ s, $\tau_2 = 300$ s. Generally, the current decayed monotonically and the apparent mass was increasing from the first moment of the step to E_2, as expected. The total mass increase observed in these experiments was much larger than that in the potential scan experiments and the analysis of p_{exp} indicates that the Faradaic efficiency ε is also higher ($p_{exp} = 13.7$ ng/mC) although still very low.

The progressive template deposition by potential steps has been carried out by pulse-deposition of GSH-templated polymer films as illustrated in Fig. 9.18 for three types of molecularly imprinted films: QC/Au/PoPD(GSH), QC/Au/PoPD(AuNP-SG), and QC/Au/MES/AuNP/PoPD(GSH). The potential pulse program used for all three syntheses was: $E_1 = 0$, $E_2 = +0.8$ V, and $E_3 = 0$, pulse width $\tau_1 = 1$ s, $\tau_2 = 300$ s. It is seen that the lowest growth rate is observed for the film PoPD(GSH) and the highest rate is observed for the film MES/AuNP/PoPD(GSH). While the higher rate of film growth for PoPD(AuNP-SG) film than for PoPD(GSH) film is most likely due to the involvement of AuNP nanoparticles which contribute strongly to the mass gain, the difference between the growth rate for PoPD(GSH) and MES/AuNP/PoPD(GSH) requires additional rationalization. There seem to be two contributing effects: (a) one associated with the enlarged surface area at the interface film-solution caused by the rough layer of the assembled AuNP in the MES/AuNP/PoPD(GSH) film, and (b) one due to the changed solution pH, from neutral to acidic medium, since the acidic environment is prompting the polymer growth.

These experiments confirm that the GSH-templated films can be grown step by step under different conditions with straightforward control of the film thickness and its conductance by a simple choice of the pulse parameters and the number of applied potential pulses. This method is also faster than the potential scanning method in which only very thin films are obtained.

9.4.2 Response of Molecularly-Templated Films to GSH Target Molecules

Three types of GSH-templated polymer film sensors have been characterized and tested for response to GSH solutions. These tests, presented in Figs. 9.19, 9.20, 9.21, include the i–t and m–t transients for the apparent mass change recorded upon injection of analyte solution. In Fig. 9.19, results obtained for a GSH-templated QC/

Fig. 9.18 Apparent mass gain recorded in consecutive cycles of a potential-step electropolymerization of a GSH-templated poly(oPD) films from 5 mM oPD solutions containing 10 mM GSH (**a–c**) and 3 nM AuNP-SG (**b**); substrate: (**a, b**) QC/Au, (**c**) QC/Au/MES/AuNP; medium: 10 mM phosphate buffer (**a**), 10 mM HClO$_4$ (**b, c**); potential program: step from $E_1 = 0$ to $E_2 = +0.8$ V vs Ag/AgCl and back to E_1, pulse duration $\tau_1 = 1$ s, $\tau_2 = 300$ s; curve numbers correspond to the cycle number (from [35], with permission from the Electrochemical Society)

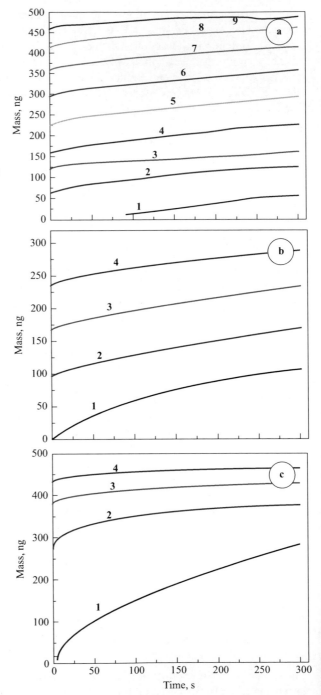

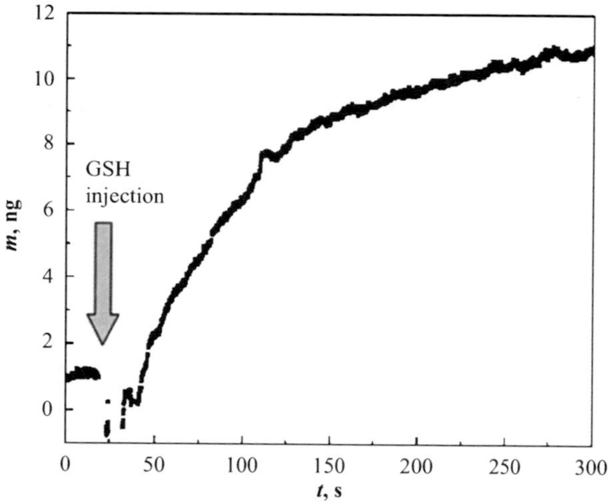

Fig. 9.19 a Chronoamperometric (1) and chronogravimetric (2) transients for the initial electropolymerization stage of a GSH-templated poly(oPD) film T1 on a QC/Au electrode from 5 mM oPD + 10 mM GSH + 10 mM $HClO_4$ solution by a potential step from $E_1 = 0$ to $E_2 = +0.8$ V vs Ag/AgCl. **b** Apparent mass vs. time response of a T1-type sensor after injection of a free GSH solution (5 mM) (from [35], with permission from the Electrochemical Society)

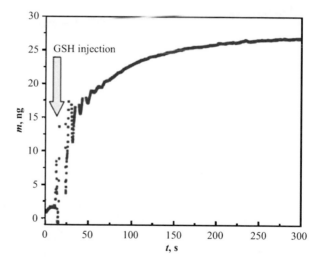

Fig. 9.20 a Chronoamperometric (1) and chronogravimetric (2) transients for the initial electropolymerization of a AuNP-SG-templated poly(oPD) film T2 on a QC/Au electrode from 5 mM oPD + 10 mM GSH + 10 mM $HClO_4$ solution by a potential step from $E_1 = 0$ to $E_2 = +0.8$ V vs Ag/AgCl. **b** Apparent mass vs. time response of a T2-type sensor after injection of a free GSH solution (5 mM) (from [35], with permission from the Electrochemical Society)

Au/PoPD(GSH) sensor are presented. The total mass of the polymer template film with embedded GSH was $\Delta m = 274$ ng ($\Delta f = 238$ Hz).

After hydrolyzing the GSH imprinted in the PoPD polymer film with 0.5 M NaOH solution, the piezosensor was tested for the response to a solution of 5 mM GSH. The time transient recorded upon injection of GSH is presented in Fig. 9.19. The total mass change $\Delta m = 10.5$ ng was observed in the 300 s recoding time. The transient represents a kinetically slow recognition process. The concentration range of sensor response, from 0.5 to 10 mM, corresponds to the GSH level in body fluids and cytosol. The reproducibility between sensors is high, ca. 90%, and with automation of the sensor fabrication, it should be further improved.

The next sensor investigated was a AuNP-SG templated QC/Au/PoPD(AuNP-SG) sensor. The results obtained for this sensor are presented in Fig. 9.20. A

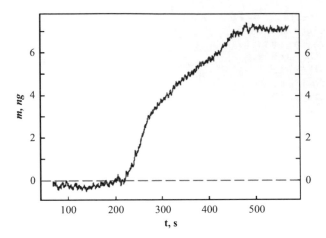

Fig. 9.21 a Chronoamperometric (1) and chronogravimetric (2) transients for the initial elec-tropolymerization of a GSH-templated poly(oPD) film T3 on a layer of AuNP-network assembled on a SAM of MES on a QC/Au electrode, from 5 mM oPD + 10 mM GSH + 10 mM HClO₄ solution by a potential step from $E_1 = 0$ to $E_2 = +0.8$ V vs Ag/AgCl.**b** Apparent mass vs. time response of a T3-type sensor after injection of a free GSH solution (5 mM) (from [35], with per-mission from the Electrochemical Society)

large total mass o the film obtained after subsequent polymerization stages was $\Delta m = 1322$ ng ($\Delta f = 1147$ Hz).

After hydrolyzing the GSH and removal of imprinted AuNP-SG in the PoPD(AuNP-SG) polymer film with NaOH solution, the piezosensor was subjected to a solution of GSH analyte. The time transient recorded upon injection of GSH is presented in Fig. 9.20. The total mass change of $\Delta m = 27$ ng was observed in the 300 s recoding time. The responsiveness of this sensor is much higher than that of the previous one (approximately three times higher) due to the high mass of the gold nanoparticle labels. This translates to a higher measurement resolution which is important for precision analyte determinations.

The process of molecular imprinting of a PoPD(GSH) film synthesized in-situ on a QC/Au/SAM/AuNP substrate is illustrated in Fig. 9.21. The total mass depos-ited was $\Delta m = 265$ ng. After the template removal from the PoPD$_{GSH}$ polymer film, the piezosensor was tested in a solution of 5 mM GSH. The time transient recorded upon injection of GSH is presented in Fig. 9.21. The total mass change $\Delta m = 7$ ng was observed. The reproducibility of this sensor fabrication (ca. 80%) was not as good as for other types of sensors tested, most likely due to the differences in the amount and packing density of AuNP in the film. However, this film design still shows a good promise since further improvements can be achieved through a better film microstructuring (*i.e.*, the improvement in the film texture).

Therefore, by highest mass gain was attained for the sensor templated with GSH-capped gold nanoparticles (design T2, QC/Au/PoPD(AuNP-SG) in Fig. 9.15b). The nanoparticle labeling enhances the nanogravimetric biosensor response because of the larger mass of the AuNP-labeled analyte.

9.5   Conclusions

The monolayer-protected gold nanoparticles interact strongly with biomarkers of oxidative stress, such as glutathione, homocysteine, and cysteine. Strong binding of these biomarkers to gold through the thiolate bonding results in an easy replacement of a self-assembled protecting monolayer on a core–shell gold nanoparticle with the biomarker SAM. By carefully controlling the solution pH, it is possible to fine-tune the biomarker-induced nanoparticle assembly mediated by interparticle zwitterionic interactions and hydrogen bonding. The surface plasmon band shift associated with the assembly can be employed for simple and inexpensive detection of the oxidative stress biomarkers. By utilizing the RELS spectroscopy, the detection of biomarker interactions with gold nanoparticles is made even more sensitive. A strong dependence of scattering cross-section on GSH concentration has been found for small AuNP (5 nm dia.). For larger AuNP (22.5 nm dia.), a dramatic reversal of this dependence has been observed, where the addition of GSH causes a sharp decrease in I_{sc}. This effect has been ascribed to the extensive GSH-cross-linking and sedimentation of AuNP networks. By carefully controlling the solution pH, the analysis of biomarkers of oxidative can be fine-tuned to distinguish GSH, Hcys, and Cys. For GSH, the strong RELS signals were observed in the low pH range (2–3.5) whereas for Hcys and Cys, the strongest RELS signals were seen in the pH range 4–5. The use of a fluorosurfactant ZONYL FSN, has enabled to differentiate RELS responses to Hcys and Cys, as the former is able to penetrate the fluorosurfactant SAM faster the latter. The functionalized gold nanoparticles have also been shown to enhance the design of molecularly-templated conductive polymer films for the detection of biomolecules. Novel designs of molecularly imprinted orthophenyl-enediamine (oPD) sensor films have been investigated for the detection of GSH. The template films were synthesized by in-situ electropolymerization of oPD in the presence of target molecules: GSH or GSH-capped gold nanoparticles. The templating has proven to be feasible and an enhancement of the templated sensor response to the GSH analyte has been obtained for imprinting GSH-capped gold nanoparticles in poly(oPD) films. Other sensor designs including a AuNP network sub-layer were also tested in an attempt to increase the interface sensitivity to the biolyte. By employing the EQCN technique, it is possible to follow each stage of the template formation and control the polymer thickness and its electrical conductivity. The molecular imprinting technique can be applied in polymer sensor designs based on biorecognition principles with piezoelectric transduction.

9.6   Materials and Methods

9.6.1   Chemicals

All chemicals used for investigations were of analytical grade purity. DL-Homocysteine   ($HS(CH_2)_3NH_2COOH$),   L-Cysteine   ($HS(CH_2)_2NH_2COOH$), tetrachloroauric(III) acid trihydrate ($HAuCl_4 \cdot 3H_2O$) with 99.9+% metals basis, D-Methionine, and L-glutathione (GSH) reduced (minimum 99%), were purchased from Sigma Aldrich Chemical Company (Atlanta, GA, U.S.A.) and used as received. ZONYL FSN-100, a fluorocarbon-ether surfactant (FES), with nominal composition $CF_3(CF_2)_m(C_2H_4O)_nCH_2OH$ and average molecular mass M = ~950 g/mol was obtained from Sigma Aldrich. Sodium citrate dihydrate (HOC(COONa) ($CH_2COONa)_2 \cdot 2H_2O$) was received from J.T. Baker Chemical Co. (Phillipsburg, NJ, U.S.A.). Sodium borohydride ($NaBH_4$) was obtained from Fisher Scientific Company. L(+) Histidine was purchased from Eastman Organic Chemicals (Rochester, NY, U.S.A.). Other chemicals were obtained from Sigma Aldrich Chemical Company. Solutions were prepared using Millipore (Billerica, MA, U.S.A.) Milli-Q deionized water (conductivity $\sigma = 55$ nS/cm). They were deoxygenated by bubbling with purified argon.

9.6.2 Apparatus

The imaging analyses of Au nanoparticles were performed using high-resolution transmission electron microscopy (HR-TEM) with Model JEM-2010 (Jeol, West Chester, PA, U.S.A.) HR-TEM instrument (200 kV). The elastic light scattering spectra were recorded using LS55 Spectrometer (Perkin Elmer) equipped with a 20 kW Xenon light source operating in 8 μs pulsing mode allowing for the use of monochromatic radiation with wavelength from 200 to 800 nm with 1 nm resolution and sharp cut-off filters: 290, 350, 390, 430, 515 nm. The dual detector system consisted of a photomultiplier tube (PMT) and an avalanche photodiode. The RELS spectra were obtained at 90° angle from the incident (excitation) light beam. The excitation beam monochromator was either scanned simultaneously with the detector beam monochromator ($\Delta\lambda = 0$) or set at a constant excitation wavelength. The UV–vis spectra were recorded using Perkin Elmer Lambda 50 Spectrophotometer in the range 400–900 nm or Ocean Optics (Dunedin, FL, U.S.A.) Model R4000 Precision Spectrometer in the range from 340 to 900 nm.

9.6.3 Procedures

The Au nanoparticles were synthesized according to the published procedure [201]. Briefly, to obtain 5 nm AuNP, a solution of $HAuCl_4$ (10 mM, 2.56 mL) was mixed with a trisodium citrate solution (10 mM, 9.6 mL), ratio 1: 3.75, and poured to distilled water (88 mL). The obtained solution was vigorously stirred and fresh cold $NaBH_4$ solution (5 mM, 8.9 mL) was added dropwise. The solution slowly turned light grey and then ruby red. Stirring was maintained for 30 min. The obtained citrate-capped core–shell Au nanoparticles (AuNP) were stored at 4°C. Their size, determined by HR-TEM imaging and UV–vis surface plasmon absorption was

5.0 nm. The concentrations of AuNP's are given in moles of particles per 1 L of solution (usually, in the nM range). Larger AuNP's were synthesized using the gallic acid procedure [202]. The RELS and UV–vis spectra for samples were obtained with 1 min of mixing of AuNP with biomolecule solutions, unless otherwise stated.

9.6.4   Calulations

Quantum mechanical calculations of electronic structures for a model fluorocarbon-ether surfactant, citric acid, cysteine and homocysteine were performed using modified Hartree-Fock methods [203, 204] with 6-31G* basis set and pseudopotentials, semi-empirical PM3 method, and density functional theory (DFT) with B3LYP functional. The molecular dynamics simulations and quantum mechanical calculations were carried out using procedures embedded in Wavefunction (Irvine, CA, U.S.A.) Spartan 6 [144]. The electron density and local density of states (LDOS) are expressed in atomic units, au^{-3}, where 1 $au = 0.529157$ Å and 1 $au^{-3} - 6.749108$ $Å^{-3}$.

Acknowledgements This work was supported by the U.S. DoD Research Program "Idea", Grant No. AS-073218.
Help of students: Kaitlin Coopersmith and Jeffrey Deeb is acknowledged.

References

1. Noble, M.; Mayer-Proschel, M.; Proschel, C., *Antioxidants Redox Signal.* **2005**, 7, 1456-1467.
2. Armstrong, D., *Oxidative Stress Biomarkers and Antioxidant Protocols.* Humana Press: Totowa, NJ, 2002.
3. Carlo, M. D.; Loeser, R. F., *Arthritis & Rheumatism* **2003**, 48, 3419-3430.
4. James, S. J.; Melnyk, S.; Jernigan, S.; Cleves, M. A.; Halsted, C. H.; Wong, D. J.; Cutler, P.; Boris, M.; Bock, K.; Bradstreet, J. J.; Baker, S. B.; Gaylor, D. W., *Am. J. Med. Genetics* **2006**, 141B, 947-956.
5. Seshadri, S.; Beiser, A.; Selhub, J.; Jaques, P. F.; Rosenberg, I. H.; D'Agostino, R. B.; Wilson, P. W. F., *N. Engl. J. Med.* **2002**, 346, 476.
6. Varadarajan, S.; Kanski, J.; Aksenova, M.; Lauderback, C.; Butterfield, D. A., *J. Am. Chem. Soc.* **2001**, 123, 5625.
7. Mills, J. L.; Scott, J. M.; Kirke, P. N.; McPartlin, J. M.; Conley, M. R.; Weir, D. G.; Molloy, A. M.; Lee, Y. J., *J. Nutrition* **1996**, 126, S756.
8. Yamamoto, Y.; Yamanashi, S., Ubiquinol/ubiquinone ratio as a marker of oxidative stress. In *Oxidative Stress Biomarkers and Antioxidant Protocols*, Armstrong, D., Ed. Humana Press: Totowa, NJ, 2002.
9. Droge, W., *Physiol. Rev.* **2002**, 82, 47-95.
10. Carmel, R.; Jacobsen, D. W., *Homocysteine in Health and Disease.* Cambridge University Press: Cambridge, U.K., 2001.
11. Jacobsen, D. W., *Arterioscler. Thromb. Vasc. Biol.* **2000**, 20, 1182-1184.
12. Refsum, H.; Ueland, P. M.; Nygard, O.; Volset, S. E., *Annu. Rev. Med.* **1989**, 49, 31.

13. Zhang, X.; Li, H.; Jin, H.; Ebin, Z.; Brodsky, S.; Goligorsky, M. S., *Am. J. Physiol. Renal. Physiol.* **2000**, 279, F671.
14. Boushey, C.; Beresford, S.; Omenn, G.; Motulsky, A., *JAMA* **1995**, 274, 1049.
15. Graham, I. M.; Daly, L.; Refsum, H.; Robinson, K.; Brattstrom, L.; Ueland, P. M., *JAMA* **1997**, 277, 1775; 5482.
16. Lentz, S. R.; Haynes, W. G., *Cleveland Clinic J. Med.* **2004**, 71, 730.
17. Welch, G. N.; Loscalzo, J., *N. Engl. J. Med.* **1998**, 338, 1042-1050.
18. Beard, K. M.; Shangari, N.; Wu, B.; O'Brien, P. J., *Mol. Cell. Biochem.* **2003**, 252, 331.
19. Bernard, S.; Enayati, A.; Redwood, L.; Roger, H.; Binstock, T., *Med. Hypotheses* **2001**, 56, 462–471.
20. Clark-Taylor, T., *Med. Hypotheses* **2003**, 62, (6), 970-975.
21. Guldener, C. V.; Robinson, K., *Semin. Thromb. Hemostasis* **2000**, 26, 313.
22. Brown, B., *Am. Acad. Ophthalmol.* **2002**, 109, 287-290.
23. Selhub, J., *Annu. Rev. Nutr.* **1999**, 19, 217–246.
24. Polidoro, G.; Ilio, C. D.; Arduini, A.; Rovere, G. L.; Federici, G., *Int. J. Biochem.* **1984**, 16, 505-509.
25. Almazzan, G.; Liu, H. N.; Knorchid, A.; Sundararajan, S.; Martinez-Bermudez, A. K.; Chemtob, S., *Free Radical Biol. Med.* **2000**, 29, 858-869.
26. Repetto, M.; Reides, C.; Carretero, M. L. G.; Costa, M.; Griemberg, G.; Llesuy, S., *Clin. Chim. Acta* **1996**, 255, 107-117.
27. Upadhya, S.; Upadhya, S.; Mohan, S. K.; Vanajakshamma, K.; Kunder, M.; Mathias, S., *Indian J. Clin. Biochem.* **2004**, 19, 80-83.
28. Clarkson, T. W., *Environ Health Perspect* **1992**, 100, 31–38.
29. Dameron, C. T.; Reese, R. N.; Mehra, R. K.; Kortan, P. J.; Carrol, M. L.; Steigerwald, M. L.; Brus, L. E.; Winge, D. R., *Nature* **1989**, 338, 596.
30. Barbas, J.; Santhanagopalan, V.; Blaszczynski, M.; Ellis, W. R.; Winge, D. R., *J. Inorg. Biochem.* **1992**, 48, 95.
31. Vatamaniuk, O. K.; Mari, S.; Lu, Y.; Rea, P. A., *J. Biol. Chem.* **2000**, 275, 31451-31459.
32. Vatamaniuk, O. K.; Bucher, E. A.; Ward, J. T.; Rea, P. A., *J. Biol. Chem.* **2001**, 276, 20817.
33. Mehra, R. K.; Winge, D. R., *J. Cell. Biochem.* **1991**, 45, 30.
34. Inouhe, M., *Braz. J. Plant Physiol.* **2005**, 17, 65-78.
35. Stobiecka, M.; Deeb, J.; Hepel, M., *Electrochem. Soc. Trans.* **2009**, 19, (28), 15-32.
36. Wang, W.; Rusin, O.; Xu, X.; Kyu, K.; Kim, K.; Escobedo, J. O.; Fakayode, S. O.; Fletcher, K. A.; Lowry, M.; Schowalter, C. M.; Lawrence, C. M.; Fronczak, F. R.; Warner, I. M.; Strongin, R. M., *J. Am. Chem. Soc.* **2005**, 127, 15949-15958.
37. Escobedo, J. O.; Rusin, O.; Wang, W.; Alptürk, O.; Kim, K. K.; Xu, X.; Strongin, R. M., Detection of Biological Thiols. In *Reviews in Fluorescence*, Springer, US: 2006; pp 139-162.
38. F.Tanaka; Mase, N.; III, C. F. B., *Chem. Comm.* **2004**, 7, 1762-1763.
39. Pacsial-Ong, E. J.; McCarley, R. L.; Wang, W.; Strongin, R. M., *Anal. Chem.* **2006**, 78, 7577-7581.
40. Agüí, L.; Peña-Farfal, C.; Yáñez-Sedeno, P.; Pingarrón, J. M., *Talanta* **2007**, 74, 412–420.
41. Lim, I. I. S.; Ip, W.; Crew, E.; Njoki, P. N.; Mott, D.; Zhong, C. J., *Langmuir* **2007**, 23, 826-833.
42. Gates, A. T.; Fakayode, S. O.; Lowry, M.; Ganea, G. M.; Marugeshu, A.; Robinson, J. W.; Strongin, R. M.; Warner, I. M., *Langmuir* **2008**, 24, 4107-4113.
43. Sudeep, P. K.; Joseph, S. T. S.; Thomas, K. G., *J. Am. Chem. Soc.* **2005**, 127, 6516-6517.
44. Wu, H. P.; Huang, C. C.; Cheng, T. L.; Tseng, W. L., *Talanta* **2008**, 76, 347-452.
45. Wasowicz, M.; Viswanathan, S.; Dvornyk, A.; Grzelak, K.; Kludkiewicz, B.; Radecka, H., *Biosens. Bioelectron.* **2008**, 24, 284-289.
46. Hepel, M.; Dallas, J.; Noble, M. D., *J. Electroanal. Chem.* **2008**, 622, 173-183.
47. Hepel, M.; Tewksbury, E., *J. Electroanal. Chem.* **2003**, 552, 291-305.
48. Hepel, M.; Tewksbury, E., *Electrochim. Acta* **2004**, 49, 3827-3840.
49. Lim, S. I.; Zhong, C. J., *Acc. Chem. Res.* **2009**, 42, 798-808.

50. Kariuki, N. N.; Luo, J.; Han, L.; Maye, M. M.; Moussa, L.; Patterson, M.; Lin, U.; Engelhard, M. H.; Zhong, C. J., *Electroanalysis* **2004**, 16, 120-126.
51. Lim, I. I. S.; Zhong, C. J., *Gold Bulletin* 40, 59-66.
52. Zhang, S.; Kou, X.; Yang, Z.; Shi, Q.; Stucky, G. D.; Sun, L.; Wang, J.; Yan, C., *Chem. Comm.* **2007**, 1816-1818.
53. Han, L.; Luo, J.; Kariuki, N.; Maye, M. M.; Jones, V. W.; Zhong, C. J., *Chem. Mater.* **2003**, 15, 29-37.
54. Zheng, W.; Maye, M. M.; Leibowitz, F. L.; Zhong, C. J., *Anal. Chem.* **2000**, 72, 2190-2199.
55. Lim, I. M. S.; Mott, D.; Ip, W.; Njoki, P. N.; Pan, Y.; Zhou, S.; Zhong, C. J., *Langmuir* **2008**, 24, 8857-8863.
56. Li, Z. P.; Duan, X. R.; Liu, C. H.; Du, B. A., *Anal. Biochem.* **2006**, 351, 18-25.
57. Rusin, O.; Luce, N. N. S.; Agbaria, R. A.; Escobedo, J. O.; Jiang, S.; Warner, I. M.; Strongin, R. M., *J. Am. Chem. Soc.* **2004**, 126, 438-439.
58. Zhang, F. X.; Han, L.; Israel, L. B.; Daras, J. G.; Maye, M. M.; Ly, N. K.; Zhong, C. J., *Analyst* **2002**, 127, 462-465.
59. Lu, C.; Zu, Y.; Yam, V. W. W., *J. Chromatogr. A* **2007**, 1163, 328-332.
60. Mocanu, A.; Cernica, I.; Tomoaia, G.; Bobos, L. D.; Horovitz, O.; M.Tomoaia-Cotisel, *Colloids Surfaces A* **2009**, 338, 93-101.
61. Wang, W.; Escobedo, J. O.; Lawrence, C. M.; Strongin, R. M., *J. Am. Chem. Soc.* **2004**, 126, 3400-3401.
62. Hostetler, M. J.; Templeton, A. C.; Murray, R. W., *Langmuir* **1999**, 15, 3782-3789.
63. Hostetler, M. J.; Wingate, J. E.; Zhong, C. J.; Harris, J. E.; Vachet, R. W.; Clark, M. R.; Londono, J. D.; Green, S. J.; Stokes, J. J.; Wignall, G. D.; Glish, J. L.; Porter, M. D.; Evans, N. D.; Murray, R. W., *Langmuir* **1998**, 14, 17-30.
64. Lu, C.; Zu, Y.; Yam, V. W. W., *Anal. Chem.* **2007**, 79, 666-672.
65. Huang, C. C.; Tseng, W. L., *Anal. Chem.* **2008**, 80, 6345-6350.
66. Anderson, M. E., *Chemico-Biological Interactions* **1998**, 111-112, 1-14.
67. Hepel, M.; Dallas, J.; Noble, M. D., *Sensors Transducers J.* **2008**, 88, 47.
68. Fang, C.; Zhou, X., *Electroanalysis* **2002**, 15, 1632-1638.
69. Zhang, S.; Kou, X.; Yang, Z.; Shi, Q.; Stucky, G. D.; Sun, L.; Wang, J.; Yan, C., *Chem. Commun.* **2007**, n/a, 1816.
70. Kou, X.; Zhang, S.; Yang, Z.; Tsung, C. K.; Stucky, G. D.; Sun, L.; Wang, J.; Yan, C., *J. Am. Chem. Soc.* **2007**, 129, 6402.
71. Ackerson, C. J.; Sykes, M. T.; Kornberg, R. D., *Proc. Natl. Acad. Sci. U.S.A.* **2005**, 102, 13383.
72. Hepel, M.; Dallas, J., *Sensors* **2008**, 8, 7224-7240.
73. Chow, E.; Hibbert, D. B.; Gooding, J. J., *Analyst* **2005**, 130, 831.
74. Takehara, K.; Aihara, M.; Miura, Y.; Tanaka, F., *Bioelectrochem. Bioenerg.* **1996**, 39, 135.
75. Takehara, K.; Aihara, M.; Ueda, N., *Electroanalysis* **1994**, 6, 1083.
76. Takehara, K.; Ide, Y.; Aihara, M., *Bioelectrochem. Bioenerg.* **1991**, 29, 113.
77. Takehara, K.; Ide, Y.; Aihara, M.; Obuchi, E., *Bioelectrochem. Bioenerg.* **1992**, 29, 103.
78. Whitesides, G. M.; Kriebel, J. K.; Love, J. C., *Science Progress* **2005**, 88, 17-48.
79. Hostetler, M. J.; Templeton, A. C.; Murray, R. W., *Langmuir* **1999**, 15, 3782.
80. Hostetler, M. J.; Wingate, J. E.; Zhong, C. J.; Harris, J. E.; Vachet, R. W.; Clark, M. R.; Londono, J. D.; Green, S. J.; Stokes, J. J.; Wignall, G. D.; Glish, J. L.; Porter, M. D.; Evans, N. D.; Murray, R. W., *Langmuir* **1998**, 14, 17.
81. Sugawara, M.; Hirano, A.; Buhlmann, P.; Umezawa, Y., *B. Chem. Soc. Jpn* **2002**, 75, 187.
82. Esplandiu, M. J.; Hagenstrom, H.; Kolb, D. M., *Langmuir* **2001**, 17, 828.
83. Wang, W. U.; Chen, C.; Lin, K. H.; Fang, Y.; Lieber, C. M., *Proc. Natll. Acad. Sci. U.S.A.* **2005**, 102, 3208-3212.
84. Gooding, J. J.; Pugliano, I.; Hibbert, D. B.; Erokhin, P., *Electrochem. Commun.* **2000**, 2, 217.
85. Szymaska, I.; Stobiecka, M.; Orlewska, C.; Rohand, T.; Janssen, D.; Dehaen, W.; Radecka, H., *Langmuir* **2008**, 24, 11239-11245.
86. Halamek, J.; Hepel, M.; Skladal, P., *Biosensors Bioelectronics* **2001**, 16, 253.

87. Pribyl, J.; Hepel, M.; Halamek, J.; Skladal, P., *Sensors Actuators B* **2003**, 91, 333.
88. Pribyl, J.; Hepel, M.; Skladal, P., *Sensors Actuators B* **2006**, 113, 900.
89. Stobiecka, M.; Cieśla, J. M.; Janowska, B.; Tudek, B.; Radecka, H., *Sensors* **2007**, 7, 1462-1479.
90. Tran, E.; Duati, M.; Whitesides, G. M.; Rampi, M. A., *Faraday Discussions* **2006**, 131, 197-203.
91. Vega, R. A.; Shen, C. K. F.; Maspoch, D.; Robach, J. G.; Lamb, R. A.; Mirkin, C. A., *Small* **2007**, 3, 1482-1485.
92. Yu, A. A.; Savas, T. A.; Taylor, G. S.; Giuseppe-Elie, A.; Smith, H. I.; Stellacci, F., *Nano Lett.* **2005**, 5, 1061-1066.
93. Hepel, M.; Dela, I.; Hepel, T.; Luo, J.; Zhong, C. J., *Electrochim. Acta* **2007**, 52, 5529-5547.
94. Luo, J.; Maye, M. M.; Kariuki, N. N.; Wang, L.; Njoki, P.; Lin, Y.; Schadt, M.; Naslund, H. R.; Zhong, C. J., *Catal. Today* **2005**, 99, 291.
95. Luo, J.; Maye, M. M.; Petkov, V.; Kariuki, N. N.; Wang, L.; Njoki, P.; Mott, D.; Lin, Y.; Zhong, C. J., *Cham. Mater.* **2005**, 17, 3086.
96. Goia, C.; Matijević, E.; Goia, D. V., *J. Mater. Res.* **2005**, 20, 1507-1514.
97. Lu, Z.; Ryde, N. P.; Babu, S. V.; Matijević, E., *Langmuir* **2005**, 21, 9866-9872.
98. Matijević, E., *Colloid Journal* **2007**, 69, 29-38.
99. Siiman, O.; Jitianu, A.; Bele, M.; Grom, P.; Matijević, E., *J. Colloid Interface Sci.* **2007**, 309, 8-20.
100. Uskoković, V.; Matijević, E., *J. Colloid Interface Sci.* **2007**, 315, 500-511.
101. Zelenev, A.; Matijević, E., *J. Colloid Interface Sci.* **2006**, 299, 22-27.
102. Suber, L.; Sondi, I.; Matijević, E.; Goia, D. V., *J. Colloid Interface Sci.* **2005**, 288, 489-495.
103. Gorantla, V. K. R.; Matijević, E.; Babu, S. V., *Chem. Mater.***2005**, 17, 2076-2080.
104. Alivisatos, A. P.; Johnsson, K. P.; Peng, X.; Wilson, T. E.; Loweth, C. J.; Bruchez, M. P.; Schultz, P. G., *Nature* **1996**, 382, 610.
105. Kamat, P. V., *J. Phys. Chem. B* **2002**, 106, 7729-2244.
106. Kariuki, N. N.; Luo, J.; Maye, M. M.; Hassan, S. A.; Menard, T.; Naslund, H. R.; Zhong, C. J., *Langmuir* **2004**, 20, (11240).
107. Westerlund, F.; Bjornholm, T., *Currenr Opinion Colloid Interface Sci.* **2009**, 14, 126-134.
108. Lu, Y.; Liu, J., *Acc. Chem. Res.***2007**, 40, 315-323.
109. Lim, I. I. S.; Chandrachud, U.; Wang, L.; Gal, S.; Zhong, C. J., *Anal. Chem.* **2008**, 80, 6038-6044.
110. Wang, C. L.; Zhang, H.; Lin, Z.; Yao, X.; Lv, N.; Li, M. J.; Sun, H. Z.; Zhang, J. H.; Yang, B., *Langmuir* **2009**, 25, 10237-10242.
111. Zhu, M. Q.; Gu, Z.; Fan, J. B.; Xu, X. B.; Cui, J.; Liu, J. H.; Long, F., *Langmuir* **2009**, 25, 10189-10194.
112. Roy, I.; Ohulchanskyy, T. Y.; Bharali, D. J.; Pudavar, H. E.; Mistretta, R. A.; Kaur, N.; Prasad, P. N., *PNAS* **2004**, 102, 279-284.
113. Matijevic, E., *Medical Applications of Colloids*. Springer Sci.: New York, 2008.
114. Tong, L.; Zhao, Y.; Huff, T. B.; Hansen, M. N.; Wei, A.; Cheng, J. X., *Adv. Mater. (Weinheim, Germany)* **2007**, 19, 3136.
115. Huff, T. B.; Tong, L.; Zhao, Y.; Hansen, M. N.; Cheng, J. X.; Wei, A., *Nanomedicine* **2007**, 2, 125.
116. Chou, C. H.; Chen, D.; Wang, C. R. C., *J. Phys. Chem. B* **2005**, 109, 11135.
117. Reynolds, R. A.; Mirkin, C. A.; Letsinger, R. L., *J. Am. Chem. Soc.***2000**, 122, 3795-3796.
118. Elghanian, R.; Storhoff, J. J.; Mucic, R. C.; Letsinger, R. L.; Mirkin, C. A., *Science* **1997**, 277, 1078-1081.
119. Liu, H.; Tian, Y.; Deng, Z., *Langmuit* **2007**, 23, 9487-9494.
120. Verma, A.; Simard, J. M.; Rotello, V. M., *Langmuir* **2004**, 20, 4178-4181.
121. Yeh, C. H.; Hung, C. Y.; Chang, T. C.; Liu, H. P.; Liu, Y. C., *Microfluid Nanofluid* **2009**, 6, 85-91.
122. Hirsch, L. R.; Jackson, J. B.; Lee, A.; Halas, N. J.; West, J. L., *Anal. Chem.* **2003**, 75, 2377-2381.

123. Chen, C. D.; Cheng, S. F.; Chau, L. K.; Wang, C. R. C., *Biosens. Bioelectron.* **2007**, 22, 926.
124. Ambrosi, A.; Castaneda, M. T.; Killard, A. J.; Smyth, M. R.; Alegret, S.; Merkoci, A., *Anal. Chem.* **2007**, 79, 5232-5240.
125. Storhoff, J. J.; Elghanian, R.; Mucic, R. C.; Mirkin, C. A.; Letsinger, R. L., *J. Am. Chem. Soc.* **1998**, 120, 1959-1964.
126. Dubertet, B.; Calame, M.; Libchaber, A. J., *Nature Biotechnology* **2001**, 19, 365-370.
127. Tsai, C. Y.; Tsai, Y. H.; Pun, C. C.; Chan, B.; Luh, T. Y.; Chen, C. C.; Ko, F. H.; Chen, P. J.; Chen, P. H., *Microsystem Technologies* **2005**, 11, 91-96.
128. Kelly, K. L.; Coronado, E.; Zhao, L. L.; Schatz, G. C., *J. Phys. Chem. B* **2003**, 107, 668-677.
129. Link, S.; El-Sayed, M. A., *J. Phys. Chem. B* **1999**, 103, 8410-8426.
130. Link, S.; Mohamed, M. B.; El-Sayed, M. A., *J. Phys. Chem. B* **1999**, 103, 3073-3077.
131. Franzen, S.; Folmer, J. C. W.; Glomm, W. R.; R.O'Neal, *J. Phys. Chem. B* **2002**, 106, 6533-6540.
132. Ungureanu, C.; Rayavarapu, R. G.; Manohar, S.; Leeuwen, T. G. v., *J. Appl. Phys.* **2009**, 105, 102032.
133. Gonzales, A. L.; Noguez, C., *J. Comput. Theor. Nanosci.* **2007**, 4, 231.
134. Levi, S. A.; Mourran, A.; Spatz, J. P.; Veggel, F. C. v.; Reinhoudt, D. N.; Moller, M., *Chem. Eur. J.* **2002**, 8, 3808.
135. Taton, T. A.; Lu, G.; Mirkin, C. A., *J. Am. Chem. Soc.* **2001**, 123, 5164-5165.
136. Taton, T. A.; Mucic, R. C.; Mirkin, C. A.; Letsinger, R. L., *J. Am. Chem. Soc.* **2000**, 122, 6305-6306.
137. Park, S. J.; Taton, T. A.; Mirkin, C. A., *Science* **2002**, 295, 1503-1505.
138. Reyolds, R. A.; Mirkin, C. A.; Letsinger, R. L., *J. Am. Chem. Soc.* **2000**, 122, 3795-3796.
139. Schroedter, A.; Weller, H., *Angew. Chem. Int. Ed.* **2002**, 41, 3218.
140. Li, Z. P.; Duan, X. R.; Liu, C. H.; Du, B. A., *Anal. Biochem.* **2006**, 351, 18.
141. Mandal, S.; Gole, A.; Lala, N.; Gonnade, R.; Ganvir, V.; Sastry, M., *Langmuir* **2001**, 17, 6262.
142. Selvakannan, P. R.; Mandal, S.; Phadtare, S.; Pasricha, R.; Sastry, M., *Langmuir* **2003**, 19, 3545.
143. Li, T.; Park, H. G.; Lee, H. S.; Choi, S. H., *Nanotechnology* **2004**, 15, S660.
144. Brelle, M.; Zhang, J. Z.; Nguyen, L.; Mehra, R. K., *J. Phys. Chem. A* **1999**, 103, 10194.
145. Maye, M. M.; Lim, I. I. S.; Luo, J.; Rab, Z.; Rabinovich, D.; Liu, T.; Zhong, C. J., *J. Am. Chem. Soc.* **2005**, 127, 1519-1529.
146. Pasternack, R. F.; Bustamante, C.; Collings, P. J.; Giannetto, A.; Gibbs, E. J., *J. Am. Chem. Soc.* **1993**, 115, 5393-5399.
147. Pasternack, R. F., *Science* **1995**, 269, (5226), 935.
148. Wang, Y. T.; Zhao, F. L.; Li, K. A.; Tong, S. Y., *Anal. Chim. Acta* **1999**, 396, 75-81.
149. Guo, Z. X.; Shen, H. X., *Anal. Chim. Acta* **2000**, 408, 177-182.
150. Wu, X.; Wang, U. Y.; Wang, M.; Sun, S.; Yang, J.; Luan, Y., *Spectrochim. Acta A* **2005**, 61, 361-366.
151. Jia, Z.; Yang, J.; Wu, X.; Sun, C.; Liu, S.; Wang, F.; Zhao, Z., *Spectrochim. Acta A* **2006**, 64, 555-559.
152. Stobiecka, M.; Deeb, J.; Hepel, M., *Biophys. Chem.* **2010**, 146, 98-107.
153. Franzen, S.; Folmer, J. C. W.; Glomm, W. R.; R.O'Neal, *J. Phys. Chem. A* **2002**, 106, 6533-6540.
154. Lim, I. I. S.; Goroleski, F.; Mott, D.; Kariuki, N.; Ip, W.; Luo, J.; Zhong, C. J., *J. Phys. Chem. B* **2006**, 110, 6673-6682.
155. Li, F.; Zu, Y., *Anal. Chem.* **2004**, 76, 1768-1772.
156. Stobiecka, M.; Hepel, M., **2010**, 350, 168-177.
157. Mirkin, C. A.; Letzinger, R. L.; Mucic, R. C.; Storhoff, J. J., *Nature* **1996**, 382, 607.
158. Skladal, P., *J. Braz. Chem. Soc.* **2003**, 14, 491.
159. Hianik, T.; Ostatna, V.; Sonlajtnerova, M.; Grman, I., *Bioelectrochem.* **2007**, 70, 127-133.
160. Tombelli, S.; Minunni, M.; Luzi, E.; Mascini, M., *Bioelectrochem.* **2005**, 67, 235-141.

161. Pividori, M. I.; Merkoci, A.; Alegret, S., *Biosens. Bioelectron.* **2000**, 15, 291-303.
162. Tassew, N.; Thompson, M., *Biophys. Chem.* **2003**, 106, 241-252.
163. Ferencova, A.; Adamovski, M.; Grundler, P.; Zima, J.; Barek, J.; Mattusch, J.; Wennrich, R.; Labuda, J., *Bioelectrochem.* **2007**, 71, 33.
164. Galandova, J.; Ovadekova, R.; Ferancova, A.; Labuda, J., *Anal. Bioanal. Chem.* **2009**, 394, 855.
165. Lim, I. I. S.; Chandrachud, U.; Wang, L.; Gal, S.; Zhong, C. J., *Anal. Chem.* **2008**, 80, 6038.
166. Ostatna, V.; Dolinnaya, N.; Andreev, S.; Oretskaya, T.; Wang, J.; Hianik, T., *Bioelectrochem.* **2005**, 67, 205-210.
167. T.Hianik; Gajdos, V.; Krivanek, R.; Oretskaja, T.; Metelev, V.; Volkov, E.; Vadgama, P., *Bioelectrochem.* **2001**, 53, 199-204.
168. Hepel, M., In *"Interfacial Electrochemistry"*, Wieckowski, A., Ed. Marcel Dekker, Inc.: New York, 1999; pp 599-630.
169. Halamek, J.; M.Hepel; Skladal, P., *Biosens. Bioelectron.* **2991**, 16, 253-260.
170. Pribyl, J.; Hepel, M.; Skladal, P., *Sensors Actuators B* **2006**, 113, 900-910.
171. Pribyl, J.; Hepel, M.; Halamek, J.; Skladal, P., *Sensors Actuators B* **2003**, 91, 333-341.
172. Greene, N. T.; Shimizu, K. D., *J. Am. Chem. Soc.* **2005**, 127, 5695-5700.
173. Levit, N.; Pestov, D.; Tepper, G., *Sensors and Actuators B* **2002**, 82, 241-249.
174. Yan, M.; Ramstrom, O., *Molecularly Imprinted Materials. Science and Technology*. Marcel Dekker: New York, 2005.
175. Priego-Capote, F.; L.Ye; Shakil, S.; Shamsi, S. A.; Nilsson, S., *Anal. Chem.* **2008**, 80, 2881-2887.
176. Ye, L.; Weiss, R.; Mosbach, K., *Macromolecules* **2000**, 33, 8239-8245.
177. Perez, N.; Whitcombe, M., *J. App. Polym. Sci.* **2000**, 77, 1855-1859.
178. Piletsky, S. A.; Subrahmanyan, S.; Turner, A. P. F., *Sensor Review* **2001**, 21, 292-296.
179. Malitesta, C.; Picca, R. A.; Ciccarella, G.; Sgobba, V.; Brattoli, M., *Sensors* **2006**, 6, 915-924.
180. Fard, S. E.; Mohammadi, A.; Abdouss, M.; Ganjali, M. R.; Norouzi, P.; Safaraliee, L., *Anal. Chim. Acta* **2008**, 612, 65-74.
181. Panasyuk, T. L.; Mirsky, V. M.; Piletsky, S. A.; Wolfbeis, O. S., *Anal. Chem.* **1999**, 71, 4609-4613.
182. Kriz, D.; Kempe, M.; Mosbach, K., *Sensors Actuators B* **1996**, 33, 178-181.
183. Sergeyeva, T. A.; Piletsky, S. A.; Brovko, A. A.; Slinchenko, E. A.; Sergeyeva, L. M.; Panasyuk, T. L.; Elskaya, A. V., *Analyst* **1999**, 124, 331-334.
184. Prasad, B. B.; Lakshmi, D., *Electroanalysis* **2005**, 17, 1260-1268.
185. Liang, C.; Peng, H.; Nie, L.; Yao, S., *Fresenius' J. Anal. Chem* **2000**, 367, 551-555.
186. Kikuchi, M.; Tsuru, N.; Shiratori, S., *Sci. Technol. Adv. Materials* **2006**, 7, 156-161.
187. Percival, C. J.; Stanley, S.; Galle, N.; Braihwaite, A.; Newton, M. I.; McHale, G.; Hayes, W., *Anal. Chem.* **2001**, 73, 4225-4228.
188. Kugimiya, A.; Takeuchi, T., *Electroanalysis* **1999**, 11, 1158-1160.
189. Yilmaz, E.; Mosbach, K.; Haupt, K., *Anal. Commun.* **1999**, 36, 167.
190. Matsuguchi, M.; Uno, T. I., *Sensors Actuators B* **2006**, 113, 94-99.
191. Tsuru, N.; Kikuchi, M.; Shiratori, S.; Kawaguchi, H., *Thin Solid Films* **2006**, 499, 380-385.
192. Stephenson, C. J.; Shimizu, K. D., *Polymer International* **2007**, 56, 482-488.
193. Tokarev, I.; Tokareva, I., *Chem. Commun.* **2006**, 3343-3345.
194. Jenkins, A. L.; Yin, R.; Jensen, J. L., *Analyst* **2001**, 126, 798-802.
195. Moreno-Bondi, M. C.; Benito-Pena, E.; Vicente, B. S.; Navarro-Villoslada, F.; Leon, M. E. d.; Orellana, O.; Aparcio, S.; Molina, L.; Kempe, M.; Fiaccabrino, G. C., *Transducers* **2003**, 2, 975-978.
196. Chen, Y.; Wang, Z.; Yan, M.; Prahl, S. A., *Luminescence* **2006**, 21, 7-14.
197. Chen, Y.; Brazier, J. J.; Yan, M.; Prahl, S. A., *Sensors Actuators B* **2004**, 102, 107-116.
198. Mahony, J. O.; Nolan, K.; Smyth, M. R.; Mizaikoff, B., *Anal. Chim. Acta* **2005**, 534, 31-39.
199. Masque, N.; Marce, R. M.; Borrull, F., *Trends Anal. Chem.* **2001**, 20, 477-486.
200. Haupt, K., *Anal. Chem.* **2003**, 75, 376A-383A.

201. Turkevich, J.; Stevenson, P. C.; Hiller, J., *Discuss. Faraday Soc.* **1951**, 11, 55-75.
202. Wang, W.; Chena, Q.; Jiang, C.; Yang, D.; Liu, X.; Xu, S., *Colloids Surfaces A: Physicochem. Eng. Aspects* **2007**, 301, 73–79.
203. Hehre, W. J.; Radon, L.; Schleyer, P. R.; Pople, J. A., *Wiley, New York* **1985**.
204. Atkins, P. W.; Friedman, R. S., *Oxford University Press, Oxford* **2004**.

Chapter 10
Stimuli-Responsive Fine Particles

Approaching Smart Drug Delivery Systems

Sergiy Minko

Abstract New generations of advanced intelligent materials and technologies will use fine particles that are hierarchically organized, multifunctional, and well-defined by size and shape, capable of programmed and controlled responses to changes in their environment or to external signals. The major long-term goal of the current research programs in the area of fine particles for biological applications is the development of novel hybrid organic–inorganic particles as active complex functional materials for applications such as smart drug delivery capsules, miniaturized biosensors for *in vivo* diagnostics, scaffolds for cell culturing and tissue engineering, microprobes and sensors for monitoring the particle's environment and smart materials for personal care and beauty.

Keywords Stimuli-responsive fine particles • Capsules • Nanogels • Core-shell particles

10.1   Introduction

New generations of advanced intelligent materials and technologies will use fine particles that are hierarchically organized, multifunctional, and well-defined by size and shape, capable of programmed and controlled responses to changes in their environment or to external signals. The major long-term goal of the current research programs in the area of fine particles for biological applications is the development of novel hybrid organic–inorganic particles as active complex functional materials for applications such as smart drug delivery capsules, miniaturized biosensors for *in vivo* diagnostics, scaffolds for cell culturing and tissue engineering, microprobes and sensors for monitoring the particle's environment and smart materials for personal care and beauty.

The synthesis of such particles is designed to address architecture of multi-level structured (hierarchically organized) colloidal particles (**MSP**) that possess a unique combination of physical and chemical properties. The engineered structure of the MSP will provide synergetic interactions across different levels of structural

S. Minko (✉)
Department of Chemistry and Biomolecular Science, Clarkson University,
8 Clarkson Avenue, SC, Box 5810, Potsdam, NY 13699, USA
e-mail: sminko@clarkson.edu

E. Matijević (ed.), *Fine Particles in Medicine and Pharmacy,*
DOI 10.1007/978-1-4614-0379-1_10, © Springer Science+Business Media, LLC 2012

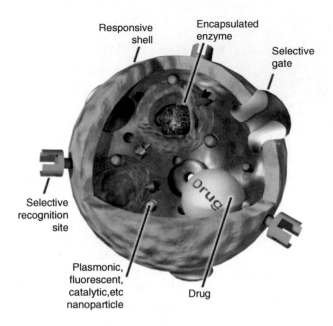

Fig. 10.1 Schematics of a complex multilevel particle designed as hierarchically organized multifunctional structures composed of functional polymers, compartments and particles [3]. For example, if a porous particle is decorated with a pH-responsive polymer, changes in the environment will affect the permeability (gating properties) of the polymer shell [4], and thus, external stimuli can be used to tune uptake or release of small molecules through the pores [5]. This hollow structure accommodates functional molecules, e.g. enzymes [6], and thus the permeability of the polymer shell can be coupled with an enzymatic process occurring in the capsule interior. The intraparticle processes could be monitored using optical effects, e.g. plasmonic coupling, etc. [7, 8]. (Reprinted with permission from Ref. [3])

organization to yield *active hybrid nanostructured* particles. The optimization of the shape and dimensions of the fine particles, and their structural organization, chemical composition, and functionalization with synthetic molecules and biomolecules will enable the development of a diverse collection of novel colloidal structures. These new complex colloidal structures (Fig. 10.1) will be dynamically changing materials that undergo reconfiguration and switching (similar to active miniaturized devices). Such intelligent colloids will be activated, manipulated and reconfigured by physical or chemical stimuli (signals) [1, 2].

The designed particles will resemble but not exactly mimic synthetic (artificial) cells in several conceptual principles: compartmentalization; combination of compartments with different functions; controlled transport of chemicals between compartments through permselective walls between the compartments; selective interactions between encapsulated molecules and structures; and exchange of chemicals and energy with surrounding environment [3, 9, 10]. The multilevel structured particles, in contrast to living cells, will not be able to reproduce themselves. However, they will behave as robust and multifunctional units capable of responsive and programmable behavior for various applications.

10.2   Synthesis of Hierarchically-Organized Fine Particles

10.2.1   Engineering of Compartmentalized Structure of Particles

Synthesis of hierarchically-organized colloidal structures generally is based on fine particles with a core–shell morphology where the core, or some components of the core, can be selectively dissolved and extracted from the particles to form hollow containers or capsules. Hollow containers obtained in this way are loaded by smaller-in-size fine particles and various molecules using two different approaches. First, the ingredients can be embedded in the core's material in the step of the fabrication of the sacrificial core. After the completion of the assembly of the shell and the dissolution of the core's material, the entrapped ingredients are liberated in the interior of the capsules. Second, the ingredients can be encapsulated by diffusion through gates (pores) in the shell. Some functional components can be built in the shell so that a variety of structures schematically shown in Fig. 10.1 can be synthesized with the multistep method.

One of exciting examples of such synthesis has been published by Kreft et al. [11]. The synthesis consists of six steps. Initially, biomolecules A (e.g., enzymes), fine particles, and magnetic particles were precipitated with $CaCO_3$. The obtained composite particles were coated with a polyelectrolyte (**PE**) shell using the layer-by-layer consecutive deposition of two oppositely charged polyelectrolytes (**LbL** method). Then, the PE shell-coated particles were used as colloidal templates for the precipitation of biomolecules B with other fine particles and $CaCO_3$. Thus, the ball-in-ball structures were synthesized. A separation in a magnetic field was used to extract the ball-in-ball particles from side products. The ball-in-ball particles were then coated again by a PE multilayer shell using the LbL method. Finally, $CaCO_3$ was selectively extracted with an ethylenediaminetetraacetic acid solution to liberate all inclusions in the inner compartment and the outer shell. The resulting structure is a capsule-within-capsule particle loaded with two different kinds of biomolecules and colloidal particles (Fig. 10.2). Two different types of biomolecules located in the different compartments (shells) are separated by a semipermeable PE membrane. Two different enzymes, glucose oxidase (GOx) and peroxidase (POx), were selected as biomolecules A and B, respectively, to demonstrate a potential application of these hierarchically organized MPS. Oxidation of glucose by GOx led to the formation of H_2O_2, which, in the presence of an electron donor, Amplex Red, was converted into the highly fluorescent resorufin by POx. After successive addition of glucose and Amplex Red to the dispersion of the particles, resorufin fluorescence occurred within a few seconds in the inner compartment and then extended into the outer compartment. This structure is an excellent platform for the development of hierarchically organized MSP where the compartments can be used to mediate a complex combination of enzymatic reactions.

Applications of polymersomes for the fabrication of the "smart" MSP have recently attracted great interest because a thin polymer wall of these vesicles resem-

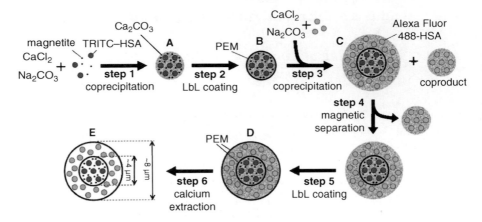

Fig. 10.2 General route for the synthesis of shell-in-shell microcapsules. *A* initial core, *B*, core–shell particle, *C* ball-in-ball particle (type I), *D* ball-in-ball particle (type II), *E* shell-in-shell microcapsule. (Reprinted with permission from Ref. [11] Wiley. 10.1002/anie.200701173)

ble a lipid membrane and can be used for the reconstruction of proteins and other biomolecules. Such MSP demonstrate a complex response, e.g., selectively gate the transport of ions and molecules across the shell, mimicking some functional behavior of living cells [12–14]. For example, multiprotein polymersomes and proteopolymersomes were used for reconstructing the cellular process of adenosine diphosphate (ATP) production using coupled reactions between a light-stimulus-driven transmembrane proton pump, bacteriorhodopsin (BR), and a motor protein, F_0F_1-ATP synthase (Fig. 10.3) [15]. The polymersomes were prepared from an amphiphilic triblock copolymer, poly(2-ethyl-2-oxazoline-*block*-dimethylsiloxane-*block*-2-ethyl-2-oxazoline (PEtOz-b-PDMS-b-PEtOz). The polymersomes had a wall thickness of about 4 nm, which was similar to a typical lipid bilayer thickness. Both BR and ATP synthases were reconstructed simultaneously into the polymersomes. BR is a light-driven proton pump that creates a proton gradient across the cell membrane via a light-induced BR photocycle. When coupled with ATP synthase, this proton gradient is used to synthesize ATP from adenosine diphosphate (ADP) and inorganic phosphate. ATP synthase maintained its rotating activity in the synthetic polymersomes and produced ATP utilizing the photoinduced proton gradient generated from BR's activity. This work demonstrated a successful biosynthesis through the coupled reactions between reconstituted transmembrane proteins in a single proteopolymersome and a molecular motor functionality in a synthetic polymer capsule. It is expected that this ATP-producing proteopolymersome will enable not only propelling biomotors, but also the development of ATP-driven nanoscale devices.

Recently, another very promising approach to the fabrication of hierarchically organized responsive particles has been attracted interest of researchers. This approach is based on templating Pickering emulsions [16–19]. For example, an o/w emulsion was stabilized with microgel particles from a poly(*N*-isopropylacryl-

Fig. 10.3 Schematic representation of proteopolymersomes reconstituted with both bacteriorhodopsin and F0F1-ATP synthase. (Reprinted with permission from [15] Copyright 2005 American Chemical Society)

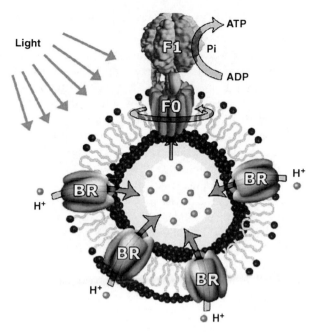

amide-co-(acrylic acid)) (poly(NIPAAm-co-AA)) random copolymer. An addition of a diblock copolymer, poly(butadiene-*block*-*N*-methyl 4-vinyl pyridinium iodide), to the oil phase helped to form a composite membrane that locked the microgel particles at the interface. The extracted material yielded a dispersion of stable hollow capsules responsive to changes in pH and temperature. Hence, this method provides an opportunity to synthesize compartmentalized particles where the shell of the hollow structure is constituted of the responsive microgel particles. The cavity in the particle and the microgel particles that surrounds the cavity could serve as separate compartments and carry different chemicals. Thus, the integrated structure could be used similarly to the capsule-within-capsule structure discussed above.

These several examples demonstrated that quite complex multifunctional properties of fine particles can be approached by engineering of the compartmentalized structure with embedded functional molecules. Building of a core–shell structure is an efficient approach for engineering MSP. Various core–shell structures were synthesized in two steps where the synthesis of particles from metals, oxides, or polymers of different sizes and shapes is followed by their functionalization with stimuli-responsive polymers. A polymeric, metallic, or metal oxide core can be synthesized by one of the known methods of fine particle fabrication. Then, the polymeric shell is formed around the core. A number of methods for the polymeric core–shell structures have been reviewed published by Hoffman-Caris [20] and Caruso [21].

The mostly used methods for the fabrication of functional core–shell structures are briefly discussed below.

10.2.2 Coacervation/Precipitation Method

The polymer coacervation denotes a process of *liquid–liquid* phase separation of a
polymer solution into a polymer-rich phase (coacervate phase) and a polymer-lean
phase (equilibrium phase) caused by changing the solution temperature, the pH in
the case of weak PEs, by adding a nonsolvent for the polymer to the solution or
by the formation of PE complexes between PEs and di- and trivalent counterions
[22]. The coacervation process leads to the formation of microscopic droplets of the
coacervate phase in the stirred liquid phase. The resulting colloidal dispersion is
usually unstable unless it is stabilized by chemical crosslinking or physical gelation
of the polymer in the coacervate phase [23]. An important advantage of the fabrication methods based on the coacervation is that stable micro- and nanoparticles can
be prepared under mild conditions without using organic solvents, surfactants, or
steric stabilizers. For an example, polymers with low critical solution temperature
(LCST) or weak PEs can be precipitated via hydrophobic collapse by gradually
changing the solution temperature or pH, respectively. Addition of a nonionic surfactant terminates the macroscopic phase separation, resulting in the formation of
colloidally stable polymer particles with low polydispersity. For example, temperature-responsive PNIPAAm [24] and pH-responsive poly(*N*-methacryloyl-L-valine)
and poly(*N*-methacryloyl-L-phenylalanine) [25] fine particles were prepared using
this approach. Obviously, the particle formation is a reversible process, suggesting a
potential application of such particles as vehicles for triggered drug delivery.

Complex coacervation is an increasingly popular approach for the fabrication of
particulate drug delivery systems, in particular for the delivery of delicate macromolecules. Stable water-soluble colloidal particles are often formed by oppositely
charged PEs mixed together under certain conditions in a ratio when a nonstoichio-
metric electrostatic complex is formed. The existing studies on complex coacervate
particles stem from the early works of Tsuchida's and Kabanov's groups [26, 27]. In
the vast majority of studies, chitosan, a naturally occurring weak polycation, and its
derivatives are used as one of the components. Besides biocompatibility, biodegrad-
ability, and low toxicity, the popularity of chitosan is explained by its mucoadhesive
properties, making it useful for transmucosal drug delivery as well as its ability
to open tight junctions between epithelial cells, enhancing the transport of macromolecules across epithelia [28]. Due to the high positive charge on its backbone,
chitosan interacts with negatively charged biomacromolecules. Colloids based on
electrostatic chitosan–DNA, chitosan–protein, and chitosan–polysaccharide complexes along with chitosan hydrogels crosslinked with polyion tripolyphosphate
(TPP) are among most studied. These colloidal systems were reviewed by Alonso
and co-workers [28, 29] and Zhang and co-workers [30] Because of the electrostatic
nature of these complexes, they are intrinsically pH- and ionic strength-responsive.
Examples of colloidal nanosystems with pH- and ionic strength-responsive release
properties include complexes formed between oppositely charged polysaccharides,
such as chitosan and carboxymethyl konjac glucomannan, and chitosan and dextran
sulfate [31–34].

Controlled chemical crosslinking of certain polymers may lead to the formation of micro- or nanoparticles by precipitation/coacervation mechanisms. For example, colloidally stable aqueous dispersions of pH-sensitive particles were prepared from chitosan via condensation reaction of amino groups of chitosan with di- and tricarboxylic acids. Depending on the acid used and the crosslinking degree, the particles with the polycation, polyanion, and polyampholyte behavior were produced by Borbely and co-workers [35, 36]. The chitosan particles *crosslink*ed with poly(ethylene glycol) dicarboxylic acid demonstrated pronounced swelling provided by the length and flexibility of the crosslinker [36]. Zhang and co-workers [37] described the fabrication of temperature-responsive hydrogel fine particles from poly(glycidilmethaclitae-co-NIPAAm) copolymer. The particles were formed in organic solvent during the crosslinking reaction of epoxy groups of glycidilmethaclitae monomer units with diamine and then transferred into an aqueous medium. Gold NPs were successfully synthesized inside the hydrogel particles in this study. The authors reported the temperature-induced swelling–deswelling and aggregation–disaggregation of the resulting nanocomposite colloids.

A simple and robust method for fabricating polyethyleneimine (PEI) nanogel particles has been recently developed using the precipitation-crosslinking method. The particles were synthesized by crosslinking PEI with dibromoethane (DBE) in organic solvents (nitromethane or dimethylformamide) [38]. Stable colloidal dispersions were obtained at elevated temperature and in solvents which provide the highest rate of the PEI alkylation reaction, steric and electrostatic mechanisms of stabilization of nanoparticles. The size of the particles swollen in water ranges from 125 to 500 nm and can be regulated by the conditions of the synthesis (solvent and concentrations of PEI and DBE). The particles respond to pH changes with a 10–15% change in diameter. Uptake and, triggered by the changes in pH, release of a anticancer drug, Rose Bengal, by the particles was associated with the electrostatic interaction of the drug with the charged gel. The encapsulating capacity of the particles with respect to the Rose Bengal was determined to be 1.56×10^{-4} M of the drug per 1 mg of the gel.

10.2.3 *Heterogeneous Polymerization*

Polymerization in heterogeneous media (emulsions and suspensions) is one of the many strategies for fabrication of colloidal particles with complex architecture. The technique has been extensively developed and has numerous applications for the synthesis of monodisperse core–shell particles and for control of the particle surface properties. Emulsion, precipitation, and dispersion polymerizations are among the mostly used synthetic approaches for preparing particles [39, 40]. The principles and possible applications of these techniques were recently reviewed by Matyjaszewski and co-workers [41]. The techniques also allow the synthesis of hybrid particles by inclusion of inorganic materials such as silica [42], alumina [43], zeolite [44], iron oxide [45, 46], and noble metal fine particles [47–50]. As recently reviewed [39, 40, 51], various types of temperature, pH, ionic strength, and light responsive particles can be designed and produced by heterogeneous polymerization processes.

For an example, PNIPAAm has long been proposed as a thermoresponsive polymer and can be produced in the form of microgel and core–shell particles using heterogeneous polymerizations. For example, thermoresponsive core–shell particles composed of a polybythilmethacrylate (PBMA) core and PNIPAAm shell were synthesized by two-stage free radical polymerization [52]. A hydrophobic, nonresponsive core of PBMA was used as a seed in the polymerization of the thermoresponsive hydrogel. Below the PNIPAAm's LCST, the shell was hydrophilic and therefore highly swollen with water. Upon raising the temperature above the LCST ($\sim 32°C$), entropically favorable hydrophobic aggregation of the polymer chains occurred, causing the hydrogel to deswell.

Temperature- and pH-responsive core–shell microgels particles composed of crosslinked PNIPAAm and poly(NIPAAm-co-AA) were synthesized via precipitation polymerization and then used as nuclei for subsequent polymerization of poly(NIPAAm-co-AA) and PNIPAAm, respectively [53]. The poly(NIPAAm-co-AA) core particles displayed a strong dependence of particle size on both temperature and pH.

A dispersion polymerization approach was applied for the synthesis of hybrid poly[NIPAAM-co-(2-hydroxyethyl methacrylate)-co-(methacrylic acid)] (PNIPAAm-co-PHEMA-co-PMAA) microgels containing magnetic Fe_3O_4 colloidal particles [54]. According to the reported procedure, first microgels particles were synthesized, and then the magnetic particles were prepared inside the microgels. The particles were found to be suitable as emulsifiers for o/w emulsions with both polar and nonpolar oils because of the particle surface activity. Both thermo and magnetic responsive particles having a magnetic core with crosslinked PNIPAAm shell were synthesized by a co-precipitation method using the ATRP method [55].

Armes' and Binks' groups have made significant advances in the synthesis of core–shell responsive latex particles [56–58]. For example, a seeded aqueous emulsion copolymerization of N,N-(dimethylamino)ethyl methacrylate (DMA) and ethylene glycol dimethacrylate (EGDMA) was conducted in the presence of polystyrene (PS) latex particles to produce a stable dispersion of core–shell latex particles, in which the shell consisted of a crosslinked poly(DMA-stat-EGDMA) overlayer [56]. Using the resulting latex particles as a pH-responsive Pickering-type emulsifier, polydisperse n-dodecane-in-water emulsions were prepared at pH 8 that could be partially broken on lowering the solution pH to 3. Armes' group also reported crosslinked, sterically stabilized responsive latexes of approximately 250 nm in diameter synthesized by emulsion polymerization of 2-(diethylamino)ethyl methacrylate (DEA) using a bifunctional oligo(propylene oxide)-based diacrylate crosslinker and a polyethylene oxide (PEO)-based macromonomer as the stabilizer at pH 9 (see [57]). These particles exhibited reversible swelling properties in water by adjusting the solution pH. At low pH, they existed as swollen microgels due to protonated amine units. Deswelling occurred above pH 7, leading to the formation of the compact latex particles.

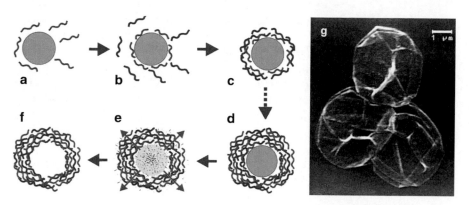

Fig. 10.4 Schematic illustration of the PE deposition process and subsequent core decomposition. The initial steps (**a–d**) involve stepwise film formation by repeated exposure of the colloids to PEs of alternating charge. The excess PE is removed by cycles of centrifugation and washing before the next layer is deposited. After the desired number of PE layers is deposited, the coated particles are exposed to 0.1 M HCl (**e**). The core immediately decomposes, as evidenced by the fact that the initially turbid solution becomes essentially transparent within a few seconds. Three additional washings with 0.1 M HCl ensure removal of the dissolved MF oligomers. Finally, a suspension of free PE hollow shells is obtained (**f**). SEM image of nine-layer [(PSS/PAH)₄/PSS] PE shells after solubilization of the MF core (**g**). The outer layer is PSS. The strong electrostatic attraction of the shells to the positively charged PEI coated glass surface leads to some spreading of the PE shell. The drying process induces a number of folds and creases. Drying, together with the topological constraints of the closed surface, results in a completely folded upper hemisphere [71]: Donath E, Sukhorukov GB, Caruso F, Davis SA, Mohwald H. Novel hollow polymer shells by colloid-templated assembly of polyelectrolytes. Angewandte Chemie-International Edition 1998;37:2202–2205. Copyright Wiley-VCH Verlag GmbH & Co. KGaA. (Reproduced with permission from Wiley)

10.2.4   LbL Method for the Fabrication of Polymeric Shell

The method of LbL deposition of oppositely charged species (organic molecules, proteins, PEs, particles, and others) was pioneered by Iler [59] and later developed by Decher [60], Fendler [61, 62], Lvov et al. [63], Rubner and co-workers [64], Laurent and Schlenoff [65], and many others [66–68]. It was successfully applied by Kunitake and co-workers [69] and Möhwald and co-workers for the fabrication of polymer shells around particulate cores [70] and transformation the core–shell particles into hollow spheres with LbL gel walls [71]. For example, they used a stepwise adsorption of oppositely charged PEs onto melanin resin templates which were later dissolved (Fig. 10.4). These capsules have mechanically unstable walls. The stability of the capsule has been improved by the deposition of inorganic particles on the inner surface of the capsule walls [72].

The LbL approach was explored by many groups for the fabrication of core–shell structures with responsive polymer shells. For example, a mesoporous silica core was used as a template to form an LbL gel capsule around the core. The mesoporous core retains in the capsule and can be used as a container for drugs and other molecules [73]. Low capacity of the mesoporous silica is the major drawback of this method. The solution to this problem has been proposed by Shi and co-workers [74]. The authors used hollow mesoporous silica particles loaded with ibuprofen and coated by LbL-gel shell.

The LbL method was also extended to the systems interacting due to the formation of hydrogen bonds. For example, polyanilin [75], PAA and PMA were combined with polyvinylpyrrolidone (PVPON) and PEO [76–78] for the fabrication of LbL films. A click chemistry approach for the LbL assembly of ultrathin pH-responsive nanocapsules has been developed by Caruso and co-workers [79]. The LbL approach was used to fabricate pH-responsive capsules from bovine serum albumin through glutaraldehyde-mediated covalent LbL assembly [80].

10.2.5 Precipitation on Colloidal Templates

If solid or polymer particles are present in a solution in which the coacervation/ precipitation process occurs, then the coacervate/precipitate phase may deposit on the particle surface producing core–shell particles. This phenomenon is referred as to *surface-controlled precipitation* and *heterocoagulation* of polymers on colloidal particles. As compared with an LbL assembly technique, the surface controlled precipitation method allows preparing relatively thick, microscopically homogeneous coatings in a single or a few steps from either charged or noncharged polymers [81]. However, this method has not received much attention because of the problems associated with the concurrent and undesired process of polymer flocculation (also referred as to *homocoagulation*) and the low colloidal stability of polymer-precipitate-coated particles [82]. Theoretical models of surface-controlled precipitation can be found in the papers of Möhwald and co-workers [82, 83]. So far, the surface controlled precipitation onto sacrificial colloidal templates (e.g., calcium carbonate particles) was used in combination with the LbL assembly as an approach for encapsulation of various delicate macromolecules (such as proteins and DNA) and poorly water-soluble drugs inside LbL capsules [83, 84].

10.2.6 Grafting to the Surface of Particles

The grafting techniques imply chemical attachment of functional polymers to the surface of colloidal particles such as latexes, silica, magnetite particles, and noble metal fine particles. Therefore, a carrier particle forms an organic or inorganic core, and it is surrounded by a polymeric shell. The responsive properties of such core–shell structures originate from the grafted polymer chains that are sensitive to changes in solvent quality, pH, ionic strength, temperature, and so forth.

"Grafting-to" and "grafting-from" are two major approaches that have been extensively investigated for covalent binding polymers to fine particles. The "grafting-to" method involves a chemical reaction of functionalized polymers with the particle surface modified by complementary functional groups. The grafting-from method involves polymerization using an initiator functionalized surface. The major advantages of the "grafting-to" method are in its simplicity, robustness, and the use of well-characterized polymers. The procedure starts from the functionalization of the particle surface using an appropriate anchoring agent. This step is followed by grafting a polymer from a solution or melt. The "grafting-to" method results in polymer brushes of a moderate grafting density because the grafting is limited by diffusion kinetics of end functionalized polymer chains through the crowded brush-like layer. The "grafting-from" method can produce tethered layers of high grafting densities because the active centers (radicals, ions, and ion pairs) of the growing chains are easily accessible for monomer molecules in the swollen brush in the course of polymerization [85, 86]. A number of polymerization techniques have been used for the fabrication of polymer brushes on the surface of fine particles, including free radical polymerization, ATRP, living anionic surface initiated polymerization, ring opening metathesis polymerization, living cationic polymerization, and hyperbranching [85, 87–89].

A grafted polymer shell may consist either of chains of a single polymer (homopolymer or random copolymer), of two or more different homopolymers (mixed brushes), or of block-copolymers.

10.2.7   Self-Assembly of Amphiphilic Block-Copolymers

Block-copolymers form various types of self-assembled structures from micelles to continuous bilayers, depending on solvent selectivity of the different blocks. Blocks with higher compatibility with the solvent are stronger swollen and exposed to the exterior of the structures, whereas less compatible blocks are densely packed in the interior of the aggregates. The packing parameter, p, is used to differentiate the type of aggregates formed by amphiphiles [90, 91], $p = v/al_c$ where v is the volume occupied by the densely packed copolymer block (e.g., hydrophobic for aqueous media); l_c is the statistical critical length normal to interface, which correlates with the counter length of the polymer chain; and a is an effective cross-sectional area per the amphiphilic block-copolymer molecule at the interface [12, 92].

Amphiphiles with p below 1/3 form spherical micelles. When p is between 1/3 and 1/2, amphiphiles form micelles, whose shape varies gradually from spherical to cylindrical (worm-like). For p values between 1/2 and 1, a gradual variation from a cylindrical micelle through vesicles (also called polymersomes [93] in the case of block copolymer vesicles) to a planar bilayer at $p = 1$ is expected. Finally, at $p > 1$, more complex systems of the inverted aggregates can form [90, 91]. For block copolymers, all mentioned parameters are much more sensitive to the properties of the surrounding medium than for the low molecular weight amphiphiles. Changes in the surrounding environment can cause subtle changes in each of the three characteristic parameters determining the packing parameter. The latter may lead to the gradual changes in aggregate dimensions or to the reorganization of the aggregate architecture and changes in its stability. Such a flexibility of block copolymers behavior reveals the great potential for gradual or sharp responses of the structures to external stimuli.

The self-assembled structures formed by block-copolymers can be also correlated with the volume fraction f ($0 < f < 1$) of one of the blocks in the copolymer molecule (see e.g. [92, 94]). A correlation was found, both experimentally and theoretically, between the fraction of the hydrophobic block in different block copolymers and their packing parameter [95, 96] expressed as $f \approx e^{-p/\beta}$ ($\beta = 0.66$) [92].

The discussion of the theories on self-assembled block copolymer architectures, experiments, and applications can be found in a series of comprehensive reviews [97–99]. The analysis of recent achievements in the field can be also found in the following references [12, 92, 100–105]. A significant number of publications reported systems which undergo changes upon the application of external stimuli. Responses of block-copolymer aggregates to pH, ionic strength, thermal, and redox stimuli are among those most commonly considered.

Self-assembled structures from block copolymers could be chemically modified and transformed into MSP of different architectures such as nanogels, capsules, and composite particles.

Block-copolymers assemblies have been also used in the fabrication of core–shell MSP using the LbL method discussed earlier in this chapter. For example, multilayers on the surface of anionic polystyrene latex particles were prepared by sequential deposition of poly(sodium styrene sulfonate) (PSS) and micelles of the partially quaternized PDMA-b-PDEA copolymer loaded with hydrophobic dye. The resulting colloids retained dye incorporated in the micelles; the micelles maintained their integrity at pH 9.3 and pH 4 [106].

Checot et al. reported on the synthesis of capsules from polymersomes based on PB-b-poly(L-glutamic acid) (PLGA) [107]. The polymersomes changed their hydrodynamic radii from 100 to 150 nm in response to changes in the pH and ionic strength of aqueous solutions. The stimuli-responsive capsules were prepared by crosslinking the polymersomes via double bonds of the PB blocks.

10.3   Mechanisms of Response

The list of practically important external stimuli used to regulate the properties of MSP includes temperature, light, mechanical force, gradient of magnetic field, electrical potential, and chemical composition of the biological environment [108–111]. The resulting demanded changes in the particles are alternations (reversible or irreversible) in their mechanical, optical, magnetic, and surface properties, and in their permeability for molecules and charge carriers. A properly engineered combination of sensitivity to external stimuli with resulting changes in the particles' properties is critically important for drug delivery capsules, capsules for diagnostics and particles-biosensors. It is important to emphasize that MSPs are seen as complex systems that could recognize signals and respond in an intelligent way. Thus, mechanisms of *high selectivity* of signal recognition, *amplification* of the signal, and *transduction* of the signal into changes in the particles' properties using energy either stored in the material or received from the external environment of the materials should be secured and combined in the engineered materials. The scenario of the MSP' response must be *programmed* and encoded in the structure of the particles. Such a complex response can be tailored by a combination of several basic mechanisms that were developed for stimuli-responsive polymeric and colloidal systems.

10.3.1   Selectivity

Mechanisms of *high selectivity* rely on molecular recognition phenomena and have been mimicked from natural systems by the conjugation of responsive polymers with biological molecules such as DNA [112, 113], enzymes [114, 115], antibodies [116], and various other proteins [117–120]. These biomolecules have a high affinity to specific biological molecules and form molecular complexes with those molecules. The complex formation typically is highly selective and characterized by large values of the equilibrium constant of the interaction.

Biorecognition events that involve DNA and proteins may cause a direct response of the particles (e.g., swelling, formation, or cleavage of junction points) or an indirect response (e.g., a shift in the critical temperature or pH of the phase transition in the responsive polymeric ingredient). Applications of enzymes refer to two major types of responsive polymeric systems: (1) polymers with enzymes (adsorbed, covalently attached, or physically entrapped) [121] and (2) polymers with fragments that are substrates for enzymes [122]. In the first type of materials, a substrate diffusing into the polymer from the surrounding aqueous medium can be biocatalytically converted into products of the enzymatic reaction. These products then interact with the polymer and cause chemical changes in the material. For example, catalytic oxidation of glucose by glucose oxidase yields gluconic acid and, consequently, results in a decrease of pH in the local environment. These changes are then used to trigger response in thin polymer films [5]. The second type of polymeric materials is typically designed in such a way that the enzyme is used as an "external stimulus" that cleaves the chemical bonds in the material. For example, this approach was used to prepare erasable polymeric structures [123].

The high costs and low stability of many biological molecules are the major drawbacks of these materials. Alternative synthetic approaches explore artificial polypeptides [124–126] and aptamers [127]. Future developments of more stable classes of molecules with recognition motifs, for example, such as polypeptoids [128], could be a promising direction for the development of intelligent MSP.

A less selective synthetic approach is based on molecular imprinted polymers [129–132]. In this case, the molecules of interest (templates) are used for the fabrication of footprints of the signaling molecules in the polymer network synthesized around the template. The cavities with the specific spatial distribution of functional groups left after removal of the templates are used as recognition sites. Binding the signaling molecules results in changes in properties of polymeric structures [133].

There are many examples of specific complex formation between synthetic polymer materials and ligands, e.g., glucose-responsive polymers with phenylboronic side groups [134]. The number of examples of highly specific recognition will be increasing because of the high demand and expectation for the development of new sensors and drug delivery devices.

In many applications for sensors and drug delivery, the concentration of biomolecules (biomarkers) or chemicals (e.g., toxins) that can trigger the response is in pico- or nanomolar range. This small concentration in many cases may be insufficient to cause changes in properties in MSP directly. Thus, low concentrations of important biomarkers may cause limitations for the range of very important biomedical applications of the MSP. This problem could be solved by the development of particles with *built-in amplification mechanisms* where a small amount of the biomarker could initiate a cascade of events resulting in the response of the particles. Practically, the demanded amplification effects should be in the range of up to several orders of magnitude. A source of energy for this cascade can be in the form of either chemical reagents stored in MSP or external sources (e.g., electrochemical processes or light-induced reactions).

10.3.2 Transduction Mechanisms

Transduction mechanisms involve conformational changes in polymer chains, or a combination of both the chemical and conformational changes, and phase transition triggered by external stimuli. These mechanisms result in substantial alternations of polymer shell or compartment's membrane properties [135–137]. The major driving force of large changes in polymer conformation is a shift in a tiny balance of the polymer chain interaction with the surrounding environment in *gels* or *melts*. This misbalance can be brought about by changes in solvent quality (changes in the temperature or chemical composition of the solvent) or by chemical changes in polymer molecules typically as a result of polymer-analogous reactions of side groups, or less frequently, as a result of the cleavage or formation of chemical bonds in the backbone of the polymer chain. The chemical changes are introduced by changes in the chemical composition of the material environment (e.g., changes in pH, ionic strength, concentration of reactive biomolecules or biomarkers) or by electrochemical or photochemical reactions. These transduction mechanisms have been intensively developed and studied in recent decades by many research groups [109–111, 137, 138] for bulk materials, thin films, molecular assemblies, and, recently, on the level of single polymer molecules [139–142].

The largest fraction of the reports on the study of swollen or melted responsive thin polymer films is based on the transduction mechanisms triggered by changes in temperature, pH, or intensity of light. The most known examples of temperature responsive systems are based on temperature-induced phase transition (sol-gel transitions) of PNIPAAM, PEO, polypropylene oxide (PPO), poly(lactic acid) homo- and copolymers, proteins, and polysaccharides in aqueous environment [143–145]. Polypeptides and polysaccharides, typically, undergo sol–gel transition upon decreasing temperature below the upper critical solution temperature (UCST) due to the coil–helix transition followed by aggregation or crystallization of helixes. These aggregated helixes form junction points in the gel. PNIPAAM undergoes a sol–gel transition upon increasing temperature above the LCST ($32°$ C for PNIPAAM) due to the dominating effect of hydrophobic interactions at elevated temperature. Examples of pH-responsive polymers refer to weak polyelectrolytes with acidic or basic functional groups (carboxylic, phosphoric, or amino functional groups). A pH-dependent ionization of weak polyelectrolytes is manifested in the swelling and shrinking of thin films as a result of an increase or decrease in the osmotic pressure inside the film with entrapped counterions [146]. Implementation of photoactive groups (such as azobenzene, spirobenzopyran, triphenylmethane, or cinnamonyl) that can undergo reversible structural changes was used for the fabrication of light responsive polymers. These functional fragments changed size or shape or formed ionic or zwitterionic spices upon irradiation [147].

Chemical changes my cause small conformational changes in polymers in the glassy state. However, these changes may result in substantial changes in the transport of charge carriers or of absorption and excitation spectra. Electroconductive polymers are typical examples of such materials where changes in the red-ox state may strongly affect the electrical conductivity and color of the thin polymer films [148].

In some cases, the demanded response must be triggered by a combination of more than one external stimulus. Depending on this combination, the scenario of the response varies. For example, release of a drug is expected if signals are received simultaneously from two biomarkers, while the presence of only one of these two biomarkers may not be sufficient to validate release. An intelligent drug delivery film could process several signals and could release the drug based on analysis of the multiple signals. Such prototype systems have been recently demonstrated on the level of proof of concept [1, 5, 149–151]. Future intelligent drug delivery systems and sensors will likely broadly use these principles of logic operations on signals received from a complex biological environment.

10.3.3   *Kinetics*

The *kinetics* of the changes in polymer materials due to external stimuli is very important for many applications. A rapid response is possible for mobile polymer chains in gels or melts. Swelling and deswelling kinetics vary from milliseconds to minutes and hours depending on the system (see, for example, [152]). In general, the kinetics of the response depend on chain dynamics, diffusivity of the signaling molecules, and kinetics of chemical changes in the responsive polymer [153]. If the response is limited by diffusion of the signaling molecules, thin films (a shell of the fine particles or walls of the capsules) have an advantage because of a much lower diffusion pass for signaling molecules compared to bulk materials. The swelling rate is inversely proportional to the square of the thinnest dimension of the gel sample [154]. However, diffusivity of molecules is slower for thin films (< 100 nm) compared to the bulk materials due to the confinement effect. For example, for the 3-nm-thick poly(4-ammonium styrenesulfonic acid) film, there is a five-orders-of-magnitude decrease in the diffusion coefficient for water molecules [155]. Chain dynamics may play an important role in systems where the local chain environment affects ionization processes. For example, a strong swelling-deswelling hysteresis loop due to the hydrophobic interactions in the poly(allylamine hydrochloride)/poly(sodium 4-styrenesulfonate) multilayer thin film was attributed to the slow kinetics of the polymer segments in hydrophobic domains formed by the polyelectrolyte backbone [156].

10.3.4   Basic Design of Polymer Structures in MSP

Polymers, that form either polymeric shell, polymeric core or a membrane to separate compartments in MSP, are typically grafted to the particle surface (capsule's walls) or envelope particles with physically or chemically cross-linked polymeric network. Thus, two major structures are used to design MSP: tethered polymer chains and thin films of polymeric networks.

Polymer chains grafted by end-functional groups form polymer brushes if the surface concentration of the grafted chains is high so that the chains overlap and experience repulsion due to the excluded volume effect in the given environment [157] In general, the thin films of polymer networks [158] and polymer brushes [137] respond to changes in their environment or to changes in the chemical structure of their side groups in a similar way. Grafted polymer chains or strands in polymer networks undergo transitions from compact coil to extended coil conformation if the solvent quality changes from poor to good, or if some molecules interact with the side groups of the polymers. This transition results in swelling and expansion of the thin film. In the case of ionizable side groups, changes in the ionization (due to changes in salt concentration or, for weak acidic or basic functional groups, due to changes in pH) result in alternation of the counter ion concentration in the thin polymer film. The counterions are entrapped inside the film, causing its strong osmotic swelling. An increase in the grafting density of chains for brushes and the density of cross-links for networks both result in a decreased swelling of the films. A combination of two or more different polymers in the same brush (mixed polymer brushes), in a block-copolymer brush [159], or in the same network (mixed network semi-[160] and interpenetrating networks [161, 162]), alternatively a patterned network [163], brings a range of additional opportunities for the regulation of the responsive properties of thin films due to rich phase behavior. In addition, there are examples of mixed brushes from polypetides [164].

For example, a mixed polymer brush demonstrates lateral phase segregation where two unlike polymers form separated domains [165]. At the same time, in a selective solvent the mixed brush can be stratified where the selected polymer is in the upmost layer while the second polymer segregates to the grafting surface [166, 167]. This kind of phase behavior extends the diversity and range of changes of different properties in the stimuli-responsive thin films.

A special kind of thin network films is a multilayer system (typically comprising tens of layers) from polyelectrolyte complexes prepared by the LbL method [168–170] or, alternatively, polymer pairs showing non-electrostatic interactions (hydrogen bonding) [171, 172]. Selecting the appropriate polymeric constituents of the multilayer and adjusting the assembly conditions can render the thin films sensitive to pH and salt concentration changes.

Stimuli-responsive thin films can be composed of complex macromolecules, e.g., multiarm block-copolymers [173], or stimuli-responsive assemblies, such as micelles or microgel particles [174, 175].

The stimuli-triggered changes of polymer thin films built into the structure of MSP will be used to regulate several important phenomena: *transport* of mass and charges across the film, in and out of compartments and the MSP, *sensing* and reporting changes in the particle's environment, *actuation* (mechanical work), and *interactions* with the environment (e.g., *wetting*, *adhesion* between different materials and tissues). An opportunity to regulate the listed properties with external stimuli is being explored for the development of various intelligent biomaterials and devices. In many cases, we are still in the stage of "*proof of concept*," since a lack of appropriate materials and level of technology are limiting factors for commercialization. Much yet should be done to approach practical solutions.

10.4   Outlook and Future Directions

There are several sound directions in the recent development of multifunctional fine particles for biomedical research and applications. The first direction is the development of MSP that respond by changing stability of colloidal dispersions for applications in a very broad area of biosensors. The second important direction is stimuli-responsive capsules that can release the cargo upon external stimuli. These capsules are interesting for delivery of drugs and contrasting agents, and for biocomposite materials (release of chemicals for various self-healing effects). Many interesting examples were demonstrated on the level of "proof of concept". The most challenging task in many cases is to engineer systems capable to work with demanded stimuli. That is not a simple task for many biomedical applications where signaling biomolecules are present in very small concentrations and a range of changes of many properties is limited by physiological conditions. An additional and well-known challenge is related to the size of capsules. Majority of publications reported microcapsules with the diameter greater than 200 nm. Many medical applications could benefit from much smaller capsules less than 50 nm in diameter. Fabrication of capsules with a narrow pore size distribution and tunable sizes could substantially improve the mass transport control. Recent developments in the stimuli-responsive porous thin gel films [176–179] could be adapted for the design of the capsule shells.

Finally, MSP could respond to weak signals, to multiple signals, and could demonstrate a multiple response. They can store energy, perform logical operations with multiple signals, absorb and consume drugs/chemicals, and synthesize and release drugs/chemicals. In other words, they could operate as an autonomous intelligent miniaturized device. The development of such MSP can be considered as a part of biomimetics inspired by living cells or logic extension of the bottom up approach in nanotechnology. Anyway, the development of the intelligent MSP faces numerous challenges related to the coupling of many functional building blocks in a single hierarchically structured MSP. These particles could find applications for intelligent drug delivery, removal of toxic substances, diagnostics and monitoring physiological functions.

References

1. Motornov, M.; Zhou, J.; Pita, M.; Gopishetty, V.; Tokarev, I.; Katz, E.; Minko, S., "Chemical transformers" from nanoparticle ensembles operated with logic. *Nano Letters* **2008**, 8, (9), 2993-2997.
2. Hinrichs, K.; Aulich, D.; Ionov, L.; Esser, N.; Eichhorn, K. J.; Motornov, M.; Stamm, M.; Minko, S., Chemical and Structural Changes in a pH-Responsive Mixed Polyelectrolyte Brush Studied by Infrared Ellipsometry. *Langmuir* **2009**, 25, (18), 10987-10991.
3. Motornov, M.; Roiter, Y.; Tokarev, I.; Minko, S., Stimuli-responsive nanoparticles, nanogels and capsules for integrated multifunctional intelligent systems. *Progress in Polymer Science* 35, (1-2), 174-211.
4. Motornov, M.; Sheparovych, R.; Katz, E.; Minko, S., Chemical gating with nanostructured responsive polymer brushes: Mixed brush versus homopolymer brush. *Acs Nano* **2008**, 2, (1), 41-52.
5. Tokarev, I.; Gopishetty, V.; Zhou, J.; Pita, M.; Motornov, M.; Katz, E.; Minko, S., Stimuli-Responsive Hydrogel Membranes Coupled with Biocatalytic Processes. *Acs Applied Materials & Interfaces* **2009**, 1, (3), 532-536.
6. Motornov, M.; Tam, T. K.; Pita, M.; Tokarev, I.; Katz, E.; Minko, S., Switchable selectivity for gating ion transport with mixed polyelectrolyte brushes: approaching 'smart' drug delivery systems. *Nanotechnology* **2009**, 20, (43), 10.
7. Tokarev, I.; Tokareva, I.; Minko, S., Gold-nanoparticle-enhanced plasmonic effects in a responsive polymer gel. *Advanced Materials* **2008**, 20, (14), 2730-2734.
8. Tokarev, I.; Tokareva, I.; Gopishetty, V.; Katz, E.; Minko, S., Specific Biochemical-to-Optical Signal Transduction by Responsive Thin Hydrogel Films Loaded with Noble Metal Nanoparticles. *Advanced Materials* **2010**, 22, (12), 1412-1416.
9. Tokarev, I.; Motornov, M.; Minko, S., Molecular-engineered stimuli-responsive thin polymer film: a platform for the development of integrated multifunctional intelligent materials. *Journal of Materials Chemistry* **2009**, 19, (38), 6932-6948.
10. Tokarev, I.; Minko, S., Stimuli-responsive hydrogel thin films. *Soft Matter* **2009**, 5, (3), 511-524.
11. Kreft, O.; Prevot, M.; Mohwald, H.; Sukhorukov, G. B., Shell-in-shell microcapsules: A novel tool for integrated, spatially confined enzymatic reactions. *Angewandte Chemie-International Edition* **2007**, 46, (29), 5605-5608.

12. Blanazs, A.; Armes, S. P.; Ryan, A. J., Self-Assembled Block Copolymer Aggregates: From Micelles to Vesicles and their Biological Applications. *Macromolecular Rapid Communications* **2009**, 30, (4-5), 267-277.

13. Li, M. H.; Keller, P., Stimuli-responsive polymer vesicles. *Soft Matter* **2009**, 5, (5), 927-937.

14. LoPresti, C.; Lomas, H.; Massignani, M.; Smart, T.; Battaglia, G., Polymersomes: nature inspired nanometer sized compartments. *Journal of Materials Chemistry* **2009**, 19, (22), 3576-3590.

15. Choi, H. J.; Montemagno, C. D., Artificial organelle: ATP synthesis from cellular mimetic polymersomes. *Nano Letters* **2005**, 5, (12), 2538-2542.

16. Lawrence, D. B.; Cai, T.; Hu, Z.; Marquez, M.; Dinsmore, A. D., Temperature-responsive semipermeable capsules composed of colloidal microgel spheres. *Langmuir* **2007**, 23, (2), 395-398.

17. Berger, S.; Zhang, H. P.; Pich, A., Microgel-Based Stimuli-Responsive Capsules. *Advanced Functional Materials* **2009**, 19, (4), 554-559.

18. Liu, H. X.; Wang, C. Y.; Gao, Q. X.; Tong, Z., Preparation of novel microcapsules by emulsion droplet template method. *Progress in Chemistry* **2008**, 20, (7-8), 1044-1049.

19. Kirnt, J. W.; Fernandez-Nieves, A.; Dan, N.; Utada, A. S.; Marquez, M.; Weitz, D. A., Colloidal assembly route for responsive colloidosomes with tunable permeability. *Nano Letters* **2007**, 7, (9), 2876-2880.

20. Hofmancaris, C. H. M., Polymers at the surface of oxide nanoparticles. *New J. Chem.* **1994**, 18, (10), 1087-1096.

21. Caruso, F., Nanoengineering of particle surfaces. *Advanced Materials* **2001**, 13, (1), 11-22.

22. Gander, B.; Blanco-Prierto, M. J.; Thomasin, C.; Wandrey, C.; Hunkeler, D., Coacervation and Phase Separation. In *Encyclopedia of pharmaceutical technology* 3rd ed.; Swarbrick, J., Ed. Informa Healthcare 2006; Vol. 1, pp 600-614.

23. Yin, X. C.; Stover, D. H., Hydrogel microspheres formed by complex coacervation of partially MPEG-grafted poly(styrene-alt-maleic anhydride) with PDADMAC and cross-linking with polyamines. *Macromolecules* **2003**, 36, (23), 8773-8779.

24. Konak, C.; Panek, J.; Hruby, M., Thermoresponsive polymeric nanoparticles stabilized by surfactants. *Colloid Polym. Sci.* **2007**, 285, (13), 1433-1439.

25. Filippov, S.; Hruby, M.; Konak, C.; Mackova, H.; Spirkova, M.; Stepanek, P., Novel pH-responsive nanoparticles. *Langmuir* **2008**, 24, (17), 9295-9301.

26. Tsuchida, E.; Osada, Y.; Sanada, K., Interaction of poly(styrene sulfonate) with polycations carrying charges in the chain backbone. *J. Polym. Sci., Part A: Polym. Chem.* **1972**, 10, (11), 3397-3404.

27. Kabanov, V. A.; Zezin, A. B., Soluble interpolymeric complexes as a new class of synthetic polyelectrolytes. *Pure Appl. Chem.* **1984**, 56, (3), 343-354.

28. Janes, K. A.; Calvo, P.; Alonso, M. J., Polysaccharide colloidal particles as delivery systems for macromolecules. *Adv. Drug Delivery Rev.* **2001**, 47, (1), 83-97.

29. Csaba, N.; Garcia-Fuentes, M.; Alonso, M. J., Nanoparticles for nasal vaccination. *Adv. Drug Delivery Rev.* **2009**, 61, (2), 140-157.

30. Liu, Z. H.; Jiao, Y. P.; Wang, Y. F.; Zhou, C. R.; Zhang, Z. Y., Polysaccharides-based nanoparticles as drug delivery systems. *Adv. Drug Delivery Rev.* **2008**, 60, (15), 1650-1662.

31. Du, J.; Sun, R.; Zhang, L. F.; Zhang, L. F.; Xiong, C. D.; Peng, Y. X., Novel polyelectrolyte carboxymethyl konjac glucomannan-chitosan nanoparticles for drug delivery. I. Physicochemical characterization of the carboxymethyl konjac glucomannan-chitosan nanoparticles. *Biopolymers* **2005**, 78, (1), 1-8.

32. Du, J.; Zhang, S.; Sun, R.; Zhang, L. F.; Xiong, C. D.; Peng, Y. X., Novel polyelectrolyte carboxymethyl konjac glucomannan-chitosan nanoparticles for drug delivery. II. Release of albumin in vitro. *J. Biomed. Mater. Res., Part B* **2005**, 72B, (2), 299-304.

33. Sarmento, B.; Ribeiro, A.; Veiga, F.; Ferreira, D., Development and characterization of new insulin containing polysaccharide nanoparticles. *Colloids Surf., B* **2006**, 53, (2), 193-202.

34. Sarmento, B.; Ribeiro, A.; Veiga, F.; Ferreira, D.; Neufeld, R., Oral bioavailability of insulin contained in polysaccharide nanoparticles. *Biomacromolecules* **2007**, 8, (10), 3054-3060.

35. Bodnar, M.; Hartmann, J. F.; Borbely, J., Preparation and characterization of chitosan-based nanoparticles. *Biomacromolecules* **2005**, 6, (5), 2521-2527.

36. Bodnar, M.; Hartmann, J. F.; Borbely, J., Synthesis and study of cross-linked chitosan-N-poly(ethylene glycol) nanoparticles. *Biomacromolecules* **2006**, 7, (11), 3030-3036.

37. Jiang, X. W.; Xiong, D. A.; An, Y. L.; Zheng, P. W.; Zhang, W. Q.; Shi, L. Q., Thermoresponsive hydrogel of poly(glycidyl methacrylate-co-N-isopropylacrylamide) as a nanoreactor of gold nanoparticles. *J. Polym. Sci., Part A: Polym. Chem.* **2007**, 45, (13), 2812-2819.

38. Lupitskyy, R.; Minko, S., Robust synthesis of nanogel particles by an aggregation-crosslinking method. *Soft Matter* 6, (18), 4396-4402.

39. Pichot, C., Surface-functionalized latexes for biotechnological applications. *Current Opinion in Colloid & Interface Science* **2004**, 9, (3-4), 213-221.

40. Pichot, C.; Elaissari, A.; Duracher, D.; Meunier, F.; Sauzedde, F., Hydrophilic stimuli-responsive particles for biomedical applications. *Macromolecular Symposia* **2001**, 175, 285-297.

41. Oh, J. K.; Drumright, R.; Siegwart, D. J.; Matyjaszewski, K., The development of microgels/nanogels for drug delivery applications. *Progress in Polymer Science* **2008**, 33, (4), 448-477.

42. Lapeyre, V.; Renaudie, N.; Dechezelles, J. F.; Saadaoui, H.; Ravaine, S.; Ravaine, V., Multiresponsive Hybrid Microgels and Hollow Capsules with a Layered Structure. *Langmuir* **2009**, 25, (8), 4659-4667.

43. Ash, B. J.; Siegel, R. W.; Schadler, L. S., Glass-transition temperature behavior of alumina/PMMA nanocomposites. *Journal of Polymer Science Part B-Polymer Physics* **2004**, 42, (23), 4371-4383.

44. Ziesmer, S.; Stock, N., Synthesis of bifunctional core-shell particles with a porous zeolite core and a responsive polymeric shell. *Colloid and Polymer Science* **2008**, 286, (6-7), 831-836.

45. Arias, J. L.; Gallardo, V.; Gomez-Lopera, S. A.; Plaza, R. C.; Delgado, A. V., Synthesis and characterization of poly (ethyl-2-cyanoacrylate) nanoparticles with a magnetic core. *Journal of Controlled Release* **2001**, 77, (3), 309-321.

46. Deng, Y. H.; Yang, W. L.; Wang, C. C.; Fu, S. K., A novel approach for preparation of thermoresponsive polymer magnetic microspheres with core-shell structure. *Advanced Materials* **2003**, 15, (20), 1729-+.

47. Kazemi, A.; Lahann, J., Environmentally Responsive Core/Shell Particles via Electrohydrodynamic Co-jetting of Fully Miscible Polymer Solutions. *Small* **2008**, 4, (10), 1756-1762.

48. Hain, J.; Schrinner, M.; Lu, Y.; Pich, A., Design of Multicomponent Microgels by Selective Deposition of Nanomaterials. *Small* **2008**, 4, (11), 2016-2024.

49. Kim, J. H.; Lee, T. R., Hydrogel-templated growth of large gold nanoparticles: Synthesis of thermally responsive hydrogel-nanoparticle composites. *Langmuir* **2007**, 23, (12), 6504-6509.

50. Feng, X. M.; Huang, H. P.; Ye, Q. Q.; Zhu, J. J.; Hou, W. H., Ag/polypyrrole core-shell nanostructures: Interface polymerization, characterization, and modification by gold nanoparticles. *Journal of Physical Chemistry C* **2007**, 111, (24), 8463-8468.

51. Pichot, C.; Taniguchi, T.; Delair, T.; Elaissari, A., Functionalized thermosensitive latex particles: Useful tools for diagnostics. *Journal of Dispersion Science and Technology* **2003**, 24, (3-4), 423-437.

52. Gan, D. J.; Lyon, L. A., Fluorescence nonradiative energy transfer analysis of crosslinker heterogeneity in core-shell hydrogel nanoparticles. *Analytica Chimica Acta* **2003**, 496, (1-2), 53-63.

53. Jones, C. D.; Lyon, L. A., Synthesis and characterization of multiresponsive core-shell microgels. *Macromolecules* **2000**, 33, (22), 8301-8306.

54. Brugger, B.; Richtering, W., Magnetic, thermosensitive microgels as stimuli-responsive emulsifiers allowing for remote control of separability and stability of oil in water-emulsions. *Advanced Materials* **2007**, 19, (19), 2973-+.

55. Frimpong, R. A.; Hilt, J. Z., Poly(n-isopropylacrylamide)-based hydrogel coatings on magnetite nanoparticles via atom transfer radical polymerization. *Nanotechnology* **2008**, 19, (17).

56. Fujii, S.; Randall, D. P.; Armes, S. P., Synthesis of polystyrene/poly[2-(dimethylamino) ethyl methacrylate-stat-ethylene glycol dimethacrylatel core-shell latex particles by seeded emulsion polymerization and their application as stimulus-responsive particulate emulsifiers for oil-in-water emulsions. *Langmuir* **2004**, 20, (26), 11329-11335.

57. Amalvy, J. I.; Unali, G. F.; Li, Y.; Granger-Bevan, S.; Armes, S. P.; Binks, B. P.; Rodrigues, J. A.; Whitby, C. P., Synthesis of sterically stabilized polystyrene latex particles using cationic block copolymers and macromonomers and their application as stimulus-responsive particulate emulsifiers for oil-in-water emulsions. *Langmuir* **2004**, 20, (11), 4345-4354.

58. Palioura, D.; Armes, S. P.; Anastasiadis, S. H.; Vamvakaki, M., Metal nanocrystals incorporated within pH-responsive microgel particles. *Langmuir* **2007**, 23, (10), 5761-5768.

59. Iler, R. K., Multilayers of colloidal particles. *J. Colloid Interface Sci.* **1966**, 21, (6), 569-594.

60. Decher, G., Fuzzy nanoassemblies: Toward layered polymeric multicomposites. *Science* **1997**, 277, (5330), 1232-1237.

61. Kotov, N. A.; Dekany, I.; Fendler, J. H., Layer-by-layer self-assembly of polyelectrolyte-semiconductor nanoparticle composite films. *J. Phys. Chem.* **1995**, 99, (35), 13065-13069.

62. Fendler, J. H., Self-assembled nanostructured materials. *Chemistry of Materials* **1996**, 8, (8), 1616-1624.

63. Lvov, Y.; Ariga, K.; Ichinose, I.; Kunitake, T., Assembly of Multicomponent Protein Films by Means of Electrostatic Layer-by-Layer Adsorption. *Journal of the American Chemical Society* **1995**, 117, (22), 6117-6123.

64. Cheung, J. H.; Fou, A F.; Rubner, M. F. In *Molecular self-assembly of conducting polymers*, 1994; Elsevier Science Sa Lausanne: 1994; pp 985-989.

65. Laurent, D.; Schlenoff, J. B., Multilayer assemblies of redox polyelectrolytes. *Langmuir* **1997**, 13, (6), 1552-1557.

66. Bertrand, P.; Jonas, A.; Laschewsky, A.; Legras, R., Ultrathin polymer coatings by complexation of polyelectrolytes at interfaces: suitable materials, structure and properties. *Macromol. Rapid Commun.* **2000**, 21, (7), 319-348.

67. Hammond, P. T., Form and function in multilayer assembly: New applications at the nanoscale. *Advanced Materials* **2004**, 16, (15), 1271-1293.

68. Caruso, F.; Spasova, M.; Susha, A.; Giersig, M.; Caruso, R. A., Magnetic nanocomposite particles and hollow spheres constructed by a sequential layering approach. *Chemistry of Materials* **2001**, 13, (1), 109-116.

69. Lvov, Y.; Ariga, K.; Onda, M.; Ichinose, I.; Kunitake, T., Alternate assembly of ordered multilayers of SiO2 and other nanoparticles and polyions. *Langmuir* **1997**, 13, (23), 6195-6203.

70. Sukhorukov, G. B.; Donath, E.; Lichtenfeld, H.; Knippel, E.; Knippel, M.; Budde, A.; Mohwald, H., Layer-by-layer self assembly of polyelectrolytes on colloidal particles. *Colloids and Surfaces a-Physicochemical and Engineering Aspects* **1998**, 137, (1-3), 253-266.

71. Donath, E.; Sukhorukov, G. B.; Caruso, F.; Davis, S. A.; Mohwald, H., Novel hollow polymer shells by colloid-templated assembly of polyelectrolytes. *Angewandte Chemie-International Edition* **1998**, 37, (16), 2202-2205.

72. Shchukin, D. G.; Sukhorukov, G. B.; Mohwald, H., Smart inorganic/organic nanocomposite hollow microcapsules. *Angewandte Chemie-International Edition* **2003**, 42, (37), 4472-4475.

73. Wang, Y. J.; Caruso, F., Enzyme encapsulation in nanoporous silica spherest. *Chemical Communications* **2004**, (13), 1528-1529.

74. Zhu, Y. F.; Shi, J. L.; Shen, W. H.; Dong, X. P.; Feng, J. W.; Ruan, M. L.; Li, Y. S., Stimuli-responsive controlled drug release from a hollow mesoporous silica sphere/polyelectrolyte multilayer core-shell structure. *Angewandte Chemie-International Edition* **2005**, 44, (32), 5083-5087.

75. Stockton, W. B.; Rubner, M. F., Molecular-level processing of conjugated polymers .4. Layer-by-layer manipulation of polyaniline via hydrogen-bonding interactions. *Macromolecules* **1997**, 30, (9), 2717-2725.

76. Sukhishvili, S. A.; Granick, S., Layered, erasable polymer multilayers formed by hydrogen-bonded sequential self-assembly. *Macromolecules* **2002**, 35, (1), 301-310.

77. Kozlovskaya, V.; Ok, S.; Sousa, A.; Libera, M.; Sukhishvili, S. A., Hydrogen-bonded polymer capsules formed by layer-by-layer self-assembly. *Macromolecules* **2003**, 36, (23), 8590-8592.

78. Zhang, Y. J.; Guan, Y.; Yang, S. G.; Xu, J.; Han, C. C., Fabrication of hollow capsules based on hydrogen bonding. *Advanced Materials* **2003**, 15, (10), 832-835.

79. Such, G. K.; Tjipto, E.; Postma, A.; Johnston, A. P. R.; Caruso, F., Ultrathin, responsive polymer click capsules. *Nano Letters* **2007**, 7, (6), 1706-1710.

80. Tong, W. J.; Gao, C. Y.; Moehwald, H., pH-responsive protein microcapsules fabricated via glutaraldehyde mediated covalent layer-by-layer assembly. *Colloid and Polymer Science* **2008**, 286, (10), 1103-1109.

81. Radtchenko, I. L.; Sukhorukov, G. B.; Mohwald, H., Incorporation of macromolecules into polyelectrolyte micro- and nanocapsules via surface controlled precipitation on colloidal particles. *Colloids Surf., A* **2002**, 202, (2-3), 127-133.

82. Sukhorukov, G. B.; Fery, A.; Brumen, M.; Mohwald, H., Physical chemistry of encapsulation and release. *Phys. Chem. Chem. Phys.* **2004**, 6, (16), 4078-4089.

83. Dudnik, V.; Sukhorukov, G. B.; Radtchenko, I. L.; Mohwald, H., Coating of colloidal particles by controlled precipitation of polymers. *Macromolecules* **2001**, 34, (7), 2329-2334.

84. Shenoy, D. B.; Sukhorukov, G. B., Microgel-Based Engineered Nanostructures and Their Applicability with Template-Directed Layer-by-Layer Polyelectrolyte Assembly in Protein Encapsulation. *Macromol. Biosci.* **2005**, 5, (5), 451-458.

85. Brittain, W. J.; Minko, S., A structural definition of polymer brushes. *Journal of Polymer Science Part a-Polymer Chemistry* **2007**, 45, (16), 3505-3512.

86. Luzinov, I.; Voronov, A.; Minko, S.; Kraus, R.; Wilke, W.; Zhuk, A., Encapsulation of fillers with grafted polymers for model composites. *Journal of Applied Polymer Science* **1996**, 61, (7), 1101-1109.

87. Luzinov, I.; Minko, S.; Tsukruk, V. V., Adaptive and responsive surfaces through controlled reorganization of interfacial polymer layers. *Progress in Polymer Science* **2004**, 29, (7), 635-698.

88. Luzinov, I.; Minko, S.; Tsukruk, V. V., Responsive brush layers: from tailored gradients to reversibly assembled nanoparticles. *Soft Matter* **2008**, 4, (4), 714-725.

89. Advincula, R. C., Surface initiated polymerization from nanoparticle surfaces. *Journal of Dispersion Science and Technology* **2003**, 24, (3-4), 343-361.

90. Israelachvili, J. N.; Mitchell, D. J.; Ninham, B. W., Theory of self-assembly of hydrocarbon amphiphiles into micelles and bilayers. *Journal of the Chemical Society, Faraday Transactions 2* **1976**, 72, (9), 1525-1568.

91. Israelachvili, J. N., *Intermolecular and Surface Forces*. Academic Press: San Diego, 1991; p 450.

92. Discher, D. E.; Ortiz, V.; Srinivas, G.; Klein, M. L.; Kim, Y.; David, C. A.; Cai, S. S.; Photos, P.; Ahmed, F., Emerging applications of polymersomes in delivery: From molecular dynamics to shrinkage of tumors. *Progress in Polymer Science* **2007**, 32, (8-9), 838-857.

93. Discher, B. M.; Won, Y. Y.; Ege, D. S.; Lee, J. C. M.; Bates, F. S.; Discher, D. E.; Hammer, D. A., Polymersomes: Tough vesicles made from diblock copolymers. *Science* **1999**, 284, (5417), 1143-1146.

94. Jain, S.; Bates, F. S., On the origins of morphological complexity in block copolymer surfactants. *Science* **2003**, 300, (5618), 460-464.

95. Won, Y. Y.; Brannan, A. K.; Davis, H. T.; Bates, F. S., Cryogenic transmission electron microscopy (cryo-TEM) of micelles and vesicles formed in water by polyethylene oxide)-based block copolymers. *Journal of Physical Chemistry B* **2002**, 106, (13), 3354-3364.

96. Srinivas, G.; Discher, D. E.; Klein, M. L., Self-assembly and properties of diblock copolymers by coarse-grain molecular dynamics. *Nature Materials* **2004**, 3, (9), 638-644.

97. Alexandridis, P.; Lindman, B., *Amphiphilic block copolymers. Self-assembly and applications*. Elsevier: Amsterdam, 2000; p 448.

98. Allen, C.; Maysinger, D.; Eisenberg, A., Nano-engineering block copolymer aggregates for drug delivery. *Colloids and Surfaces B-Biointerfaces* **1999**, 16, (1-4), 3-27.

99.   Forster, S.; Antonietti, M., Amphiphilic block copolymers in structure-controlled nanomaterial hybrids. *Advanced Materials* **1998**, 10, (3), 195-+.
100. Discher, D. E.; Ahmed, F., Polymersomes. *Annual Review of Biomedical Engineering* **2006**, 8, 323-341.
101. Meng, F. H.; Zhong, Z. Y.; Feijen, J., Stimuli-Responsive Polymersomes for Programmed Drug Delivery. *Biomacromolecules* **2009**, 10, (2), 197-209.
102. Onaca, O.; Enea, R.; Hughes, D. W.; Meier, W., Stimuli-Responsive Polymersomes as Nanocarriers for Drug and Gene Delivery. *Macromolecular Bioscience* **2009**, 9, (2), 129-139.
103. Mackay, J. A.; Chilkoti, A., Temperature sensitive peptides: Engineering hyperthermia-directed therapeutics. *International Journal of Hyperthermia* **2008**, 24, (6), 483-495.
104. Ni, P., Nanoparticles comprising pH/temperature-responsive amphiphilic block copolymers and their applications in biotechnology. In *Colloidal Nanoparticles in Biotechnology*, Elaissari, A., Ed. John Wiley & Sons: Hoboken, New Jersey, 2007; pp 31-63.
105. Riess, G., Micellization of block copolymers. *Prog. Polym. Sci.* **2003**, 28, (7), 1107-1170.
106. Addison, T.; Cayre, O. J.; Biggs, S.; Armes, S. P.; York, D., Incorporation of Block Copolymer Micelles into Multilayer Films for Use as Nanodelivery Systems. *Langmuir* **2008**, 24, (23), 13328-13333.
107. Checot, F.; Lecommandoux, S.; Klok, H. A.; Gnanou, Y., From supramolecular polymersomes to stimuli-responsive nano-capsules based on poly(diene-b-peptide) diblock copolymers. *European Physical Journal E* **2003**, 10, (1), 25-35.
108. Chen, G. H.; Hoffman, A. S., Graft-copolymers that exhibit temperature-induced phase-transitions over a wide-range of pH. *Nature* **1995**, 373, (6509), 49-52.
109. Qiu, Y.; Park, K., Environment-sensitive hydrogels for drug delivery. *Advanced Drug Delivery Reviews* **2001**, 53, (3), 321-339.
110. Gil, E. S.; Hudson, S. A., Stimuli-reponsive polymers and their bioconjugates. *Progress in Polymer Science* **2004**, 29, (12), 1173-1222.
111. Chilkoti, A.; Dreher, M. R.; Meyer, D. E.; Raucher, D., Targeted drug delivery by thermally responsive polymers. *Advanced Drug Delivery Reviews* **2002**, 54, (5), 613-630.
112. Yurke, B.; Lin, D. C.; Langrana, N. A., Use of DNA nanodevices in modulating the mechanical properties of polyacrylamide gels. In *Lecture notes in computer science*, Carbone, A.; Pierce, N. A., Eds. Springer-Verlag Berlin: BERLIN, 2006; Vol. 3892, pp 417-426.
113. Li, Z.; Zhang, Y.; Fullhart, P.; Mirkin, C. A., Reversible and chemically programmable micelle assembly with DNA block-copolymer amphiphiles. *Nano Letters* **2004**, 4, (6), 1055-1058.
114. Miyata, T.; Uragami, T.; Nakamae, K., Biomolecule-sensitive hydrogels. *Advanced Drug Delivery Reviews* **2002**, 54, (1), 79-98.
115. Ulijn, R. V., Enzyme-responsive materials: a new class of smart biomaterials. *Journal of Materials Chemistry* **2006**, 16, (23), 2217-2225.
116. Miyata, T.; Asami, N.; Uragami, T., A reversibly antigen-responsive hydrogel. *Nature* **1999**, 399, (6738), 766-769.
117. Cutler, S. M.; Garcia, A. J., Engineering cell adhesive surfaces that direct integrin alpha(5) beta(1) binding using a recombinant fragment of fibronectin. *Biomaterials* **2003**, 24, (10), 1759-1770.
118. Rosso, F.; Marino, G.; Giordano, A.; Barbarisi, M.; Parmeggiani, D.; Barbarisi, A., Smart materials as scaffolds for tissue engineering. *Journal of Cellular Physiology* **2005**, 203, (3), 465-470.
119. Stayton, P. S.; Shimoboji, T.; Long, C.; Chilkoti, A.; Chen, G. H.; Harris, J. M.; Hoffman, A. S., Control of protein-ligand recognition using a stimuli-responsive polymer. *Nature* **1995**, 378, (6556), 472-474.
120. Hoffman, A. S. In *Bioconjugates of intelligent polymers and recognition proteins for use in diagnostics and affinity separations*, 2000; Amer Assoc Clinical Chemistry: 2000; pp 1478-1486.

121. Hassan, C. M.; Doyle, F. J.; Peppas, N. A., Dynamic behavior of glucose-responsive poly(methacrylic acid-g-ethylene glycol) hydrogels. *Macromolecules* **1997**, 30, (20), 6166-6173.

122. Patel, K.; Angelos, S.; Dichtel, W. R.; Coskun, A.; Yang, Y. W.; Zink, J. I.; Stoddart, J. F., Enzyme-responsive snap-top covered silica nanocontainers. *Journal of the American Chemical Society* **2008**, 130, (8), 2382-2383.

123. Gopishetty, V.; Roiter, Y.; Tokarev, I.; Minko, S., Multiresponsive Biopolyelectrolyte Membrane. *Adv. Mater.* **2008**, 20, (23), 4588-4593.

124. Rezania, A.; Healy, K. E., Biomimetic peptide surfaces that regulate adhesion, spreading, cytoskeletal organization, and mineralization of the matrix deposited by osteoblast-like cells. *Biotechnology Progress* **1999**, 15, (1), 19-32.

125. Pakalns, T.; Haverstick, K. L.; Fields, G. B.; McCarthy, J. B.; Mooradian, D. L.; Tirrell, M., Cellular recognition of synthetic peptide amphiphiles in self-assembled monolayer films. *Biomaterials* **1999**, 20, (23-24), 2265-2279.

126. Wang, C.; Stewart, R. J.; Kopecek, J., Hybrid hydrogels assembled from synthetic polymers and coiled-coil protein domains. *Nature* **1999**, 397, (6718), 417-420.

127. Liu, J. W.; Lu, Y., Smart nanomaterials responsive to multiple chemical stimuli with controllable cooperativity. *Adv. Mater.* **2006**, 18, (13), 1667-1671.

128. Kirshenbaum, K.; Barron, A. E.; Goldsmith, R. A.; Armand, P.; Bradley, E. K.; Truong, K. T. V.; Dill, K. A.; Cohen, F. E.; Zuckermann, R. N., Sequence-specific polypeptoids: A diverse family of heteropolymers with stable secondary structure. *Proceedings of the National Academy of Sciences of the United States of America* **1998**, 95, (8), 4303-4308.

129. Ramstrom, O.; Ansell, R. J., Molecular imprinting technology: Challenges and prospects for the future. *Chirality* **1998**, 10, (3), 195-209.

130. Ansell, R. J.; Kriz, D.; Mosbach, K., Molecularly imprinted polymers for bioanalysis: Chromatography, binding assays and biomimetic sensors. *Current Opinion in Biotechnology* **1996**, 7, (1), 89-94.

131. Haupt, K., Imprinted polymers - Tailor-made mimics of antibodies and receptors. *Chemical Communications* **2003**, (2), 171-178.

132. Byrne, M. E.; Park, K.; Peppas, N. A., Molecular imprinting within hydrogels. *Advanced Drug Delivery Reviews* **2002**, 54, (1), 149-161.

133. Tokareva, I.; Tokarev, I.; Minko, S.; Hutter, E.; Fendler, J. H., Ultrathin molecularly imprinted polymer sensors employing enhanced transmission surface plasmon resonance spectroscopy. *Chem. Commun.* **2006**, (31), 3343-3345.

134. Lapeyre, V.; Ancla, C.; Catargi, B.; Ravaine, V., Glucose-responsive microgels with a core-shell structure. *Journal of Colloid and Interface Science* **2008**, 327, (2), 316-323.

135. Balazs, A. C.; Singh, C.; Zhulina, E.; Chern, S. S.; Lyatskaya, Y.; Pickett, G., Theory of polymer chains tethered at interfaces. *Prog. Surf. Sci.* **1997**, 55, (3), 181-269.

136. Peppas, N. A.; Huang, Y.; Torres-Lugo, M.; Ward, J. H.; Zhang, J., Physicochemical, foundations and structural design of hydrogels in medicine and biology. *Annual Review of Biomedical Engineering* **2000**, 2, 9-29.

137. Minko, S., Responsive polymer brushes. *Polym. Rev.* **2006**, 46, (4), 397-420.

138. Luzinov, I.; Minko, S.; Tsukruk, V. V., Adaptive and responsive surfaces through controlled reorganization of interfacial polymer layers. *Prog. Polym. Sci.* **2004**, 29, (7), 635-698.

139. Minko, S.; Roiter, Y., AFM single molecule studies of adsorbed polyelectrolytes. *Cur. Opin. Colloid Interf. Sci.* **2005**, 10, (1-2), 9-15.

140. Roiter, Y.; Minko, S., Single Molecule Experiments at the Solid-Liquid Interface: In Situ Conformation of Adsorbed Flexible Polyelectrolyte Chains. *J. Am. Chem. Soc.* **2005**, 127, (45), 15688-15689.

141. Roiter, Y.; Jaeger, W.; Minko, S., Conformation of single polyelectrolyte chains vs. salt concentration: Effects of sample history and solid substrate. *Polymer* **2006**, 47, (7), 2493-2498.

142. Roiter, Y.; Minko, S., Adsorption of polyelectrolyte versus surface charge: in situ single-molecule atomic force microscopy experiments on similarly, oppositely, and heterogeneously charged surfaces. *Journal of Physical Chemistry B* **2007**, 111, (29), 8597-8604.

143. Jeong, B.; Kim, S. W.; Bae, Y. H., Thermosensitive sol-gel reversible hydrogels. *Advanced Drug Delivery Reviews* **2002**, 54, (1), 37-51.
144. Schmaljohann, D.; Beyerlein, D.; Nitschke, M.; Werner, C., Thermo-reversible swelling of thin hydrogel films immobilized by low-pressure plasma. *Langmuir* **2004**, 20, (23), 10107-10114.
145. Schmaljohann, D., Thermo- and pH-responsive polymers in drug delivery. *Advanced Drug Delivery Reviews* **2006**, 58, (15), 1655-1670.
146. Rühe, J.; Ballauff, M.; Biesalski, M.; Dziezok, P.; Gröhn, F.; Johannsmann, D.; Houbenov, N.; Hugenberg, N.; Konradi, R.; Minko, S.; Motornov, M.; Netz, R. R.; Schmidt, M.; Seidel, C.; Stamm, M.; Stephan, T.; Usov, D.; Zhang, H., Polyelectrolyte Brushes. *Adv. Polymer Sci.* **2004**, 165, 79-150.
147. Keller, P.; Li, M. H., Stimuli-responmsive polymer vesicles. *Soft Matter* **2009**, 5, (5), 927-937.
148. Kang, E. T.; Neoh, K. G.; Tan, K. L., Polyaniline: A polymer with many interesting intrinsic redox states. *Prog. Polym. Sci.* **1998**, 23, (2), 277-324.
149. Tam, T. K.; Zhou, J.; Pita, M.; Ornatska, M.; Minko, S.; Katz, E., Biochemically controlled bioelectrocatalytic interface. *J. Am. Chem. Soc.* **2008**, 130, (33), 10888-10889.
150. Zhou, J.; Tam, T. K.; Pita, M.; Ornatska, M.; Minko, S.; Katz, E., Bioelectrocatalytic System Coupled with Enzyme-Based Biocomputing Ensembles Performing Boolean Logic Operations: Approaching "Smart" Physiologically Controlled Biointerfaces. *ACS Applied Materials & Interfaces* **2008**, 1, (1), 144-149.
151. Motornov, M.; Zhou, J.; Pita, M.; Tokarev, I.; Gopishetty, V.; Katz, E.; Minko, S., Integrated Multifunctional Nanosystem from Command Nanoparticles and Enzymes. *Small* **2009**, 5, (7), 817 - 820.
152. Jiang, G. Q.; Baba, A.; Ikarashi, H.; Xu, R. S.; Locklin, J.; Kashif, K. R.; Shinbo, K.; Kato, K.; Kaneko, F.; Advincula, R., Signal enhancement and tuning of surface plasmon resonance in Au nanoparticle/polyelectrolyte ultrathin films. *J. Phys. Chem. C* **2007**, 111, (50), 18687-18694.
153. Hinsberg, W.; Houle, F. A.; Lee, S. W.; Ito, H.; Kanazawa, K., Characterization of reactive dissolution and swelling of polymer films using a quartz crystal microbalance and visible and infrared reflectance spectroscopy. *Macromolecules* **2005**, 38, (5), 1882-1898.
154. Tanaka, T.; Fillmore, D. J., Kinetics of swelling of gels. *J. Chem. Phys.* **1979**, 70, (3), 1214-1218.
155. Vogt, B. D.; Soles, C. L.; Lee, H. J.; Lin, E. K.; Wu, W. L., Moisture absorption and absorption kinetics in polyelectrolyte films: Influence of film thickness. *Langmuir* **2004**, 20, (4), 1453-1458.
156. Itano, K.; Choi, J. Y.; Rubner, M. F., Mechanism of the pH-induced discontinuous swelling/deswelling transitions of poly(allylamine hydrochloride)-containing polyelectrolyte multilayer films. *Macromolecules* **2005**, 38, (8), 3450-3460.
157. Brittain, W. J.; Minko, S., A structural definition of polymer brushes. *J. Polym. Sci. Part A. Polym. Chem.* **2007**, 45, (16), 3505-3512.
158. Toomey, R.; Freidank, D.; Ruhe, J., Swelling behavior of thin, surface-attached polymer networks. *Macromolecules* **2004**, 37, (3), 882-887.
159. Brittain, W. J.; Boyes, S. G.; Granville, A. M.; Baum, M.; Mirous, B. K.; Akgun, B.; Zhao, B.; Blickle, C.; Foster, M. D., Surface rearrangement of diblock copolymer brushes - Stimuli responsive films. In *Surface- Initiated Polymerization Ii*, Springer-Verlag Berlin: Berlin, 2006; Vol. 198, pp 125-147.
160. Ju, H. K.; Kim, S. Y.; Lee, Y. M., pH/temperature-responsive behaviors of semi-IPN and comb-type graft hydrogels composed of alginate and poly (N-isopropylacrylamide). *Polymer* **2001**, 42, (16), 6851-6857.
161. Zhang, J.; Peppas, N. A., Synthesis and characterization of pH- and temperature-sensitive poly(methacrylic acid)/poly(N-isopropylacrylamide) interpenetrating polymeric networks. *Macromolecules* **2000**, 33, (1), 102-107.

162. Aoki, T.; Kawashima, M.; Katono, H.; Sanui, K.; Ogata, N.; Okano, T.; Saku-
      rai, Y., Temperature-responsive interpenetrating polymer networks constructed with
      poly(acrylic acid) and poly(N,N-dimethylacryl-amide). *Macromolecules* **1994**, 27, (4),
      947-952.
163. Hu, Z. B.; Chen, Y. Y.; Wang, C. J.; Zheng, Y. D.; Li, Y., Polymer gels with engineered
      environmentally responsive surface patterns. *Nature* **1998**, 393, (6681), 149-152.
164. Yang, C. T.; Wang, Y. L.; Yu, S.; Chang, Y. C. I., Controlled Molecular Organization of
      Surface Macromolecular Assemblies Based on Stimuli-Responsive Polypeptide Brushes.
      Biomacromolecules **2009**, 10, (1), 58-65.
165. Minko, S.; Muller, M.; Usov, D.; Scholl, A.; Froeck, C.; Stamm, M., Lateral versus per-
      pendicular segregation in mixed polymer brushes. *Phys. Rev. Lett.* **2002**, 88, (3), Art. No.
      035502.
166. Minko, S.; Luzinov, I.; Luchnikov, V.; Muller, M.; Patil, S.; Stamm, M., Bidisperse mixed
      brushes: Synthesis and study of segregation in selective solvent. *Macromolecules* **2003**, 36,
      (19), 7268-7279.
167. Usov, D.; Gruzdev, V.; Nitschke, M.; Stamm, M.; Hoy, O.; Luzinov, I.; Tokarev, I.; Minko,
      S., Three-dimensional analysis of switching mechanism of mixed polymer brushes. *Macro-
      molecules* **2007**, 40, (24), 8774-8783.
168. Sukhishvili, S. A., Responsive polymer films and capsules via layer-by-layer assembly.
      Current Opinion in Colloid & Interface Science **2005**, 10, (1-2), 37-44.
169. Sui, Z. J.; Schlenoff, J. B., Phase separations in pH-responsive polyelectrolyte multilayers:
      Charge extrusion versus charge expulsion. *Langmuir* **2004**, 20, (14), 6026-6031.
170. Wong, J. E.; Richtering, W., Layer-by-layer assembly on stimuli-responsive microgels.
      Current Opinion in Colloid & Interface Science **2008**, 13, (6), 403-412.
171. Quinn, J. F.; Johnston, A. P. R.; Such, G. K.; Zelikin, A. N.; Caruso, F., Next generation,
      sequentially assembled ultrathin films: beyond electrostatics. *Chemical Society Reviews*
      2007, 36, (5), 707-718.
172. Kharlampieva, E.; Sukhishvili, S. A., Hydrogen-bonded layer-by-layer polymer films.
      Polym. Rev. **2006**, 46, (4), 377-395.
173. Lupitskyy, R.; Roiter, Y.; Tsitsilianis, C.; Minko, S., From Smart Polymer Molecules to
      Responsive Nanostructured Surfaces. *Langmuir* **2005**, 21, (19), 8591-8593.
174. Webber, G. B.; Wanless, E. J.; Butun, V.; Armes, S. P.; Biggs, S., Self-organized monolayer
      films of stimulus-responsive micelles. *Nano Letters* **2002**, 2, (11), 1307-1313.
175. Webber, G. B.; Wanless, E. J.; Armes, S. P.; Tang, Y. Q.; Li, Y. T.; Biggs, S., Nano-anem-
      ones: Stimulus-responsive copolymer-micelle surfaces. *Adv. Mater.* **2004**, 16, (20), 1794-
      1798.
176. Gopishetty, V.; Roiter, Y.; Tokarev, I.; Minko, S., Multiresponsive Biopolyelectrolyte Mem-
      brane. *Advanced Materials* **2008**, 20, (23), 4588-4593.
177. Tokarev, I.; Minko, S., Multiresponsive, Hierarchically Structured Membranes: New, Chal-
      lenging, Biomimetic Materials for Biosensors, Controlled Release, Biochemical Gates, and
      Nanoreactors. *Advanced Materials* **2009**, 21, (2), 241-247.
178. Tokarev, I.; Orlov, M.; Minko, S., Responsive polyelectrolyte gel membranes. *Advanced
      Materials* **2006**, 18, (18), 2458-+.
179. Orlov, M.; Tokarev, I.; Scholl, A.; Doran, A.; Minko, S., pH-Responsive thin film mem-
      branes from poly(2-vinylpyridine): Water vapor-induced formation of a microporous struc-
      ture. *Macromolecules* **2007**, 40, (6), 2086-2091.

Chapter 11
Medicine in Reverse

Colloid Technologies for Combating Chemical Overdoses

Richard Partch, Adrienne Stamper, Evon Ford, Abeer Al Bawab and Fadwa Odeh

Abstract The axiom "all life is dependent on chemicals" is true but the chemicals must be the right kind and in the right proportions for healthy growth. A second truth is that "external or internal contact with chemicals in overdose concentrations may result in either discomfort, disease or in death". Often the very therapeutic chemicals (medicines by any other name) created to reduce or eliminate one type of malady result in complications elsewhere in the body when taken accidently in too high dosage. Furthermore, overdoses of illegitimate drugs and exposure to biotoxins damage organs and may lead to death. Examples of some molecules commonly overdosed are shown in Fig. 11.1. This chapter gives fundamental and applied information on oil-in water microemulsion and on functionalized carrier particle colloids designed to selectively bind and physiologically deactivate *in vivo* several commonly overdosed chemicals. EKG and NMR techniques have been employed to obtain important diagnostic information.

Keywords Chemical overdoses • Chemical overdose remediation • Medicine in reverse • Microemulsions • Functionalized carrier particles • Surfactants • Charge transfer complexes • Pi-pi aromatic binding • Donor-acceptor interactions • NMR chemical shifts • EKG effects of chemicals • Silica • Silation reactions • Pluronic surfactants • Bupivacaine • Amitriptyline • Cocaine • Ricin • Atomic force microscopy • Cardiotoxicity • Biotoxin

11.1 Introduction

The axiom "all life is dependent on chemicals" is true but the chemicals must be the right kind and in the right proportions for healthy growth. A second truth is that "external or internal contact with chemicals in overdose concentrations may result in

R. Partch (✉)
Chemistry Department and Center for Advanced Materials Processing, Clarkson University, Potsdam 13699-5814, NY, USA
e-mail: partch@clarkson.edu

E. Matijević (ed.), *Fine Particles in Medicine and Pharmacy,*
DOI 10.1007/978-1-4614-0379-1_11, © Springer Science+Business Media, LLC 2012

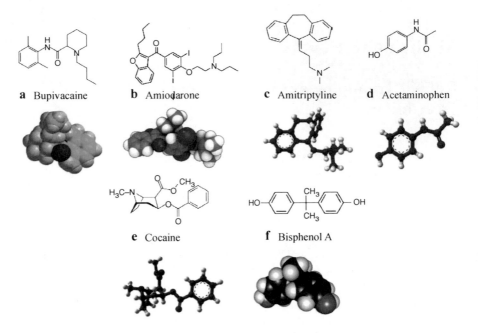

Fig. 11.1 Representative molecules which cause undesirable reaction if overdosed. **a** local anesthetic, **b** antiarrythmic, **c** antidepressant, **d** analgesic, **e** alkaloid, **f** estrogenic plasticizer

either discomfort, disease or in death". Often the very therapeutic chemicals (medicines by any other name) created to reduce or eliminate one type of malady result in complications elsewhere in the body when taken accidently in too high dosage. Furthermore, overdoses of illegitimate drugs and exposure to biotoxins damage organs and may lead to death. Examples of some molecules commonly overdosed are shown in Fig. 11.1.

One method of reducing complications from overdoses employs ionic attraction of the chemical to the surface of a microemulsion oil droplet, with subsequent absorption of the molecule into the oil. The other concept involves adsorption, binding the chemical to ligands attached to the surface of functionally modified carrier nanoparticles. Figure 11.2 depicts two types of colloids capable of achieving the two phenomena [1].

This chapter summarizes the synthesis and physicochemical characterization of the microemulsion and spherical nanoparticle species, the proposed mechanisms by which each type of dispersed phase lowers overdosed therapeutic or cocaine concentration in blood, analytical data on the efficiency of phase transfer of toxin out of normal saline, blood plasma and whole blood, and efficacy of the microemulsions when employed *in vivo* in rats [2–4]. Furthermore, evidence is presented on application of both atomic force microscopic (AFM) and diffusion nuclear magnetic resonance (NMR DOSY) analyses to verify that the concepts have merit for predicting the structure of a new inhibitor for Ricin biotoxin.

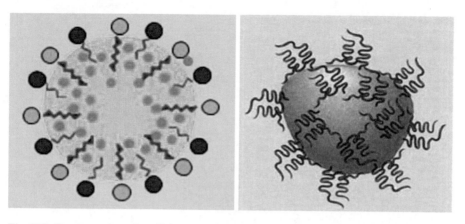

Fig. 11.2 Representation of a colloidal oil droplet (*yellow*) in water stabilized by surfactants (*blue* and *purple*) and showing absorbed overdosed chemical (*orange*) (*left*); and of a carrier particle with chemical adsorbing functions attached (*blue*) for binding overdosed chemical (*right*)

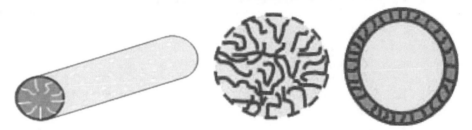

Fig. 11.3 Representation of internally functionalized nanotube, of microgel and of core–shell microemulsion colloids

The research has been interdisciplinary and international in scope between anesthesiologists, chemical and materials engineers and chemists at Clarkson University, the University of Florida [5], Kyungwon University, Korea [6] and the University of Jordan, Amman [7].

Evaluation of microgels [8], core–shell microemulsions [9] and of nanotubes with ligands [10] for remediation of overdoses are outside the scope of this chapter (Fig. 11.3).

11.2   Colloidal Oil-in-Water Microemulsion Studies

11.2.1   *General Discussion on Emulsions*

Emulsions and microemulsions abound in nature and in the synthetic chemical world. They are widely employed in, e.g., food processing, paint manufacture, and

Fig. 11.4 Detail of oil-
in-water microemulsion
structure showing oil (*yellow*), Pluronic and fatty acid
surfactants, and partitioning
of drug/toxin molecules
(*orange*) between the bulk
water phase (*grey*) and oil
droplet

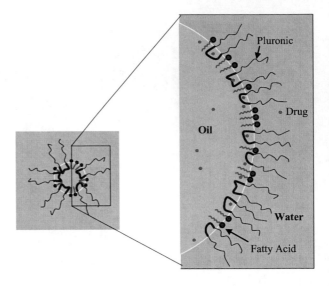

drug delivery. Most are composed of droplets of oil dispersed in water but the reverse is becoming more widely studied as new applications demand. In either case, their stability is enhanced by the addition of one or more surfactants, molecules which have both hydrophobic and hydrophilic regions that transcend the interface between droplet and bulk medium (Fig. 11.4).

Time honored research on the preparation of emulsions and microemulsions, and their applications for drug **delivery** are well documented by Shah [11–13], and others [14]. That information is seminal to the present quest for "Medicines in Reverse", biocompatible oil-in-water microemulsions for *in vivo* **removal** of commonly overdosed molecules shown in Fig. 11.1 from free circulation in blood.

An oil-in-water emulsion differs from a microemulsion by the size of the dispersed oil droplets, the latter being controlled by the type and ratio of the chemical components. Typically, oil droplets in an emulsion are greater than 200 nm in diameter and scatter visible light, causing the dispersion to have a milky, opaque appearance. A multicomponent oil-in-water microemulsion having droplets less than 100 nm in diameter is usually optically isotropic, clear and thermodynamically stable. That is, the oil and water components do not separate over time. The examples shown in Fig. 11.5 are soybean oil–water mixtures having the same oil–water ratio prepared with different surfactants.

Several combinations of non-ionic and ionic surfactants and oils were used in the present work to obtain preliminary data on the preparation and stability of microemulsions and their possible employment as absorbants of toxins. Proof of concept has been achieved using the relatively biocompatible components ethyl butyrate ($LD_{50} > 250$ mg/kg in dogs) as the oil, Pluronic ($LD_{50} \sim 10$ g/kg in dogs) and fatty acid co-surfactants.

"Pluronic" is a generic name for a series of triblock copolymers composed of different ratios of more hydrophobic propylene oxide (PPO) monomers flanked on each side by more hydrophilic ethylene oxide (PEO) monomers [15–18]. Several commercially available analogues of Pluronics are shown in Fig. 11.6.

Fig. 11.5 Appearance of an emulsion versus microemulsion and comparison of oil droplet sizes and surface areas

	Emulsion	Microemulsion
Particle Size:	~ 400 nm	~ 30 nm
Surface Area:	~ 15 m²/ml	~ 215 m²/ml

Fig. 11.6 Structures of some commercially available Pluronic surfactants. PEO refers to polyethylene oxide and PPO refers to polypropylene oxide segments having repeat units of X and Y in length

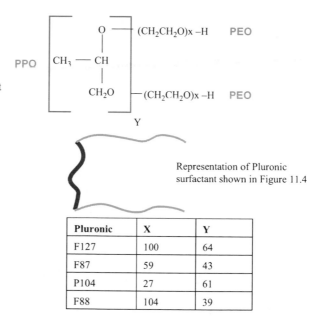

Representation of Pluronic surfactant shown in Figure 11.4

Pluronic	X	Y
F127	100	64
F87	59	43
P104	27	61
F88	104	39

These amphiphilic molecules aggregate and form micelles when mixed with water. However, with oil droplets present Pluronic molecules assemble at the interface between the oil core and water, with the hydrophilic PEO termini extending into the water phase.

11.2.2   Quantitative Data on Drug Absorption by Microemulsions

Experimental results reveal that 10 nM Pluronic F-127 is the most effective for the phase transfer of bupivacaine from water, blood plasma or whole blood into ethyl butyrate microemulsion droplets. The fundamental reason why F-127 Pluronic iso-

Fig. 11.7 Extraction of
1 mM bupivacaine in normal
saline by Pluronic 127 MEs
EB = ethyl butyrate; *SC, SL,
SM* = capric, lauric, myristic
fatty acids; *AOT* = succinate
salt

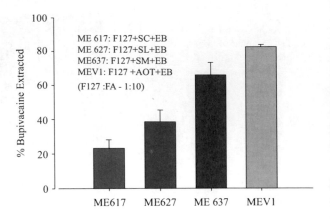

Fig. 11.8 Extraction of
1 mM Amitriptyline in
normal saline by microemulsions having incremental
compositions

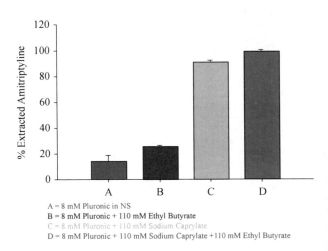

A = 8 mM Pluronic in NS
B = 8 mM Pluronic + 110 mM Ethyl Butyrate
C = 8 mM Pluronic + 110 mM Sodium Caprylate
D = 8 mM Pluronic + 110 mM Sodium Caprylate +110 mM Ethyl Butyrate

mer participates in better phase transfer of the drug into the microemulsion compared to other derivatives is not known.

As indicated in Figs. 11.2 and 11.4, fatty acid co-surfactants are also used in the
optimized formulation of the microemulsions. A fatty acid co-surfactant having a
chain length at least eight carbons long not only helps stabilize the dispersed phase
in blood but enhances phase transfer of drug from blood into the ethyl butyrate
droplet (Figs. 11.7 and 11.8). Comparison studies on the efficiency of extraction
of amitriptyline versus bupivacaine as a function of the type of ionic co-surfactant
reveal that longer chain carboxylic acid salts greatly increase bupivacaine partitioning from normal saline. Also, extraction efficiency of toxin from blood into the
microemulsion droplets is increased by increasing the concentration of the ionic co-
surfactant in microemulsions; and, depending on the chemical system, amitriptyline
extraction is more sensitive to this variable (Fig. 11.9).

Extraction of bupivacaine from normal saline was also evaluated using biocompatible microemulsions composed of ethyl butyrate oil, fatty acid salts and

Fig. 11.9 Effect of sodium laurate concentration on extraction of 1 mM drug in normal saline into Pluronic-stabilized oil-in-water microemulsion

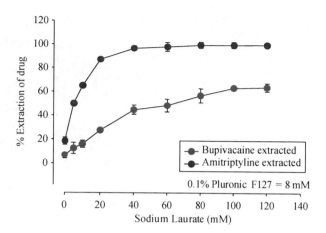

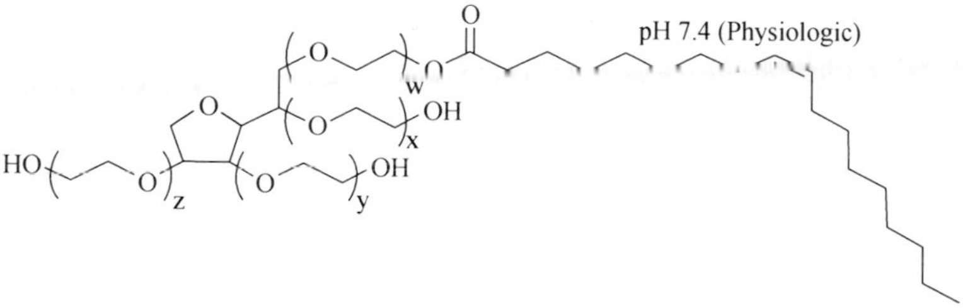

Fig. 11.10 General structure of Tween surfactant derivatives. For Tween-80, $X+Y+W+Z = 80$

polyethylene glycol surfactants commonly referred to as Tween (Fig. 11.10) and PEG.

Chromatography data show that increasing Tween-80 from 0.07 to 0.20 M, while holding the PEG, sodium caprylate and ethyl butyrate concentrations constant, results in an increase in drug extracted from 7 to 22%. Using the same chemicals but increasing PEG concentration from 0.0 to 0.5 M reduces by 60% the amount of bupivacaine extracted.

Replacing fatty acid salts with the more organophilic dioctyl sulfosuccinate ester salt surfactant (AOT) enhances bupivacaine extraction even more (Fig. 11.7).

In summary, the concepts and data presented in Sects. 11.2.1 and 11.2.2 are proof that a microemulsion can efficiently remove two of the therapeutic drugs of concern from saline. An optimized composition for extraction of bupivacaine from normal saline at pH 7.4 contains 9 mM Pluronic F127, 36 mM sodium laurate, and 155 mM ethyl butyrate [19]. The slight turbidity upon addition of butyrate oil quickly dissipates during stirring. Dynamic light scattering and photon correlation spectroscopy was employed to determine oil droplet size in the microemulsions. The diameters depend on the amount and type of surfactants and range from 11 to 40 nm with an average polydispersity index of 0.251 nm. These small sizes are compatible for recirculation in humans without disruption [20].

Fig. 11.11 Attenuation of bupivacaine-induced QRS interval prolongation of isolated rat heart

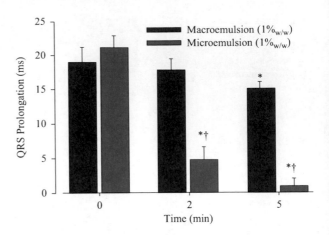

11.2.3 Electrocardiogram Data on Success of Microemulsions to Reduce Cardiotoxin Activity

The optimized microemulsion was next tested for ability to restore a toxin-induced decreased rate of heart beat, otherwise referred to as an increased QRS time interval, of an isolated rat heart. In these experiments, carried out by team members Morey and Dennis at Shands Medical School, University of Florida [5], the heartbeat was continually monitored while the being infused first with normal saline doped with bupivacaine and then with the optimized microemulsion defined at the end of Sect. 11.2.2 above.

The experiment was repeated using an emulsion having larger oil droplets compared to those in a microemulsion. As shown in Fig. 11.11, the emulsion is much less effective in restoring the heartbeat to normal even after 5 min.

Thromboelastography and cell lysis experiments using whole blood verified that the microemulsions showing better drug removal properties did not cause excessive thrombosis or rupture of blood cells.

Two variations of experiments were carried out to determine if microemulsions ME617 and ME627 (see Fig. 11.7 for compositions) could reverse *in vivo* cardiac arrest induced by amitriptyline in rats. When ME617 was administered 5 min prior to infusion of the drug, the undesired percent change increase in QRS interval was not as great as when infusing normal saline. This experiment served to demonstrate that ME617 functioned *in vivo* as an antidote for the cardiotoxicity of amitriptyline. More relevant to a real-life drug overdose senario is the fact that the drug-induced increase in the QRS interval is greatly reduced within 30 min after ME627 is administered (Fig. 11.12). This latter sequence of administering chemicals into the blood stream indicates that the microemulsions being developed may serve as antidotes for persons previously overdosed on amitrityline, bupivacaine or related toxins [21].

Fig. 11.12 Intravenous treatment of oral amitriptyline (190 mg/kg) poisoning in rats using normal saline (*black*) or microemulsion (*red*) to lower the QRS interval

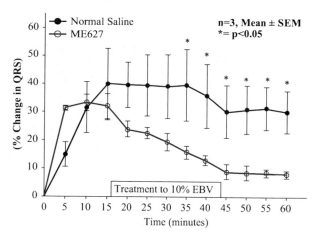

Fig. 11.13 Extraction of 1,000 uM amitriptyline in saline by 10% v/v ME617 at different pH values

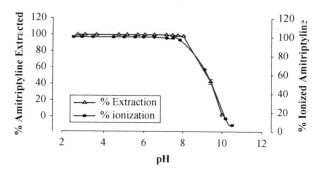

11.2.4 Effect of pH on Drug Absorption by Microemulsions

Several of the target drugs included in this study for removal from blood by microemulsions are molecules containing amino groups (Fig. 11.1) and are administered as therapeutics in the form of water soluble hydrochloride salts. However, due to the hydrophobic aromatic ring portions of their structures the protonated forms of the molecules assume surfactant-like properties [22].

At physiological pH of 7.4 an equilibrium is established between protonated and unprotonated amine functionalities. The neutral unprotonated forms of the molecules are lipophilic and are more susceptible to partitioning into the microemulsion oil droplets than the ionic protonated form. Amitriptyline has a pK_a of 9.4 and is 99% protonated at pH 7.40; bupivacaine has a pK_a of 8.1 and is 83% protonated at the same pH [22]. It is interesting that these rather small percentage differences play a major role in the amount of each of these drugs that can be removed from normal saline by a microemulsion. Chromatography data show that while amitriptyline extraction from saline declines from 100% at pH 6.5 to near 0% when the saline is at pH 10 (Fig. 11.13), the amount of bupivacaine extracted increases with increasing pH.

Fig. 11.14 Computer model
of bupivacaine structure

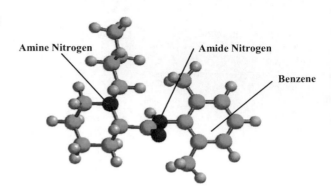

Fig. 11.15 Proposed interaction of amitriptyline with microemulsion droplet. Core oil, *yellow*; hydrophilic region of Pluronic surfactant, *green*

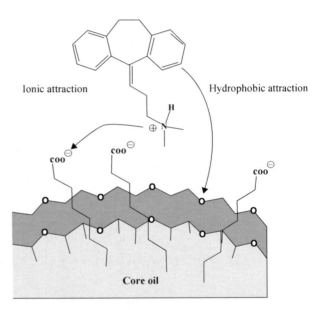

The data show that at physiological pH 99% amitriptyline undergoes phase transfer from saline into the ME617 microemulsion. This suggests that steric differences in molecular structures may play a role in the phase transfer processes. Close examination of the molecular structures of amitriptyline and bupivacaine (Fig. 11.1) reveals that in the former therapeutic the hydrophobic aromatic portion is sterically voluminous and the amine portion is much less hindered. In bupivacaine, however, the amine nitrogen atom is positioned deep inside a cage-like hydrocarbon region, a steric hindrance factor that controls the pK_a and ability of the drug to partition into oil (Fig. 11.14).

The above described molecular features of amitriptyline and bupivacaine allow a prediction why amitriptyline hydrochloride salt undergoes partitioning into microemulsion oil droplets at or below physiological pH, and why bupivacaine does so to a much lesser degree. Figure 11.15 is a hypothetical drawing of the approach of the protonated form of amitriptyline to the surface of a microemulsion droplet. Data in Figs. 11.8 and 11.9 shows that fatty acid salts present in the co-surfactant shell around core ethyl butyrate droplets greatly enhance amitriptyline absorption. This

Fig. 11.16 Effect of amitriptyline concentration on proton chemical shifts in ME617. *EO* = ethylene oxide; *PO* = propylene oxide repeat units; *SC* = sodium caprylate; *EB* = ethyl butyrate. The greater the shifts the more interaction the drug has with the designated protons

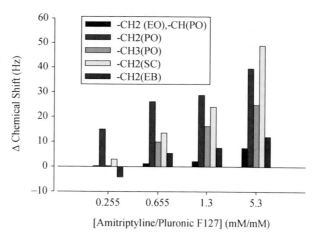

[Amitriptyline/Pluronic F127] (mM/mM)

suggests that electrostatic attraction between the co-surfactant carboxylate groups protruding into the aqueous medium around an oil droplet, and the sterically unencumbered protonated amine group of amitriptyline, is one factor favoring partitioning of amitriptyline into the oil droplet at lower pH. Hydrophobic interaction between the aromatic portion of the molecule, the propylene oxide repeat units of the Pluronic and the oil in the droplet also may be a factor.

The interfacial transport of bupivacaine into oil droplets is not controlled to the same extent by the amine/protonated amine ratio. Hydrophobicity of the unprotonated form of bupivacaine at pH 7.40 may be the controlling molecular feature for extraction because the amine group in protonated bupivacaine is sterically buried inside hydrocarbon moieties and partially shielded from the carboxylate head groups.

11.2.5 NMR Evaluation of Drug Interacton with Microemulsions

Nuclear Magnetic Resonance (NMR) spin relaxation measurements is a technique that measures changes in chemical shifts of protons positioned on molecules in solution as they diffuse [23]. The technique was employed to measure the time-dependent interaction of protonated amitriptyline with ME617 microemulsion oil droplets. The chemical shift intensities in Fig. 11.16 show that the protons associated with the propylene oxide portion of the triblock Pluronic surfactant are magnetically altered first at low amitriptyline concentration. This data may be interpreted by referring to Fig. 11.4. It suggests that amitriptyline diffuses quickly through the hydrophilic ethylene oxide repeat units (green region in Fig. 11.15) more external to the oil droplet, then associates with the more hydrophobic propylene oxide units embedded in the surface of the oil droplet. With increasing concentration of the drug the protons on the alpha carbon of the fatty acid (yellow bar in Fig. 11.16) are affected. At the highest amitriptyline concentration the chemical shift data show magnetic interaction with propylene oxide repeat unit protons and those of the core ethyl butyrate molecules (blue bar); however, interaction with the fatty acid protons (yellow) dominate. Overall, the data suggest that the postulated interactions for amitriptyline with the colloidal oil droplets shown in Fig. 11.15 may be correct. If the NMR data has been correctly interpreted, it appears that the oil droplet core, the ethyl butyrate, serves mostly as a final reservoir in removing amitriptyline from saline.

11.3   Selective Absorption of Drugs by Microemulsions

In the section above it has been amply demonstrated that oil-in-water microemulsions successfully absorb representative chemicals commonly overdosed. The ethyl butyrate-based oil droplets used in those studies cannot differentiate aromatic versus aliphatic chemicals. This fact resulted in evaluating how a different oil might be used that could show selectivity in absorption. The choice reduced to making a microemulsion with oil having a benzene ring as part of its structure. The fundamental reason for this choice is explained in the next section.

11.3.1   Π–Π Charge Transfer Concepts

Charge transfer π–π complexes between aromatic rings have been a well-known interaction since the early discoveries made by Benesi and Hildebrand [24], through the pioneering work of Briegleb [25] and the valence-bond, and molecular orbital descriptions by Mulliken [26], and Dewar and Lepley [27], respectively. Π–π interactions are one of the principal noncovalent forces governing the fundamental process of molecular recognition between molecules in biology and chemistry [28–40]. For example, charge transfer interactions between benzene rings influence structures of proteins, DNA, host–guest complexes and self-assembled supramolecular architectures. The diversity and significance of charge transfer complex formation in chemical and biological recognition has been illustrated in recent reviews [28, 29].

Aromatic π–π interactions are driven mostly by electrostatic forces between two rings with relatively small binding energies in the range 2–5 kcal/mol [41–43]. Such interactions are enhanced when the p orbitals of one ring are enriched in electron density (donor), and the p orbitals of the other ring are deficient (acceptor). Figure 11.17 shows the influence of the two types of substituents on NMR proton chemical shifts in each of the donor and acceptor molecules measured separately. The arrows pointing away from the benzene ring on the left and towards attached functional groups indicate depletion of p orbit electron density from the ring. The functional groups on the ring on the right have the opposite effect. Chemical shifts for the protons on unsubstituted benzene are shown in the center.

Fig. 11.17 Influence of electron withdrawing (*left* molecule) and electron donating (*right* molecule) substituents on chemical shifts of protons on the benzene rings. The *ring* in the center is unsubstituted benzene

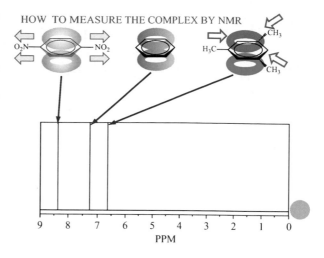

Table 11.1 Hammett substituent effects on aromatic ring electron density

Electron-donating substituents	Electron-withdrawing substituents
$-CH_3,-OH,-OCH_3,-NH_2$	$-NO_2,-CN,-CF_3,-CO_2Et$

The efficiency of electron density removal from or donation to a ring for different functional groups can be estimated from Hammett substituent constant values [44, 45]. The $-NO_2$ groups are very strong electron withdrawing groups while $-CH_3$ groups are electron donating (Table 11.1).

11.3.2   NMR Chemical Shift Concepts for Π–Π Complexes

NMR spectroscopy is a proven technique for evaluating charge-transfer binding between aromatics [41, 42]. Before mixing the two aromatic compounds suspected of undergoing π–π interaction, the chemical shift values for the protons on each ring are measured (Fig. 11.17). When acceptor and donor molecules are mixed the chemical shifts of the protons on the acceptor aromatic ring are compared to the shifts before mixing. Π–π complexation causes the protons on the acceptor ring to be shifted upfield to the right on the parts per million (ppm) scale because of electron density transferred to the acceptor ring from the donor ring (Fig. 11.18).

An upfield chemical shift for acceptor nuclei results from a change in the magnetically induced ring current and stacking interactions between the aromatic rings [46, 47]. Hanna and Ashbaugh developed a mathematical expression to evaluate NMR data similar to the Benesi-Hildebrand equation using UV–vis data for correlating π–π complexes formed between 7,7,8,8-tetracyanoquinodimethane (TCNQ) and various aromatic donors [24, 41]. The equations are valid if the donor:acceptor

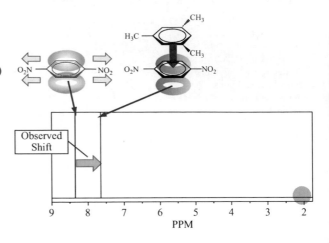

Fig. 11.18 Observed upfield chemical shift of protons on the acceptor ring (*green*) when a π–π complex forms with a donor (*right* molecule)

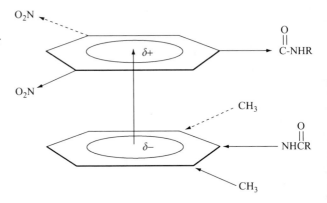

Fig. 11.19 Typical donor–acceptor π–π charge transfer complex structure. The lower ring is a mimic of bupivacaine donor

concentration ratio is large (e.g. >20). Figure 11.19 depicts the interaction between an acceptor dinitroaromatic and the donor aromatic portion of bupivacaine.

NMR chemical shifts have been reported even for π–π complexes of antimalarials with heme species [48, 49], for paraquat probes of certain aspects of polypeptide and protein conformations [50], and for charge transfer complex formation between trinitrobenzene and a structural analog of the important neurochemical dopamine [51].

11.3.3   NMR Data for Binding Overdosed Chemicals

In the work summarized in this chapter, bupivacaine (BPC) (Fig. 11.14) and its molecular mimics 2,6-dimethylaniline (DMA) and 2,6-dimethylacetanilide (DMAc) were evaluated as donors for π–π complexation with the series of acceptors. Figure 11.20 identifies acceptors (A) as well as the donor (D) synthesized. As an ex-

$$Ar-X + RY \longrightarrow \text{Amide Products}$$

Compound	AR	X	RY	Z
DNB (A)	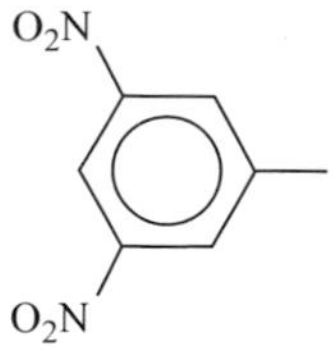 (benzene ring with O$_2$N and O$_2$N substituents)	COCl	CH$_2$ NH$_2$	CONHCH$_3$
EDNB (A)	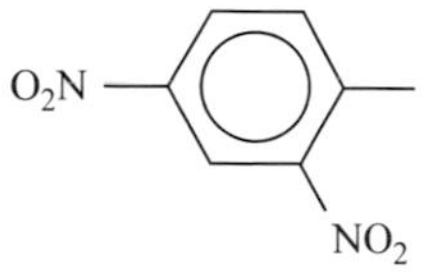 (benzene ring with O$_2$N and NO$_2$ substituents)	COCl	CH$_3$ CH$_2$ NH$_2$	CONHCH$_2$CH$_3$
EDNS (A)	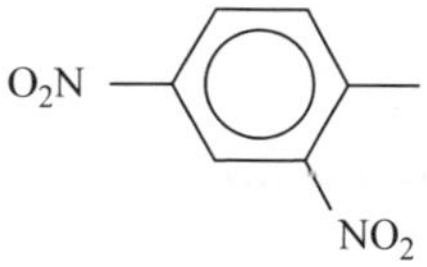 (benzene ring with O$_2$N and NO$_2$ substituents)	SO$_2$Cl	CH$_3$ CH$_2$ NH$_2$	SO$_2$NHCH$_2$CH$_3$
MPFB (A)	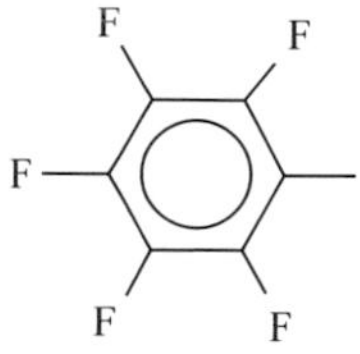 (benzene ring with F, F, F, F substituents)	COCl	CH$_3$ NH$_2$	CONHCH$_3$
DMAc (D)	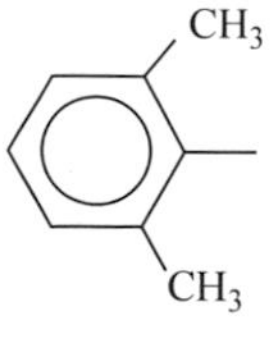 (benzene ring with CH$_3$ and CH$_3$ substituents)	NH$_2$	CH$_3$ COCl	NHCOCH$_3$

Fig. 11.20 Structures of acceptor (*A*) and donor (*D*) molecules synthesized in this study

ample, EDNB was prepared from 2,4-dinitrobenzoyl chloride reacting with ethyl-amine.

The protons on the rings of the acceptor molecules N-methyl-3,5-dinitrobenza-mide (DNB) and N-ethyl-dinitrobenzenesulfonamide (EDNS), when mixed with DMAc or BPC, exhibit upfield shifts as shown in Table 11.2. It is noted in Table 11.2 that all acceptor ring protons have upfield shifts but the magnitude of the shifts for protons H3, H5 and H6 attached to different positions on the ring of EDNS or DNB are not equal.

This can be understood in terms of their positions relative to the electron with-drawing groups in the acceptor molecule. The stacking of one ring over another may

Table 11.2 Observed changes in the chemical shifts of N-ethyl-2,4-dinitrobenzamide and N-ethyl-2,4-dinitrobenzenesulfonamide when complexed with 2,6-dimethylacetanilide (DMAc) and bupivacaine (BPC) in chloroform-d

Donor	Observed chemical shift changes (ppm)					
	Benzamide (DNB)			Sulfonamide (EDNS)		
	H3	H5	H6	H3	H5	H6
DMAc	0.2496	0.3242	0.3663	0.2252	0.3014	0.2514
BPC	0.3531	0.3686	0.2903	–	–	–

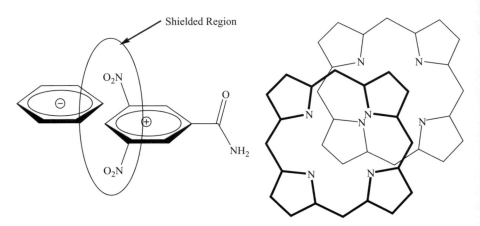

Fig. 11.21 Examples of offset stacking of acceptor–donor π–π complexed aromatics. Monocyclic aromatics (*left*); porphyrins (*right*)

not be perfectly vertical and symmetrical and the same for the induced magnetic influence on the shared rings currents (Fig. 11.21) [27, 49].

11.3.4 Use of Fluorinated Acceptor Aromatics in Microemulsions to Achieve Selective Aromatic Drug Absorption

Recalling the title of Sect. 11.3 of this chapter, it was of interest to select an aromatic molecule for use *in vivo* as oils in microemulsions to evaluate differential π–π binding between two or more overdosed donor chemicals.

Recent reviews report that fluorine-containing compounds have enjoyed a rich history in bioorganic and medicinal chemistry [52, 53]. Replacement of fluorine atoms for protons in therapeutics has been widely studied as a result of their diverse functions, ranging from being isosteric and isoelectronic replacements for hydroxyl groups, to enhancing the metabolic stability of drugs and modifying electronic and physical properties (for example, lipophilicity, acidity, and steric hindrance).

The precedent that fluorine substitution into molecules intended for use in the human body retains biocompatibility encouraged our use of 1,3-bis(trifluoromethyl) benzene (BTFMB) and N-methylpentafluorobenzamide (MPFB) as acceptor oils in microemulsions. Fluorine atoms are more electronegative than carbon or proton atoms and have a Hammett substituent constant indicating they are capable, like nitro groups ($-NO_2$), of withdrawing electron density from a benzene ring to which they are attached.

The bupivacaine mimic DMAc (Fig. 11.20) and cocaine (Fig. 11.1) were used as donors in separate experiments. As expected, all proton and fluorine atoms on the two acceptor molecule rings exhibit upfield shifts, signifying that both DMAc and cocaine act as π–π donor molecules and enter into π–π complex formation with the fluorinated acceptors.

As with the differently positioned protons in the acceptors shown in Table 11.2, the magnitude of the shifts are not equal for all of the fluorine or proton atoms in different positions in the BTFMB and MPFB acceptor molecules. For example, in BTFMB acceptor there are three different positions on the ring where protons are located. They are positions 2 (1H), 4 and 6 (2 H's) and 5 (1H).

Microemulsion studies for demonstrating selective removal of one chemical over another from saline were carried out using a mixture of both DMAc and co-caine donor molecules in the same solution. Importantly, for the first time a smart microemulsion has been prepared which has selective absorption for one bioactive molecule over another in a mixture. Using BTFMB as the oil rather than ethyl bu-tyrate, and Pluronic and fatty acid co-surfactants for dispersion stabilization, the dispersion shows remarkable selectivity for extraction of bupivacaine over cocaine (Fig. 11.22) [54].

It is proposed that the increased partitioning of bupivacaine from normal saline over that of cocaine into BTFMB oil in the microemulsion is due at least in part to cocaine's benzene ring having less negative polarity and donor ability than the ring of bupivacaine. The acyl carbonyl attached to the benzene ring in cocaine has elec-tron withdrawing capability according to Hammett (Table 11.1), while the methyl groups on the bupivacaine ring are electron donating (Fig. 11.23).

11.3.5   Π–Π *Concepts for Proposed Ricin Biotoxin Inhibitors*

Ricin powder is a lethal protein from Castor beans that has been used at very low microgram/kilogram levels for assassination of humans. Robertus has been one of several leaders in elaborating the structure of Ricin, how it functions to kill cells, and determining what structural features an inhibitor must have [55–58]. It is com-posed of two peptide chains, RTA and RTB, with the former protein hosting the active site which has two exposed tyrosine amino acid molecules across opposite sides of the opening (Fig. 11.24). Furthermore, Robertus has defined an "iterative crystallographic inhibitor search" algorithm and from it has shown that a derivative

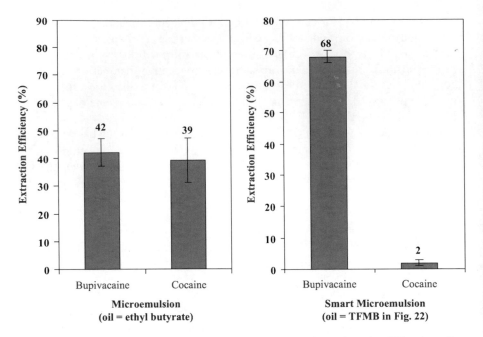

Fig. 11.22 Microemulsion selectivity for extraction of bupivacaine and cocaine differs depending on oil core in the droplets (Fig. 11.15)

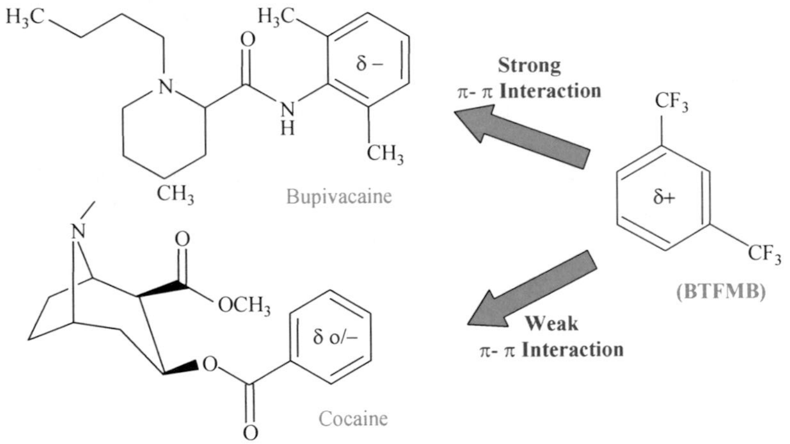

Fig. 11.23 Selective π–π interaction between a smart oil (BTFMB) used for microemulsions and different π donors

of the aromatic heterocyclic amine pteridine shows reasonable inhibitor activity [55].

A key connection between the discovery that pteridine derivatives enter and partially close off the binding site, and the discussion in this chapter about π–π

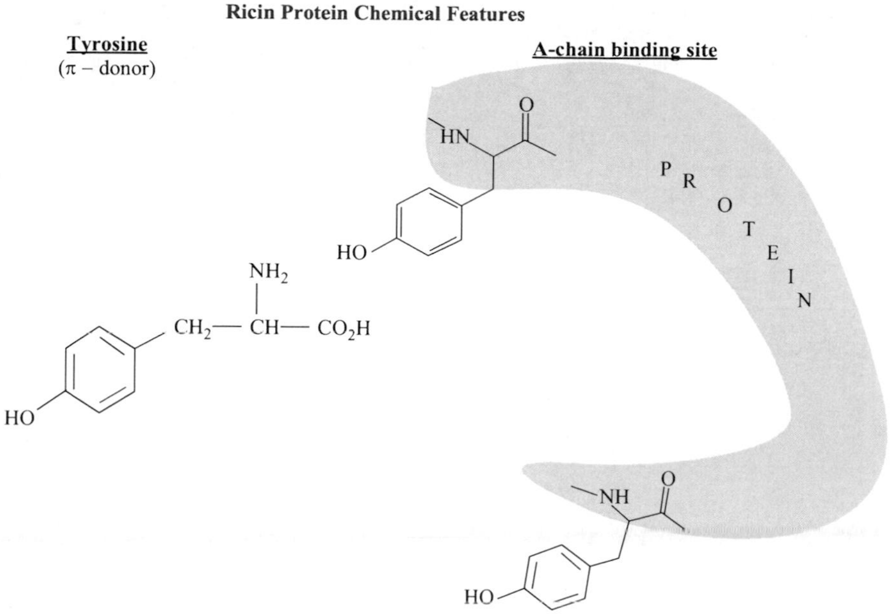

Ricin Protein Chemical Features

Fig. 11.24 Ricin RTA protein binding site showing the exposed tyrosine amino acid residues

complex formation between acceptor and donor aromatics, is that there is evidence that the heterocycle inhibitor interacts with the toxin near the two exposed tyrosine residues in the opening of the binding site. It may that the pteridine derivative is attracted to and held in position by π–π interaction between the aromatic portion of the heterocycle and a tyrosine molecule. However, the inhibitor activity is claimed to be due mostly to multivalent hydrogen bonding with amino acid residues in the distant back region of the binding site (Fig. 11.25) [59–61].

11.3.5.1   Measurement of Π–Π Bonding Using AFM

As an extension of the work reported by the authors of this chapter on π–π concepts for binding overdosed chemicals, NMR and atomic force microscopy (AFM) experiments have been carried out yielding encouraging results that π–π concepts may lead to an improved inhibitor for Ricin.

Demonstrated in sections above is that upfield NMR chemical shifts for acceptor protons occur when acceptor–donor complexes form. Relevant to binding an acceptor to Ricin, it was of interest to evaluate such shifts when acceptor DNB was exposed to tyrosine and related donors. Table 11.3 data shows that only proton B shown on the DNB molecule is shifted upfield when any of the four donors are present. This suggests that the stacking of the two aromatic rings in the complex is asymmetrical as in Fig. 11.21 (left).

Fig. 11.25 Two representations of interaction of a pteridine derivative with amino acids in the binding site of Ricin RTA protein. Tyrosine amino acid residues (TYR) in left structure. Pteridine hydrogen bonding in right structure

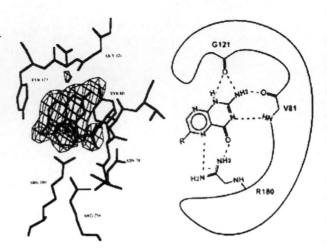

Table 11.3 Chemical shift data for the various protons on acceptor DNB when the four different donors are present. Std. Dev. ±0.03%

	Donor Molecules (Chemical Shifts are in ppm)			
Acceptor Protons	p–Cresol	L–Tyrosine-HCl	4–aminophenol	4–methylanisole
A	N/A	−0.00318	−0.01109	−0.06860
B	0.09758	0.06150	0.02184	0.10940
C	−0.01124	−0.16259	−0.03620	−0.12611
D	0.03421	0.00168	−0.03066	−0.14311

Atomic Force Microscopy (AFM) was developed in 1986 and is well known for its ability to examine the topography of a surface on the nanometer scale in 3-dimensions [62]. Another interesting aspect of AFM is using a force probe to measure the long-range and adhesive forces between an AFM tip and a substrate. The microscope is sensitive to the order of a few picoNewtons of force and can easily detect electrostatic, Van der Waals, hydrogen bonding, and other molecular forces between molecules. The tip and the substrate can be chemically modified in order to perform

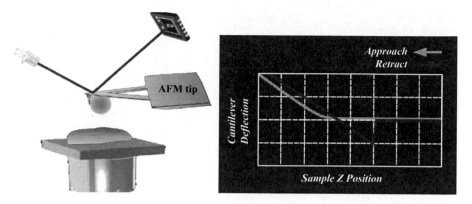

Fig. 11.26 AFM apparatus configuration (*left*) and representative graph of standard AFM adhesion force curves. *Orange* and *red lines* are for cantilever tip approach and retraction from a substrate, respectively

such experiments. A standard adhesion force curve between the cantilever deflection and sample Z position can be seen in Fig. 11.26 [63].

In Fig. 11.26 graph the orange line is showing the tip approaching the substrate surface from right to left. No interaction is evident between the tip and the substrate on the approach until the tip is a few angstroms from the surface. Then, as shown, there is repulsion between the molecules on the tip and the molecules on the substrate due to the contact electrostatic interactions between the tip and substrate. This effect can be seen by the upward curve of the approach line from right to left. This region of the force curve is known as the contact regime.

The interesting part of a force curve is during the retraction of the tip from the substrate after they have been in contact. In the force curve shown, the red line is showing the retraction of the tip from the substrate from left to right. Once the tip has come out of the contact regime the curve dips sharply downward showing that the cantilever tip is adhering to the substrate. When the tip detaches from the substrate there is a break in the curve and the line returns back to the natural resting position of the cantilever. For statistical accuracy force curve data is collected many times and the adhesion between the molecules on the tip and the molecules on the substrate can be calculated.

11.3.5.2 AFM Force Measurement Data on Donor–Acceptor Pairs

The AFM technique is useful for evaluating the attractive forces between a π-acceptor DNB and 3,5-dimethylacetanilide (DMAc), the latter being a donor molecule like tyrosine in the opening of the active site of Ricin. The adhesion calculated in this study directly correlates to the stacking of aromatic rings and the interaction of π–π complexes formed.

In 1999, Clear and Nealey investigated the influence of solvent polarity on adhesion using AFM [64]. In their study hydrophobic methyl groups were attached to

Fig. 11.27 Force curves
showing solvent effects on
hydrophobic interactions
between molecules on an
AFM tip and on the substrate

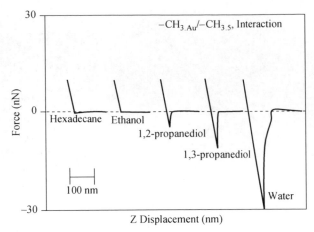

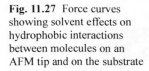

Fig. 11.28 Synthesis scheme for the attachment of acceptor 3,5-dinitrobenzoyl chloride (DNB) to
a polyaminostyrene particle (PS) mounted on an AFM cantilever tip

both the AFM tip and substrate. Force analyses were performed in solvents: water,
1,3-propanediol, 1,2-propanediol, ethanol, and hexadecane. According to curves
shown in Fig. 11.27, as the solvent polarity increases, the adhesion between the
functional groups on the tip and substrate also increases.

This study showed that hydrophobic attraction between the methyl functional
groups on a tip and substrate could be "cancelled out" by using an apolar solvent.
This work provides useful information for future experiments by the authors on
examining π–π interactions between tyrosine moieties and a known π acceptor molecule.

11.3.5.3 AFM Force Measurement Data on DMA–DNB Complex

The force of interaction between an acceptor–donor π–π complex of aromatic rings
was first determined using the pair of molecules DNB and DMAc (Fig. 11.20). Each
donor and receptor were covalently attached to either the AFM cantilever tip or the
substrate that the tip was going to approach and retract from. Example chemistries
employed for these attachments are shown in Figs. 11.28 and 11.29.

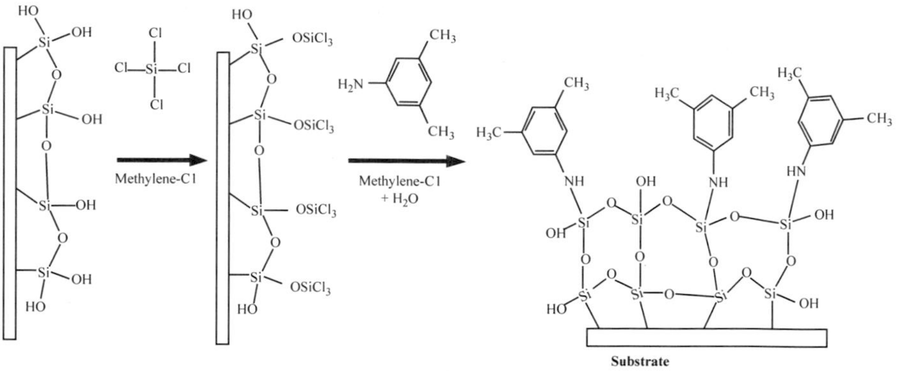

Fig. 11.29 Two step modification for attachment of dimethylaniline (DMA) to the AFM silica substrate

Both infrared spectroscopy and thermogravimetric analyses verified that the desired conversions were successful.

The results for the force analysis studies performed with modified/unmodified PS tips vs. modified/unmodified silica substrates in an aqueous environment are shown in Table 11.4. The experiment was performed using a fluidic cell that was filled with 1 mM KNO_3. The adhesion force of 66 pN measured in water was reduced to 30 pN when measurements were made in octanol solvent. By comparison with the curves in Fig. 11.27, the lower value may be the force of adhesion due solely to π–π interaction between the donor and acceptor molecules when in contact with each other.

This work provides useful information for future experiments by the authors on examining π–π interactions between tyrosine-containing peptides and a π acceptor molecule with the goal of identifying a new and more powerful inhibitor for Ricin.

11.3.5.4 Diffusion NMR Studies for Evaluating Interaction of a Π Acceptor with Tyrosine of Ricin Peptide

Sections 11.3.5.1–11.3.5.4 has established that π–π interaction forces between acceptor and donor aromatic rings can be obtained using the AFM technique. In parallel, NMR diffusion ordered spectral measurements (DOSY) yield information corroborating conclusions obtained from AFM experiments.

NMR-DOSY gives information on the rate of diffusion of molecules in a solvent, a process that is controlled by the degree of molecular association of the chemical species being monitored [65]. Using the pulse gradient spin echo sequence, different angles of radio frequency are used to study the Brownian motion of molecules. To study the interactions of two molecules, the diffusion coefficient of one molecule is measured and then the diffusion coefficient is measured of the two molecules

Table 11.4 Force analysis summary of modified/unmodified AFM tips and substrates performed in 1 mM KNO$_3$

Tip	Substrate	Force Interaction
O$_2$N— [benzene ring] —NO$_2$, X	[benzene ring], X	Adhesion, 66pN's
CH$_2$CH$_2$NH$_2$	[benzene ring], X	Adhesion
O$_2$N— [benzene ring] —NO$_2$, X	O$_2$N— [benzene ring] —NO$_2$, X	Repulsion
CH$_2$CH$_2$NH$_2$	O$_2$N— [benzene ring] —NO$_2$, X	Repulsion
O$_2$N— [benzene ring] —NO$_2$, X	Si-OH	Repulsion
CH$_2$CH$_2$NH$_2$	Si-OH	Repulsion

mixed together. If the diffusion coefficient decreases for molecules in the mixture, there is strong association between the two.

In the context of the overall theme of Sect. 11.3.5 of this chapter, the DOSY technique has been employed to determine if tyrosine forms π–π complexes with a known acceptor like dinitrobenzamide (DNB).

Each of the chemical species, tyrosine, DNB and the desired π–π complex, have different molecular weights and therefore should have different rates of diffusion. The experimental data in Table 11.5 supports the conclusion that not only monomeric tyrosine acts as a donor to complex with DNB, but so does tyrosine residue in a tripeptide derivative. The acceptor DNB in solution without a donor present has a larger diffusion coefficient than when any of the three donors shown in Table 11.5 are in the solution. This information may contribute to the discovery of a new Ricin inhibitor incorporating a strong π acceptor ring for π–π interaction with one of the two tyrosine residues in the opening of the binding sight (Figs. 11.24 and 11.25). Powerful multivalent binding may occur if a pendant group is attached to the accep-

Table 11.5 Diffusion coefficients as measured by the NMR DOSY technique for solutions of DNB π acceptor without and with added π acceptors. Std. Dev. $=\pm5\%$

Sample in DMSO	Diffusion coefficient (m^2/s)
DNB alone	1.465×10^{-9}
DNB+Tyrosine (1:20)	4.684×10^{-11}
DNB+Bupivacaine (1:20)	5.757×10^{-10}
DNB+Leu-Tyr-Leu (1:20)	1.636×10^{-10}

Decrease in D_i indicates molecular association

tor ring capable of hydrogen bonding to other amino acid residues in another region of the sight [59–61].

11.4 Functionalized Colloidal Particle Studies

11.4.1 Surface Modification Chemistry

A plethora of technical publications over the past several decades reveal that surface modification of solid core particles of several compositions, shapes and sizes, using all types of chemical reagents and reaction techniques, is now well understood and applied to enhance the properties of many powders used in manufacturing and medicine [66–61]. One aspect of the present work is to attach molecular units to the surface of carrier particles which, when introduced into the blood stream, will complex with and deactivate overdosed toxins (Refer to Fig. 11.2) In other words, to have controlled **removal** of drug/toxin molecules rather than controlled **release**.

A wide range of both inorganic and organic solid carrier particles exist to which π acceptor moieties can be attached. Inorganic examples are oxides of aluminum, calcium, cerium and silicon. An example of an organic carrier particle is oligochitosan. Both core and surface compositions should be biocompatible and the receptors must be able to differentiate overdosed toxin molecules from ones having similar structures but needed for normal human health. If the modified carrier particles are less than 10 nm in diameter they may be able to pass through the kidney carrying the toxin with them during urination. Alternatively, carrier particles too large to be removed by normal body function may be composed of biodegradable material of which smaller degradation units carrying toxin may be eliminated.

Both silica and oligochitosan carrier particles have been successfully used to establish proof-of-concept that π–π charge-transfer chemistry for binding toxic molecules has merit. Classic Stober method chemistry was employed to prepare several sizes of nearly monodisperse spherical silica particles (Fig. 11.30) [92, 93]. In addition, higher surface area and more porous silica particles were prepared by the sol–gel method by incorporating pore templating molecules into the reaction mixture and subsequently removing them by solvolysis or thermal treatment [94–98]. An example experiment of the latter type consisted of adding bupivacaine or its mimics 3,5-dimethyl aniline or N-acetylated derivative into the sol–gel reaction.

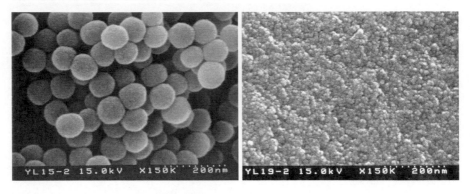

Fig. 11.30 Scanning electron micrographs of two sizes of colloidal silica synthesized by the Stober sol–gel method. Both photos are at 150 K magnification

After particle formation the entrained organics were removed from the silica carrier particles by thermal treatment in air at 300° C. Their specific surface area was greater than 1,000 m^2 g^{-1}.

The surfaces of silica particles have many exposed silanol–OH groups which undergo facile reaction with, for example, reagents having carboxylic acid chloride functional groups [99–101]. Thus, preparation of the desired functionalized silica carrier particles was easily accomplished by refluxing 8 nm diameter silica in a nonhydroxylic solvent with excess π acceptor reagents such as dinitro or bis-trifluoromethyl substituted benzoyl chlorides. This process yielded silica particles having numerous π receptor groups covalently attached to the colloidal silica (Figs. 11.30 and 11.31 [2–4]).

Attachment of π receptor groups to colloidal oligochitosan having reactive amino and hydroxyl groups (Fig. 11.32) has also been achieved [102–106]. Specifically, oligochitosan having average molecular weight of 1,150 Da and containing 8% water was dissolved in DMSO containing 2,4-dinitrobenzenesulfonyl chloride and stirred at 20°C for 48 h. Addition of ethanol caused the oligochitosan particles having 300–500 nm diameters functionalized with the π acceptor to precipitate. The π acceptors thus attached showed good ability to bind π donors such as bupivacaine or its mimics dimethylaniline and dimethylacetanilide to the particles [3, 4].

Preliminary results prove that silica carrier particles functionalized with a π acceptor as shown in Fig. 11.31 are very efficient in removing bupivacaine hydrochloride dissolved in normal saline.

HPLC analysis showed that the silica particles not functionalized with a π acceptor removed no drug from the liquid containing from 1,000 to 30,000 μM bupivacaine. However, 0.05 and 0.10% w/v derivatized nanoparticles exhibited removal of the drug in amounts of 1,800 and 4,000 μM, respectively when the original bupivacine was 10,000 μM or more (Fig. 11.33).

The proposed mechanism of removal is depicted in Fig. 11.34. The preliminary results do not allow differentiation as to whether 1 or 2 point interaction of the drug with the particles occurs. If the number of π acceptor units attached to a particle,

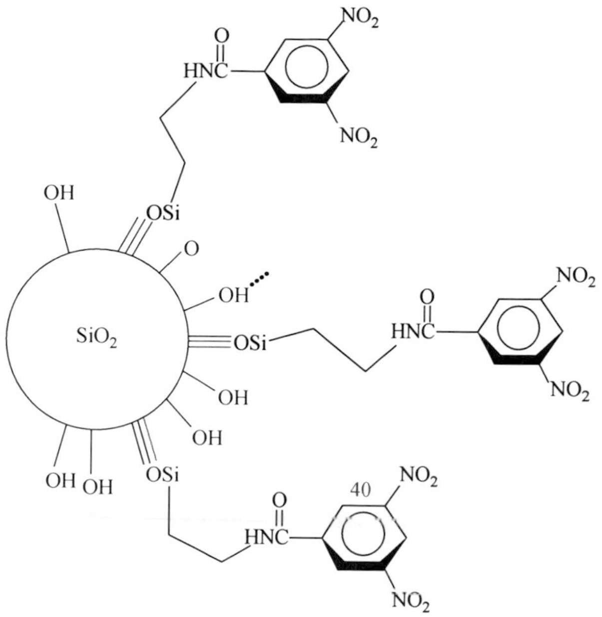

Fig. 11.31 Synthetic procedure for preparing silica modified with DNB-type π acceptors

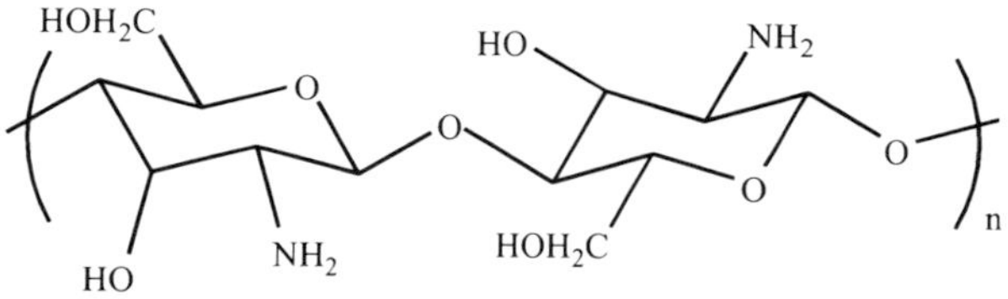

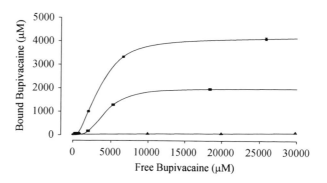

Fig. 11.32 Chemical structure of repeat unit of oligochitosan

Fig. 11.33 Effectiveness of removal of bupivacaine hydrochloride from normal saline by silica (▲), and by two concentrations of functionalized silica (■, ●) represented by Fig. 11.31

and their freedom of movement on the end of the tethers linking them to the silica allows the basic nitrogen of a drug to diffuse close to and hydrogen bond with the residual silanol groups, the 2 point interaction should be facilitated and improve binding efficiency [59–61]. Referring to Figs. 11.14 and 11.5 it may be that the less

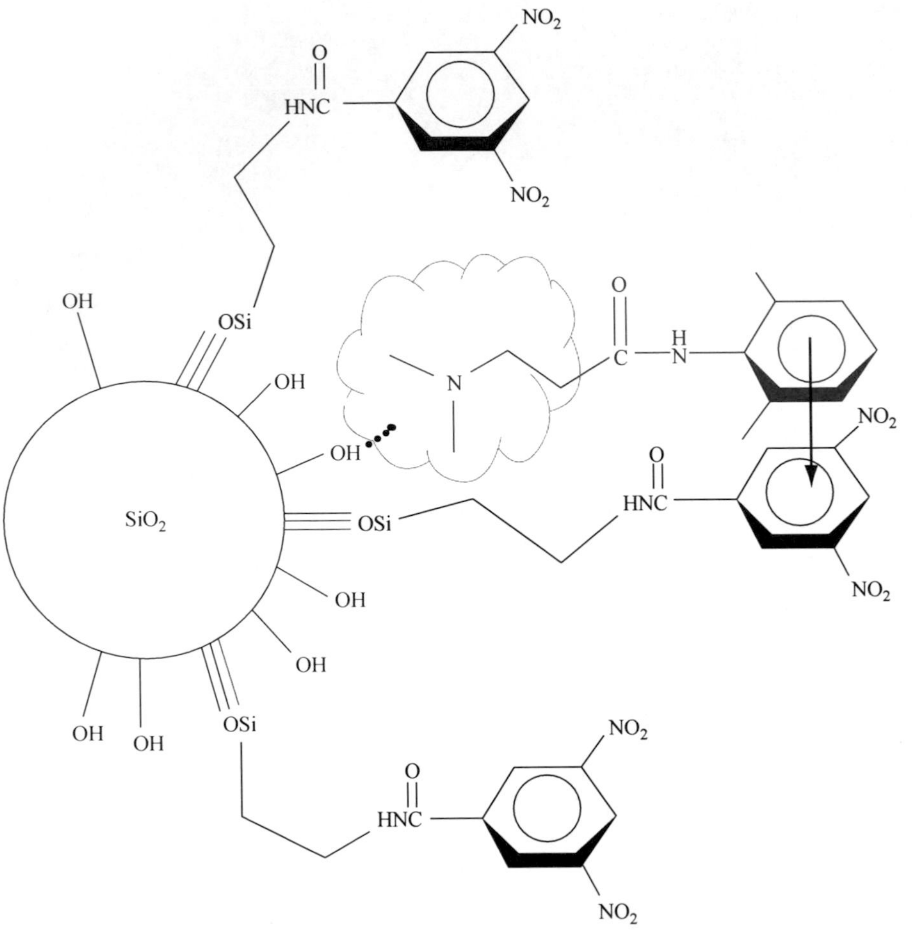

Fig. 11.34 Surface chemical features of a nanoparticle showing covalently attached π acceptor aromatic rings complexed in two possible ways to the π donor aromatic rings of bupivacaine

sterically encumbered amino group of amitriptyline could reach within hydrogen bonding distance to the Si–OH groups. The amino group of bupivacaine is surrounded by considerable hydrocarbon structure and may be hindered from entering that crowded space.

11.5   Summary and Path Forward

The experimental results presented in this chapter show that the concept of π–π complex formation between acceptor and donor aromatic rings applies to the theme "Medicine in Reverse". Three types of injectable and biocompatible dispersed phases have been prepared which show promise as antidotes for a selection of commonly overdosed drugs. The oil-in-water microemulsions deplete overdoses of amitriptyline, bupivacaine and even cocaine through a combination of electrostatic and hydrophobic action. A smart microemulsion has been prepared using a π acceptor oil that can differentiate between strong and weak π donor molecules. Colloidal oligochitosan and silica nanoparticles, each derivatized with π acceptor dinitroaromatics, bind the same drugs efficiently and rapidly while their underivatized precursors do not.

Proof-of-concept that the research may lead to clinically viable remedies for some drug overdoses comes from the described results of *in vitro* and *in vivo* experiments. Preliminary experiments show that the cardiotoxicity of at least drugs **a**, **c** and **e** drugs in Fig. 11.1 is successfully reversed when the microemulsion or oligochitosen phases are infused into excised rat hearts or whole animals.

And, preliminary results not described above reveal that compounds **d** and **f** in Fig. 11.1 have donor properties and can be bound by complexation with an acceptor.

Finally, the results obtained both by AFM and NMR DOSY experiments hint that through further study a new and more active inhibitor of Ricin biotoxin may be forthcoming by incorporating both π–π complexation and hydrogen bonding concepts into the molecule.

The path forward by the interdisciplinary team will be to (1) further improve the compositions of the microemulsions to make them more "smart" regarding drug adsorption, (2) adapt the π acceptor π donor concept for use with all three types of dispersed phases, (3) test the effectiveness of the phases to reverse overdose effects of other than the three cardiotopxic therapeutics, (4) obtain fundamental information on interaction of molecules **d** and **f** in Fig. 11.1, and even methamphetamine, with the dispersed phases using NMR techniques and ^{13}C-enriched drugs [107], and (5) enhance the biocompatibility of the optimized formulations.

References

1. For general discussions on colloids in medicine see Matijevic, E. ed., *Medical Applications of Colloids,* 1st Ed., Springer Pub., 2008; and the multiple authored issue of *MRS Bull,* **2009**, *34*, 406-464.
2. Partch, R., Powell, E., Lee, Y-H.,Varshney, M., Shah,D., Baney, R., Lee, D., Dennis, D., Morey, T. and Flint, J., *Finely Dispersed Particles*, A. Spasic and J-P. Hsu, Eds, **2005**, *28*, 813-832.
3. Lee, D., Flint, J., Morey, T., Dennis, D., Partch, R., and Baney, R., *J. Pharm. Sci.* **2004**, *94*, 373-381.
4. Ford, E., Lee, D., Flint, J., Morey, T., Dennis, D., Partch, R. and Baney, R., *Int. J. Nanomedicine* **2007**, *2*, 1-11.
5. MD/PhD's D. Dennis and T. Morrey, Shands Medical School; Professors R. Baney and B. Moudgil, Material Science and Engineering; Professor D. Shaw, Department of Chemical Engineering; Visiting Professor M. Varshney, Hamdara University, India.
6. Professor Young H. Lee, Department of Chemistry.

7. Dr. Abeer Al Bawab, Faculty of Science and Director of Nanotechnology Center Abeer Al Bawab and Dr. Fadwa Odeh, Chemistry Department.
8. Somasundaran, P., Liu, F. and Gryte, C., *Science and Engineering* **2003**, *89*, 235-241.
9. Underhill, R., Jovanovic, A., Carino, S., Varshney, M., Shah, D., Dennis, D., Morey, T. and Duran, R., *Chem. Mater.* **2002**, *14*, 4919-25.
10. Mitchel, D., Lee, S., Trofin, L., Li, N., Nevanen, T., Soderlund, H. and Martin, C., *J. Am. Chem. Soc.* **2002**, *124*, 11864-65.
11. Bagwe, R., Kanicky, J., Palla, B., Patanjali, P. and Shah, D., *Critical Reviews in Therapeutic Drug Carrier Systems* **2001**, *18*, 77-140.
12. Palla, B., Shah, D., Garcia-Casillasa, P. and Matutes-Aquino, J., *Journal of Nanoparticle Research* **1999**, *1*, 215-226.
13. Shah, D.O. (Ed.), *Micelles, Microemulsions, and Monolayers*, Marcel Dekker Inc., New York, **1998**, 1-610.
14. Sjoblom, J.; Lindberg, R.; and Friberg, S., *Adv. Colloid Interface Sci.* **1996**, *65*, 125- 87.
15. Scottmann, T., *Current Opinions Colloid Interface Sci.* **2002**, 7, 57-65.
16. Jakobs, B., Scottmann, T., Strey, R., Allgaier, J., Willner, L. and Richter, D., *Langmuir* **1999**, *15*, 6707.
17. Friberg, S., Mortensen, M. and Neogi, P., *Sep. Sci. Technol.* **1985**, *20*, 285.
18. Orringer, E., Casella, J., Ataga, K., Koshy, M., Adams-Graves, P., Luchtman-Jones, L., Wun, T., Watanabe, M., Shafer, F., Kutlar, A., Abboud, M., Steinberg, M., Adler, B., Swerdlow, P., Terregino, C., Saccente, S., Files, B., Ballas, S., Brown, R., Wojtowicz-Praga, S. and Grindel, J., *JAMA* **2001**, *286*, 2099-3010.
19. Varshney, M., Morey, T., Shah, D., Flint, J., Moudgil, B., Seubert, C. and Dennis, D., *J. Am. Chem. Soc.* **2004**, *126*, 5108-12.
20. Allemann, E., Gurny, R. and Dekker, E., *Eur. J. Pharm. Biopharm.* **1993**, *39*, 173.
21. Junquera, E.; Romero, J. and Aicart, E., *Langmuir* **2001**, *17*, 1826-32; Pancrazio, J., Kamatchi,,G., Pentel, P. and Benowitz, N., Tricyclic , *Med. Toxicol.* **1986**, *1*, 101-121; Roscoe, A. and Lynch, C., *J. Pharmacol. Exp. Ther.* **1998**, 284, 208-214; Marshall, J. and Forker, A., *Am. Heart J.* **1982**, *103*, 401-414.
22. Strichartz, G.; Sanchez, V.; Arthur, G.; Chafetz, R. and Martin, D., Fundamental Properties of Local Anesthetics. II, Anesth. Analg., **1990**, *71*, 158-70.
23. Sanders, J. and Hunter, B., *NMR Spectroscopy*, 2n **1997** OxfordUniversityPress.
24. Benesi H and Hildebrand J., *J. Am. Chem. Soc.* **1949**, *71*: 2703.
25. Briegleb G. *Electronen-donator-acceptor-komplexe*. **1961**, Berlin. Springer-Verlag; Chujo, Y. and Tamaki R., *MRS Bull,* **2001**, May, 389.
26. Mulliken R., *J. Phys. Chem.* **1952**, 56: 801.
27. Dewar,M and Lepley, A., *J. Am. Chem. Soc.* **1961**, *83*, 4560.
28. Meyer, E., Castellano, R and Diederich F., *Angew. Chem. Int. Ed.* **2003**, *42*, 1210.
29. Hunter, C., *Chem. Soc. Rev.* **1994**, *23*, 101-109.
30. Boal, A. and Rotello, V., *J. Am. Chem. Soc.*, **2000**, *122,* 734
31. Stinson, S., Chiral Cravings, *Chem. Eng. News,* **2001**, Dec. *10*, pg 35-38.
32. Brunsveld, L., Folmer, B. and Meijer, E., *MRS Bull* **2000**, *25*, 49-53; Arnold, C. , *Chem. Eng. News,* **2008**, July 21, 48-50; Ritter, S., *Chem. Eng. News*, **2008**, April, 21, 50-51.
33. Rouhi, A., *Chem. Eng. News* **2001**, July 30, 46-49.
34. Goodnow, T., Reddington, M., Stoddart, F. and Kaifer, A., *J. Am. Chem. Soc.* **1991**, *113*, 4335-37.
35. Wasset, W., Ghobrial, N. and Agami, S., *Spectrochim. Acta* **1991**, *47A*, 623-27.
36. Fesik, S., Medek, A., Hajduk, P. and Mack, J., *J. Am. Chem. Soc.* **2000**, *122*, 1241.
37. Regeimbal, J., Gleiter, S., Trumpower, B., Yu, C-A., Diwakar, M., Ballou, D. and Bardwell, J., *Proc. Nat. Acad. Sci.* **2003**, *100*, 13779-84.
38. Adams, H., Blanco, J., Chessari, G., Hunter, C., Low, C., Sanderson, J. and Vinter, J., *Chem. Eur. J.* **2001**, *7*, 3494-3503.
39. Sinnokrot, M., Valeev, E. and Sherrill, C., *J. Am. Chem. Soc.* **2002**, 124, 10887-893; Rao, C., *J. Am. Chem. Soc.*, **2010**, DOI: 10.1021/ja100190p.

40. Ramraj A. and Hunter, I., *J. Chem. Inf. Model.* **2010**, *50*, 585-588.
41. Hanna, M., and Ashbaugh, A., *J. Phys. Chem.* **1964**, *68*, 811; Kotov N. and Westenhoff, S., *J. Am. Chem. Soc.*, **2002**, *124*, 2448.
42. Foster, R. and Fyfe, C., *Proc. Nucl. Magn. Reson. Spectrosc.* **1969**, *4*, 1-9; Aida, T., Yamamoto, Y., Fukushima, T., *Science,* **2006**, *314*, 1761 , 1829-43.
43. Neusser, H. and Krause, H., *Chem. Rev.* **1994**, 94, 1829-1843; Imanishi, Y. and Higashimura, T., *Macromolecules,* **1977**, *10*, 125-130.
44. Jaffe, H., *Chem. Rev.* **1953**, *53*, 191-261; Teixidor, F. and Vinas, C., *J. Am. Chem. Soc.* **2005**, *127*, 10158.
45. Hansen, O., *Acta Chem. Scand.* **1962**, *16*, 1593-1600.
46. Ishida, T., Ibe, S. and Inoue, M., *J. Chem. Soc. Perkin Trans.*, **1984**, *2*, 297.
47. Hunter, C., Lawson, K. and Perkins, J. et al., *J. Chem. Soc. Perkin Trans.* **2001**, *2*, 651.
48. Constantinidis, I., and Satterlee, J., *J. Am. Chem. Soc.* **1988**a, *110*, 927.
49. Constantinidis, I. and Satterlee, J., *J. Am. Chem. Soc.* **1988**b, *110*, 4391; Hunter, C. and Sanders, J., *J. Am. Chem. Soc..* **1990**, *112*, 5525-5534.
50. Verhoeven, J., Verhoeven-Schoff, A., Masson A., et al., *Helv. Chim. Acta,* **1974**, *57*, 2503; http://www.cis.rit.edu/htbooks/nmr/inside.htm
51. Dust, J., Can. *J. Chem.* **1992**, 70, 151-57.
52. Meyer, E., Castellano, R., and Diederich, F., 2003 *Angew. Chem. Int. Ed* **2003**, *42*, 1210; Ritter, S., Chem. Eng. News **2007**, September 10, 36-37; Borman, S., Chem. Eng. News **2005**, 40-42.
53. Thayer, A., 2006 *Chem. & Eng. News,* **2006**, 84(23): 15.
54. Henry, C., *Chem. Eng. News*, **2004**, June 28, 11.
55. Robertus, J., Yan, X., Ernst, S., Monzingo, A., Worley, S., Day, P., Hollis, T. and Svinth, M., *Toxicon.* **1996**, *11/12*, 1325-1334.
56. Yan, X., Hollis,T., Svinth, M., Dat, P., Monzingo, A., Milne,G., Robertus, J., *J. Mol. Biol.* **1997**, *266*, 1043-49.
57. Miller, D., Ravikumar, K., Shen, H., Suh, J-K., Kerwin, S., Robertus, J., *J. Med. Chem.* **2002**, *45*, 90-98
58. Robertus, J., Monzingo, A., *Mini-Reviews in Med. Chem.* **2004**, *4*, 483-492.
59. Fan, E., Zhang, Z., Minke, W., Hou, Z., Verlinde, C., Hol, W., *J. Am. Chem. Soc.,* **2000**, *122*, 2663-2664.
60. Kitov, P., Sadowska, J., Mulvey, G., Armstrong, G., Ling, H., Panu, N., Read, R. and Bundle, D., *Nature,* **2000**, *403*, 669-672
61. Borman, S., Chem. Eng. News **2000**, October 9, 48-53.
62. Frisbie, C., Rozsnyai, L., Noy, A, Wrighton, M. and Leiber, C.M., *Science,* **1994**, *265*, 2072; Zhang, X., Moy, V., *Biophysical Chemistry,* **2003**, *104*, 271-278
63. http://stm2.nrl.navy.mil/how-afm/how-afm.html.
64. Clear, S. and Nealey, P., *J. Coll. Int. Sci..* **1999**, *213*, 238-250.
65. Viel, S., Mannina, L. and Segre, A., *Tetrahedron,* **2002**, *43*, 2515-2519; see also Ludwig, C., Michiels, P., Wu, X., Kavanagh, K., Pilka, E., Jansson, A., Opperman, U., Gunther, U., *J. Med. Chem.*, **2008**, *51*, 1-3.
66. Keklikian, L. and Partch, R., *Colloids and Surfaces* **1989**, *41*, 327-37.
67. Nishida, Y., Iso, M., Matsuoka, M. and Partch, R., *Adv. Powder Technol.* **2004**, *15*, 247-61.
68. Huang, C-L., Partch, R. and Matijevic, E., *J. Coll. Interface Sci.* **1995**, *170*, 275-83.
69. Avella, M., Martuscelli, E., Raimo, M., Partch, R., Gangolli, S. and Pascucci, B., *J. Mat. Sci.* **1997**, *32*, 2411-16.
70. Dilsiz, N., Partch, R., Matijevic, E. and Sancaktar, E., *J. Adhesion Sci. Technol.* **1997**, *11*, 1105-18.
71. Partch, R., *Materials Synthesis and Characterization*, Perry, D. Ed., **1997**, Plenum Press, New York, 1-17.
72. Josef, I., Abu-Reziq, R., Avnir, D., *J. Am. Chem. Soc.* **2008**, *130*, 11880.
73. Zhang, Y., Iijima, S., *Science* **1998**, *281*, 973.
74. Pita, M., Tatz. E., *J. Am. Chem.* **2008**, *130*, 36-37.

75. Bao, J., Chen, W., Liu, T., Zhu, Y., Jin, P., L. Wang., Liu, J. and Li, Y., *ACS Nano,* **2007**, *1*, 293-298.
76. Nakamura, E., *Science,* **2007**, DOI: 10.1126/science.1138690
77. Weissleder, R., Kelly, K., Sun, E., Shtatland, T. and Josephson, L., *Nature Biotechnology,* **2005**, *23*, 1418-1423.
78. Riveros, T., Lo, R., Liu, X., Valdez, A., Lozano, M. and Gomez, F., *American Laboratory,* **2010**, *42*, 11-19.
79. Lin, H. and Wang, N., *J. Am. Ceram. Soc.,* **2007**, DOI: 10.1111/j.1551-2916.2007.02132x.
80. Paulson, J., *Blood*, **2010**, *115*, 4778.
81. Shukla, N., Bartel, M. and Gellman, A., *J. Am. Chem. Soc.*, **2010**, *132*, 8575-8580.
82. Vairaprakash, P., Ueki, H., Tashiro, K. and Yaghi, O., *J. Am. Chem. Soc.*, **2011**, *133*, 759-761.
83. Zhang, J., Su, D., Blume, R., Schogl, R., Wang, R., Yang, X. and Gajovic, A., *Angew, Chem. Int. Ed.*, **2010**, *49*, 8640-8644.
84. Chermont, Q., Chaneac, C., Pelle, F., Maitrejean, S., Jolivet, J-P., Gourier, D., Bessodes, M. and Scherman, D., *Proc. Natl Acad. Sci*, **2007**, *104*, 9266-9271.
85. Prencipe, G., Tabakman, S., Welsher, K., Liu, Z., Goodwin, A., Zhang, L., Henry, J.and Dai, H., *J. Am. Chem. Soc.,* **2009**, 131-4783-4787.
86. Lee, N. and Hyeon, T., *Proc. Natl. Acad. Sci.,* **2010**, DOI: 10.1073/pnas.1016409108.
87. Sokolov, I., Kievsky, Y. and Kaszpurenko, J., *Small,* **2007**, *3*, 419-423.
88. Huang, X., E-Sayed, I., Qian, W. and El-Sayed, M., *J. Am. Chem. Soc.,* **2006**, *128*, 2115-2120.
89. Lal, S. *Acct. Chem. Res.*, *41*, 1842-1851.
90. Hoffman, R., *Nat. Rev. Cancer*, **2005**, *5*, 796-806
91. Kamiya, H., Yoshida, M. and Mitsui, M., *Mat Res Soc Symp Proc*, **1998**, *501*, 241- 246.
92. Iler, R., *The Chemistry of Silica*, **1979**, Wiley.
93. Brinker, J. and Scherer, G., *Sol-Gel Science*, **1989**, Academic Press.
94. Dickey, F., *Proc. Natl. Acad. Sci.* **1949**, *35*, 227-29.
95. Raman, K., Anderson, M. and Brinker, C., *Chem. Mater.* **1996**, *8*, 1682-1701.
96. Makote, R., and Collinson, M., *Chem. Mater.* **1998**, *10*, 2440-45.100.
97. Zimmermann, C., Partch, R. and Matijevic, E., *Colloids and Surfaces* **1991**, *57*, 177-182.
98. Katz, A. and Davis, M., *Nature* **2000**, *403*, 286-88.
99. Product Catalogue, Gelest Co.
100. Wang, H-Y., Ph.D. Thesis, Clarkson University, **2000**.
101. Peri, J. and Hensley, A., *J. Phys. Chem.* **1968**, *72*, 2926-33.
102. Pancrazio, J.J., Kamatchi,G.L., Pentel, P. and Benowitz, N., *Med. Toxicol*, **1986**, 101-121.
103. Roscoe, A. and Lynch, C., *J. Pharmacol. Exp. Ther.* **1998**, *284*, 208-214.
104. Marshall, J., and Forker, A., *Am. Heart J.* **1982**, *103*, 401-414; Hawley, A., Illum L. and Davis, S., *Pharm. Res.*, **1997**, *14*, 657-661
105. Guo, L., Liu, G., Hong, R-Y and Li, H-Z., *Marine Drugs*, **2010**, *8*, 2212-2222.
106. Tomczak, M., Slocik, J., Stone, M. and Naik, R., *MRS Bull*, **2008**, *33*, 519-523.
107. Patist, A., Kanicky, J., Shukla, P. and Shah, D., *J. Colloid Interface Sci.* **2002**, *245*, 1-15.

Index